Sorting

A complete list of titles in this series appears at the end of this volume.

Sorting
A Distribution Theory

HOSAM M. MAHMOUD
The George Washington University

A Wiley-Interscience Publication
JOHN WILEY & SONS, INC.
New York • Chichester • Weinheim • Brisbane • Singapore • Toronto

This book is printed on acid-free paper. ∞

Published simultaneously in Canada.

For ordering and customer service, call 1-800-CALL-WILEY.

Library of Congress Cataloging-in-Publication Data is available.
ISBN 0-471-32710-7

Printed in the United States of America

10 9 8 7 6 5 4 3 2 1

إهداء

إلى فريدة رهنورد محمود

مع حبى و تقديرى

Contents

Preface

Putting data in order is an intellectual activity that is perhaps one of the oldest problems of applied mathematics. The wall of the corridor of Abydos Temple in Egypt is padded with a chronologically ordered list (Gallery of the List of Kings). Dating back to around 1250 B.C., this display allegedly lists the Pharaohs who had preceded Siti I (modern research proves that the list makes some false historical claims; a few rulers of Egypt before Siti I are omitted from that list!) The Inakibit-Anu Babylonian tablets, dating back to 200 B.C., contain a sorted list of what seems to be several hundred records, of which only a little over 100 records are preserved. Alphabetical ordering of words dates at least as far back as Western dictionaries. Over the millennia people have been yearning to find fast efficient sorting algorithms. With the appearance of computers in the twentieth century, sorting occupied a central position among computing problems for its numerous business and administrative potentials.

This is a book about sorting methods with a general focus on their analysis. We concentrate on well-known algorithms that are widely used in practice. Of course, to the basic skeleton of any such algorithm small improvements have been proposed. We present these algorithms in their simplest form, even when a modification is known to improve them. This choice is motivated by a desire to make the material lucid. While they may affect lower-order terms, the modifications usually do not change the dominant asymptotic behavior. The analysis techniques are as varied as the sorting algorithms themselves, giving rise to a fascinating variety of methods, ranging from standard treatment such as elementary probability theory, combinatorial counting, and graph-theoretic methods, to instances of more modern techniques such as martingales, Poissonization, Wasserstein's metric space, and the Mellin transform.

We discuss a variety of standard sorting methods that can be readily implemented on a conventional computer. The book takes the following algorithmic view. When a *deterministic* algorithm runs on a set of data of size n, certain computing resources are committed. Among standard resources are the running time and the amount of memory needed to support the algorithm. A suitable combination of these computing resources may be referred to as the *cost*. Running the deterministic algorithm on the same data set will always give the same results. However, running the deterministic algorithm on a different data set of size n may result in different cost. Putting a reasonable probability measure on the set of inputs of size n renders the cost a random variable. Interpreting costs as random variables provides some understanding of the behavior of the algorithm over the variety of data sets the algorithm may face. Probabilistic analysis addresses the natural questions one usually asks about random variables. What are the average, variance, higher moments, and exact distributions? In many cases it may be prohibitive to get exact answers. Even when such answers are obtained they may give little insight into the algorithm's behavior. Exact answers often take the form of multifolded sums of products or similar complicated forms. Insight is gained by simplifying such expressions asymptotically, that is, finding a simple representation of the leading terms in the form of elementary functions of n, when n gets large. Asymptotics then provide quick ways of estimating how the mean and other moments grow with n, and how probabilities change. Limiting distributions give ways to approximate probabilities from standard distributions that may be well known and tabulated. A question of interest to the practitioner is how large n should be before one can consider limit distributions as reliable approximations. This is a question of finding rates of convergence. We give a few examples of rates of convergence in the book. It is fascinating to find that some phenomena in sorting algorithms may have essential periodic fluctuations in leading asymptotic terms or in rates of convergence to such terms.

The issue we consider is how a deterministic algorithm may behave when it is presented with *random data*. This is to be distinguished from an area of algorithmics that deals with *randomized algorithms*, where the algorithm itself may follow different paths to obtain the same result, or even produce different results when it is run repeatedly on the same data set. Such variability is intrinsic to a randomized algorithm as the algorithm makes its decisions based on random outcomes (like generating random numbers or flipping coins). We touch in passing on the area of randomized algorithms in a few exercises, but it is not the main focus of this book.

The book is intended to be used as a reference by computer professionals, scientists, mathematicians, and engineers. The book may also be used for teaching. The material is accessible to first-year graduate students in fields like computer science, operations research, or mathematics. At least a portion of each chapter is accessible to advanced undergraduates. A reading course or a special topics seminar may be based on the book. If used in this fashion, a one-year course in probability and a one-year course in algorithms are recommended.

The developments that led to this book are numerous. Many research contributions were made toward a distribution theory of sorting algorithms over the past quarter century. These contributions are mostly published in highly specialized research outlets such as scientific journals and conference proceedings. The aim of the book is to organize this massive and esoteric body of research that is accessible only to a few specialized researchers into a coherent, well-founded distribution theory and make it accessible to a wider audience by clarifying and unifying the research and bringing it down a notch to be understood by students and other less-specialized interested readership. This is accomplished via the slower-paced presentation of a textbook, the patient construction of a logical framework for sorting, and careful explanation that appeals to intuition as closely as possible. It is hoped that the manuscript is written in a lively style, starting at the basic principles and building from the ground up, integrating in the process some of the most modern techniques that have succeeded in cracking some long-standing open problems in sorting, like the distributions associated with QUICK SORT.

The book is organized as follows. It opens with a chapter on the general area of complete and partial sorting (identification of order statistics). The opening chapter sets the tone for the entire monograph: broad meaning, motivation, and applications. The opening sends a message to the reader about the level of complexity of the material and what is expected as background. The opening chapter outlines general methodology (analytic, probabilistic, and combinatorial methods used in later chapters of the book).

The rest of the book dedicates one chapter for every standard sorting algorithm with a careful explanation of the mechanics of the algorithm both as code, for the specialized computer science audience, and verbal description, for the broader readership of scientists, mathematicians, and engineers. A typical chapter delves into the domain of analysis. A broad array of classical and modern techniques converge hand in hand to substantiate a distribution theory of sorting.

Each chapter provides exercises. Chapter 1 is a collection of peripheral topics that are somewhat disparate; the exercises on each topic of this chapter come at the end of the topic. Each of the remaining chapters has exercises at the end of the chapter. The exercises touch on peripheral concepts and even instill new material in small doses. The exercises vary in complexity from the elementary, only reinforcing the basic definitions, to the very challenging, bordering on the threshold of recently understood problems.

HOSAM M. MAHMOUD

Acknowledgments

Professionals who have made contributions to the study of sorting are too many to be listed individually. I am indebted to the algorithmics community at large. However, in particular I am indebted to the participants of the seminar series *Average-Case Analysis of Algorithms*. Special thanks are in order to Philippe Flajolet, Rainer Kemp, Helmut Prodinger, Robert Sedgewick, and Wojciech Szpankowski, the founding fathers of the seminar series.

The encouraging remarks of Philippe Flajolet, Donald Knuth, and Robert Sedgewick were inspiring. Parts of the book were written while the author was on sabbatical leave at Institut National de Recherche, Rocquencourt, France and Princeton University, New Jersey, U.S. Both places have provided an excellent research atmosphere. The hospitality of Philippe Flajolet, Mireille Régnier, and Robert Sedgewick, who made these visits possible, is greatly acknowledged. Hsien-Kuei Hwang, Philippe Jacquet, Janice Lent, Guy Louchard, Uwe Rösler, Robert Smythe, and Wojciech Szpankowski took interest in earlier drafts of the manuscript and identified several improvements.

The common wisdom of many friends not specialized in the field has been a great inspiration. Many have contributed to the production of the book by providing moral support and unwittingly giving motivation. A few friends not specialized in algorithmics have asked questions that brought to the focus of my attention what the intelligent consumer from other fields may expect from the producers in the field of algorithmics. Talaat Abdin, Kareem and Sherif Hafez, and Christian Hesse may not be aware that they have made contributions to this book. Last but not least, I thank my wife Fari for being an inexhaustible source of patience and understanding.

1

Sorting and Associated Concepts

1.1 SORTING

Sorting a list of items is a very mundane task. The term generally refers to arranging the items of the list in ascending or descending order according to some ordering relation. Sometimes this activity is desired for its own sake such as the case in ranking Olympic contenders by sorting their scores in an event, or the scores of a class to determine letter grades. Sorting is often intended to facilitate searching, another fundamental activity in data processing. Imagine how difficult it would be to consult a dictionary on the meaning of a word, if the dictionary is not organized in some order to guide our search. With relative ease we can find a word in a dictionary or a person's phone number in a directory when its entries are kept in the standard alphabetical order.

Sorting is a more convoluted task than searching. Generally it is more expensive and time consuming. Nonetheless, we view it as a long-term investment—we take the time and effort to sort a file once to repeatedly search quickly in it for items.

The general problem of sorting can be stated as follows. Say we have a list of items

$$X_1, X_2, \ldots, X_n$$

that we intend to rearrange in increasing order. Our goal is to design efficient algorithms to create a permutation

$$X_{(1)}, X_{(2)}, \ldots, X_{(n)}$$

where $X_{(i)}$ is the ith smallest (or ith *order statistic*) among the items on our list. The term *smallest* refers to relative smallness according to an ordering relation often denoted by $\le$. For example, $X_1, X_2, \ldots, X_n$ may be real numbers and $\le$ is the usual arithmetic comparison. The final sorted list in this case will be $X_{(1)}, X_{(2)}, \ldots, X_{(n)}$, with $X_{(1)} \le X_{(2)} \le \ldots \le X_{(n)}$, that is, the numbers arranged in increasing order.

Another way to view the situation is the following. Note that $X_{(1)}$ may turn out to be any of the n numbers—there is an index i_1 so that $X_{(1)} = X_{i_1}$. Generally, the same is true for any other order statistic: $X_{(j)} = X_{i_j}$, for some index i_j. We can think of sorting as an attempt to come up with a map

$$\Pi_n : \{1, 2, \ldots, n\} \to \{1, 2, \ldots, n\}, \tag{1.1}$$

where $\Pi_n(1) = \pi_1, \Pi_n(2) = \pi_2, \ldots, \Pi_n(n) = \pi_n$, with the integers $\pi_1, \ldots, \pi_n$ being all distinct, and

$$X_{\pi_1} \le X_{\pi_2} \le \ldots \le X_{\pi_n}. \tag{1.2}$$

In other words, $\pi_1, \ldots, \pi_n$ is a permutation of the integers $\{1, 2, \ldots, n\}$, with X_{π_j} being the jth smallest item. We shall use the notation $\Pi_n = (\pi_1, \ldots, \pi_n)$ to represent such a permutation. Permutations can be represented in a number of ways. A standard two-line representation helps in visualizing where the numbers go. This representation consists of a top row of indexes and a bottom row of corresponding values; under each index i, we list π_i. For example, the permutation $\Pi_8 = (7\ 6\ 8\ 3\ 1\ 2\ 5\ 4)$ can also be represented as

$$\begin{pmatrix} 1 & 2 & 3 & 4 & 5 & 6 & 7 & 8 \\ 7 & 6 & 8 & 3 & 1 & 2 & 5 & 4 \end{pmatrix}.$$

Producing the sorted list is an act of actual rearrangement. The task of sorting n elements may effectively be accomplished without moving any elements (if so desired) by specifying the permutation Π_n of the construction (1.2) as a separate list. For in this case the knowledge of $\pi_1, \ldots, \pi_n$ can effectively be used to obtain any order statistic. For example, we can print the still-unsorted list $X_1, \ldots, X_n$ in sorted order by printing

$$X_{\pi_1}, X_{\pi_2}, \ldots, X_{\pi_n}.$$

EXERCISES

1.1.1 Sort each of the following lists in "ascending" order:

(i) 2.6 1.3 7.1 4.4 3.4 2.9.

(ii) 60 33 17 84 47 29 71 56.

(iii) Jim Beatrice Philip Alice Johnny Elizabeth.

1.1.2 For each list in the previous exercise, construct a permutation to effectively sort the list without changing its composition.

1.1.3 Interpret the mapping (1.1) when the list contains repeated elements.

1.2 SELECTION

Another problem of practical interest is that of the selection of some order statistics either ordered by rank or as a set known to have some predesignated ranks but not necessarily in order. An example of finding an ordered group of ranked order statistics is common in competitive sports. It is often the case that from the scores obtained by the competing athletes the top three contenders with the highest three scores (in order) are to be selected for medals.

To emphasize the distinction between sorting and selection, sorting is sometimes referred to as *complete sorting*, whereby all the elements are moved to their correct positions (or indexes are created to indicate these positions). By contrast, selection is referred to as *partial sorting*, as the selection process imparts only partial information on the correct position of some elements.

Statistics based on a few order statistics (with known orders) abound in the literature of statistical inference. For example, weighted averages of tertiles and quartiles[1] of a sample $X_1, X_2, \ldots, X_n$, are in common use in estimating the center of the distribution from which the sample was gathered. Two such statistics with different weights are Gastwirth's estimator

$$G_n = \frac{3}{10} X_{(\lfloor \frac{n}{3} \rfloor + 1)} + \frac{2}{5} X_{(\lfloor \frac{n}{2} \rfloor)} + \frac{3}{10} X_{(n - \lfloor \frac{n}{3} \rfloor)},$$

and Tukey's tri-mean

$$T_n = \begin{cases} \frac{1}{8}\left(X_{(\frac{n}{4})} + X_{(\frac{n}{4}+1)}\right) + \frac{1}{2} X_{(\frac{n}{2})} & \\ \qquad + \frac{1}{8}\left(X_{(\frac{3n}{4})} + X_{(\frac{3n}{4}+1)}\right), & \text{if } n \text{ is a multiple of } 4; \\ \frac{1}{4} X_{(\lfloor \frac{n}{4} \rfloor)} + \frac{1}{2} X_{(\lfloor \frac{n}{2} \rfloor)} + \frac{1}{4} X_{(\lceil \frac{3n}{4} \rceil)}, & \text{otherwise.} \end{cases}$$

The book by Andrews, Bickel, Hampel, Huber, Rogers, and Tukey (1972) gives many other statistical applications involving only a few order statistics. Efficient algorithms for finding quantiles are needed for quick computation of these statistics. Many other applications in statistics need such algorithms to find few order statistics, typically less than six, for a statistical design.

Another flavor of the problem of finding order statistics may request finding a group of unranked order statistics without necessarily knowing the relative ranks in the group. For example, a university that intends to admit 1000 students may want to find the 1000 students with the top 1000 scores (not necessarily in order) among the applicants whose number typically exceeds 10000 each academic season. An example from statistics where an unordered group of order statistics is required is the α-trimmed mean. The *α-trimmed mean* is a statistic that throws out the upper and lower α proportion of the (sorted) data deeming them unreliable outliers, with $0 < \alpha < 1$. The α-trimmed mean is given by

$$\frac{1}{n - 2\lfloor \alpha n \rfloor} \sum_{i=\lfloor \alpha n \rfloor + 1}^{n - \lfloor \alpha n \rfloor} X_{(i)}.$$

[1]An α-quantile of a sample having no duplicates is a real number so that proportion α of the data is less than it; the rest are greater. When n is an exact multiple of 3, the $\frac{1}{3}$- and $\frac{2}{3}$ quantiles are called tertiles. The proportions $\frac{1}{3}$ and $\frac{2}{3}$ of the data are interpreted respectively as $\lfloor \frac{n}{3} \rfloor$ and $\lfloor \frac{2n}{3} \rfloor$ of the points when n is not a multiple of 3. With a similar interpretation of "fourth," "half," and "three-fourths," the $\frac{k}{4}$-quantiles are called quartiles, for $k = 1, 2, 3$ and the $\frac{1}{2}$-quantile is called the sample median.

We may design an algorithm to identify the two bounding order statistics $X_{\lfloor \alpha n \rfloor}$ and $X_{n-\lfloor \alpha n \rfloor+1}$ and in the process place all intermediate order statistics in the list between positions $(\lfloor \alpha n \rfloor + 1), \ldots, (n - \lfloor \alpha n \rfloor)$ in any order, not necessarily sorted. For the computation of the α-trimmed mean we need not sort these intermediate order statistics since we are only seeking their average.

Algorithms for finding a predesignated number of order statistics (ordered or unordered) are generally called *selection algorithms*. Sorting itself is the special case of a selection algorithm, when the selection algorithm finds all order statistics.

One can find order statistics by sorting a data file. The item at position i is then the ith order statistic. This may be much more work than necessary; it is often the case that a more efficient algorithm can be designed for the direct selection of specified order statistics as we shall see. Several sorting algorithms discussed in this book can be adapted for general purpose selection.

EXERCISES

1.2.1 Write an algorithm to find the maximum element in a list. How many comparisons does your algorithm make to pick the largest number in a list of n numbers?

1.2.2 Write an algorithm to simultaneously find the maximum and minimum elements in a list of n elements using no more than $3n/2$ comparisons.

1.3 JARGON

A sorting algorithm is *comparison based* if it sorts an input (a collection of data items called *keys*) by making a series of decisions relying on comparing pairs of the input data. (Many algorithms in this book are comparison based.) Nonetheless, there are sorting algorithms that are not comparison based. For instance, RADIX SORT, an algorithm discussed in a later chapter, bases its decisions on another operation different from comparing pairs of data—this operation is the extraction of digits of the input keys; RADIX SORT makes decisions according to these digits. RADIX SORT is a digital non-comparison-based sorting method.

Broadly speaking, sorting algorithms fall in two categories. The class of *naive* sorting algorithms, the ones that occur first in a natural way to most people, is a class of algorithms with $\Theta(n^2)$ running time[2] when they sort most inputs of size n and that tend to have an average running time of order $\Theta(n^2)$. On close examination we realize that naive algorithms do not effectively use information gained from the earlier stages of the algorithm and may tend to perform redundant or repetitious tasks. Careful parsimonious design that avoids redundancy can introduce a substantial reduction in the order of magnitude of the running time of an algorithm, leading to an algo-

[2]For definition of the asymptotic symbols Θ, O, o, $\sim$, and Ω see the appendix.

rithm in the second class of parsimonious algorithms. We shall see shortly that the best comparison-based algorithm must take at least $\Omega(n \ln n)$ time[3] on some input of size n. That is, the best comparison-based sorting algorithm must take $\Omega(n \ln n)$ time on its worst case of size n, or comparison-based sorting is inherently bounded from below by $\Omega(n \ln n)$ for some input. There are practical algorithms that do take $O(n \ln n)$ time on all inputs. A *parsimonious* sorting algorithm is one that possesses an average running time that is of the order $\Theta(n \ln n)$. A parsimonious algorithm performs on most inputs as the best comparison-based sorting algorithm would on its worst-case instance.

Parsimonious algorithms are often recursive and use the paradigm of *divide-and-conquer*, whereupon the list to be sorted is split into two or more segments according to some splitting criterion, then the parts which are typically smaller than the input list itself are recursively attacked individually. The process continues until very small lists are considered (typically of size 1 or 0) for which the sorting task is trivial or has a directly obvious answer. The solution to the original sorting problem is then assembled by combining the solutions of the parts.

A sorting algorithm is called *stable* if equal elements in the raw data maintain their relative position in the final sorted list. For example, if $X_5 = X_{22} = X_{47}$ are the only keys equal to 18 in an unsorted list of integers, a stable sorting algorithm may move X_5 to position 3, X_{22} to position 4, and X_{47} to position 5. Stability is desired when satellite information is present alongside the primary keys. For example, suppose we have an alphabetically sorted list of passengers out of an international airport on some day. When a stable sorting algorithm is applied to the flight number field of the records, the alphabetical order is not disturbed within one flight; we get the day's activity sorted by flight numbers, and within each flight the passenger names will be sorted alphabetically.

A desirable feature of a sorting algorithm is the capability to finish the job without creating large secondary data structures to hold copies of the data at some stage. Several algorithms discussed in this book are capable of sorting using only a chief host container of data, such as an array. When performing on n data items, the sorting task is accomplished by moving data around within the container, that is, making swaps via a few additional intermediate swap variables (typically $O(1)$ extra space). Recursive sorting algorithms may set up a hidden stack of small size, typically averaging to $O(\ln n)$ additional space. An algorithm that sorts a list of n items "in place" without allocating data structures of size comparable to the list it is sorting—that is, the secondary storage is no more than $o(n)$ on average, is said to sort *in situ*.

EXERCISES

1.3.1 Study the algorithm of Figure 1.6 first. Is this algorithm comparison based? Is it stable? Does it operate in situ?

[3]See the appendix for the convention for the base of the logarithm.

1.4 ALGORITHMIC CONVENTIONS

We specify many algorithms in this book in pseudocode. It is essentially similar to standard conventions of a pedagogical programming language, except that we occasionally take the liberty of relaxing the syntax when convenient or more readable. For example, exponential expressions like 2^k are easier to read than the usual representation by double asterisks or computation via logarithms. Most of such algebraic symbolism is represented within pseudocode in standard mathematical notation. For example, ceils and floors are inserted in our pseudocode and division in the form of a two-level fraction, as in

$$\frac{a+b}{c+d},$$

is preferred to the parenthesized expression $(a + b)/(c + d)$, because the former is more readable and uses fewer symbols.

We also give a brief verbal explanation for each algorithm to make the material accessible to nonprogramming readership of scientists, mathematicians, and engineers.

The reader is assumed to be familiar, at least conceptually, with basic data structures such as arrays and linked lists. Data to be sorted will most of the time be assumed to be already loaded in a typical array structure like the following:

global A: **array** $[1\,..\,n]$ **of real**;

As an algorithm progresses with its tasks we shall often need to specify a particular stretch (subarray) on which the algorithm operates next. Such a subarray will typically extend from a lower index ℓ, to an upper index u. The convenient notation $A[\ell\,..\,u]$ will refer to this subarray (and to its content).

Arrays and subarrays will be depicted as vertical columns with lower indexes near the top. Sometimes it is more convenient for the typography to depict an array horizontally. In horizontal representation the lower indexes will be on the left.

Many sorting algorithms perform by swapping array elements. In the background of sorting algorithms that need to swap data we shall assume the presence of a procedure

$$swap(x, y)$$

that interchanges the contents of the two variables x and y. A typical call in the context of array sorting will be

call $swap(A[i], A[j])$

to interchange the data at positions i and j. Generally, the background *swap* procedure will be assumed to be capable of handling whatever data type is being sorted.

EXERCISES

1.4.1 Implement the sorting algorithm discussed in Section 1.7 on a computer.

1.4.2 Implement a *swap* procedure on a computer to interchange the content of two variables.

1.5 ORDER

Central to the analysis of sorting and selection is the notion of partial order. We shall briefly review the basic concepts of order. Let A and B be two sets. The *Cartesian product* $A \times B$ is the set of all pairs $\{(a, b) \mid a \in A \text{ and } B \in B\}$; an element of $A \times B$ is an *ordered pair*. A relation R from A to B is a subset of $A \times B$. A relation from A to A is simply called a relation *on* A and is thus a subset of $A \times A$. Let R be a relation on A. If $(x, y) \in R$ we say that x and y are *related* by R and in infix notation we write $x \; R \; y$.

A relation $\leq$ (commonly called less than or equal to) on a set A is a *partial order* if the relation is:

(i) *Reflexive*, that is, $a \leq a$, for all $a \in A$.

(ii) *Antisymmetric*, that is, $a \leq b$ and $b \leq a$ imply $a = b$, for all $a, b \in A$.

(iii) *Transitive*, that is, $a \leq b$ and $b \leq c$ imply $a \leq c$, for all $a, b, c \in A$.

Two related elements a and b are said to be *comparable* (either $a \leq b$, or $b \leq a$). The notation $a \leq b$ is called a *true comparison* when $a \leq b$ is in the partial order. For the comparable pair $a \leq b$, we use expressions like a is less than or equal to b, a has a rank lower than b, a is below b, or a precedes b in the partial order, and so on. Similarly, we often use expressions like b is larger or has rank higher than a, b is above a, or follows a in the partial order. When two elements a and b are not comparable, that is, neither $a \leq b$ nor $b \leq a$ is in the partial order, we say a and b are *incomparable*. The set A and the ordering relation $\leq$ together are called a *partially ordered set* denoted by $(A, \leq)$, also called a *poset*. A true comparison is equivalent to an ordered pair and the partial order may be represented as a set of true comparisons.

A partial order on the universe A is a *total order*, when every pair of the elements of the universe are comparable. For example, the set of integers and the usual $\leq$ arithmetic comparison is a total order.

The following example illustrates several definitions discussed in the preceding paragraphs. Suppose we have the universe $\{X_1, \ldots, X_8\}$. The set of true comparisons

$$Q = \{X_i \leq X_i \mid i = 1, \ldots, 8\} \cup \{X_1 \leq X_2, X_1 \leq X_3, X_1 \leq X_4, X_1 \leq X_5,$$
$$X_2 \leq X_3, X_2 \leq X_4, X_2 \leq X_5, X_3 \leq X_5, X_6 \leq X_7\} \qquad (1.3)$$

is a partial order on the universe and is a subset of the total order induced by indexing, that is, the ordering relation $X_i \le X_j$ whenever $i \le j$. If it is not known whether a total indexing order exists, the partial order Q gives only partial information on the relative ranking of some elements.

The goal of a comparison-based sorting algorithm on a finite totally ordered set is to discover the total order. The algorithm does that by asking questions in steps. Every step reveals more information and a larger partial order is attained. The sorting process starts with the empty partial order, as there is no prior information available, then goes on to build nested posets. The algorithm terminates when enough information is gathered to construct the total order.

The set of *minima* in a poset $\mathcal{P} = (A, \le)$ is the set

$$\{a \in A \mid \text{the only element in } A \text{ that is } \le a \text{ is } a\};$$

a member in the set of minima is a *minimum* or *minimal element*. We similarly define the set of *maxima* of $\mathcal{P}$ as

$$\{b \in A \mid \text{the only element } x \in A \text{ satisfying } b \le x \text{ is } b\};$$

a member in the set of maxima is a *maximum* or *maximal element*.

Often a poset is very conveniently represented pictorially by a diagram of the objects of the poset with an arrow going from a to b for any true comparison $a \le b$ between two different elements $a \ne b$. (The self-loops corresponding to reflexivity are usually not shown but implicitly understood.) As a visual aid to the essence of the relation, the minima are drawn at the bottom of the diagram. Higher-ranked elements usually appear in the drawing above lower-ranked elements of the poset. Figure 1.1 shows the partial order Q of display (1.3).

With every minimal element of a poset $(A, \le)$ there are associated *chains* of the partial order, where each chain is formed by the following procedure. If $a_1 \in A$ is a minimum, the chains starting at a_1 are formed by first formally writing the string a_1. Whenever a_k is at the rightmost end of a string $a_1 \le a_2 \le \ldots \le a_k$, and a_k is below $b_1, \ldots, b_j$ in the partial order, the string is cloned into j strings; the symbol $\le$ is added at the right end of each clone followed by one of the elements $b_1, \ldots, b_j$;

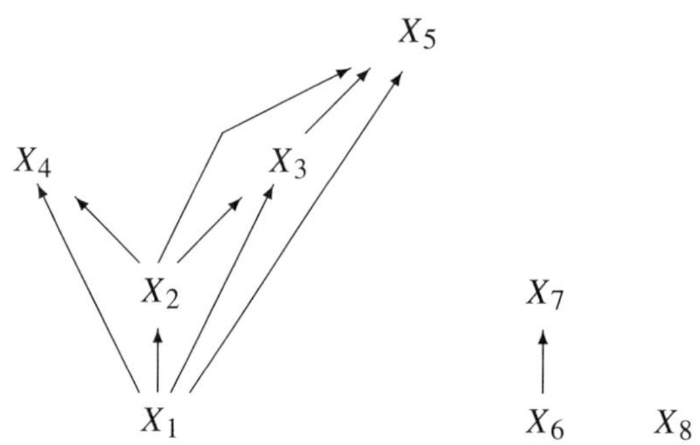

Figure 1.1. A poset.

this step in the process yields

$$a_1 \le a_2 \le \ldots \le a_k \le b_1,$$
$$a_1 \le a_2 \le \ldots \le a_k \le b_2,$$
$$\vdots$$
$$a_1 \le a_2 \le \ldots \le a_k \le b_j.$$

The rules apply recursively to every clone until it is no longer possible to add any elements. Strings that are obtained by removing symbols from longer strings are then discarded. Each remaining formal string is a chain that starts at a minimal element and ends at a maximal element of the partial order. Together, the chains reflect how much order is known. A total order has only one chain.

Finally, suppose A is a subset of some universal set S. The set A is said to be *compatible* with a partial order $(S, \le)$, if whenever $a \le b$, and $a \in A$ then $b \in A$. Informally, if a is a member of a chain and $a \in A$, then all the elements of the chain higher than a all the way up to the maximal element of that chain are in A. None of the elements of A is known to be below an element from the complement set $S - A$.

To fix the ideas, let us consider again the partial order Q of display (1.3). The set of minima of Q is $\{X_1, X_6, X_8\}$. The set of maxima is $\{X_4, X_5, X_7, X_8\}$. The elements X_2 and X_5 are comparable; but X_1 and X_7 are incomparable. The chains of Q are $X_1 \le X_2 \le X_3 \le X_5$, $X_1 \le X_2 \le X_4$, $X_6 \le X_7$, and X_8. The set $\{X_2, X_3, X_4, X_5, X_8\}$ is compatible with Q, but the set $\{X_2, X_5, X_6\}$ is not.

The following lemma is important in deriving lower bounds on sorting and selection.

Lemma 1.1 *A comparison-based algorithm for finding the jth order statistic must implicitly determine the set of $j - 1$ items of lower ranks. Consequently the set of $n - j$ items of higher ranks is also determined.*

Proof. We prove the lemma by contradiction. Suppose we have an algorithm that finds the jth smallest in the data set $\{X_1, \ldots, X_n\}$ of distinct elements, without determining the sets $\{X_{(1)}, \ldots, X_{(j-1)}\}$ of smallest $j - 1$ order statistics and $\{X_{(j+1)}, \ldots, X_{(n)}\}$ of largest $n - j$ order statistics. There are indexes k, ℓ, and m such that the algorithm does not know whether $X_{(k)} = X_\ell \le X_{(j)} = X_m$, or vice versa. No chain of the partial order discovered by the algorithm contains both X_ℓ and X_m; the existence of such a chain would make X_ℓ and X_m comparable. We can then construct an ordering of the input, starting with a chain containing X_m. In this chain let $a \le X_m$ be the last element that is known to be below X_ℓ, and let b be the first element known to be above X_ℓ such that $X_m \le b$; one or both of the elements a or b may not exist. In a manner consistent with the partial order, we can complete a linear order by first inserting X_ℓ anywhere between a and b. (If a does not exist and b does, we can place X_ℓ anywhere below b. Similarly, if b does not exist and a does, we can place X_ℓ anywhere above a. If both do not exist we are at liberty to put X_ℓ

anywhere as we completely lack any information on its relative rank in the chain.) Specifically, we insert X_ℓ next to X_m, which is presumed to be at position j. The rest of the elements not already in the chain are then added in any way consistent with the partial order. Transposing X_ℓ and X_m changes the position of X_m in the listing, but remains to be an ordering consistent with the partial order. This uncertainty of the position of X_m, supposedly the jth smallest in the final linear order, contradicts the assumption that the algorithm has identified the jth order statistic. ■

EXERCISES

1.5.1 Let the set $A = \{2, 3, 4, \ldots\}$ be partially ordered by divisibility—for $x, y \in A$, the ordered pair (x, y) is in the partial order if x divides y. What are the minima of A? What are the maxima of A? What are the chains of the poset? Is the set of odd numbers in A compatible with the partial order?

1.6 BINARY TREES

The *binary tree* is a hierarchical structure of *nodes* (also called *vertices*) that underlies many sorting and other combinatorial algorithms. The nodes of the structure are arranged in *levels*. The level count begins at 0. The only node at level 0 is called the *root node*. A node may have up to two nodes above it (viewed as *children*) that are put on the next level—a node (also viewed as a *parent*) may have no children, one left child, one right child, or two children (one left and one right). The children of a node are joined to their parent by links called *edges* or *branches*. It is customary to represent the tree by a drawing in which nodes are black bullets or circles of various sizes to illustrate different purposes, and edges are straight lines. The *depth* of a node in the tree is the node's distance from the root, that is, its level.

Historically trees are drawn upside down with the root at the top of the drawing. We adopt this convention in this book. If a tree has n nodes we say it is of *size* or *order n*. Figure 1.2 shows an example of a binary tree of size 8. We shall continue to use this historical convention; nevertheless, the language of the text will more

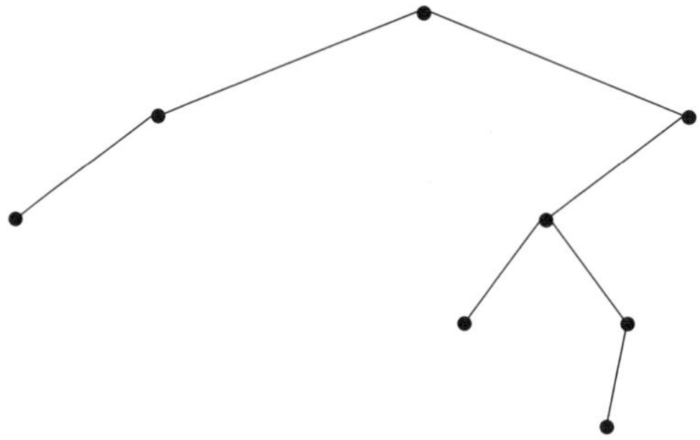

Figure 1.2. A binary tree on 8 nodes.

closely comport with the growth direction of natural trees—the root is at the bottom and tracing a path starting from the root is a process of moving "up" in the tree or climbing it, etc.

It is sometimes useful to think of the binary tree structure in terms of the following inductive definition. A binary tree is either empty, or has two substructures (called *subtrees*) distinguished as a left subtree and a right subtree, with the left and right subtrees both being binary trees. The collection of nodes in the subtree rooted at a node (excluding the node itself) is referred to as its *descendants*. By a parallel terminology, the collection of nodes encountered by climbing down a path from a node to the root of the whole tree is the node's *ancestors* or *predecessors*.

Often the parent-child links in a binary tree are given an orientation with sense leading from parent to child. Thus each node has only one link leading to it, or the so-called *indegree* is 1 for each node (except the root, whose indegree is 0). Each node may have up to two edges emanating out of it leading to its children. The number of edges coming out of a node is called the *outdegree* of the node. The outdegree of any node in a binary tree is at most 2. Level 1 can have up to two nodes; level 2 can have up to 4 nodes, and so forth. The maximum number of nodes that can appear on level i is 2^i. If level ℓ has 2^ℓ nodes, we say that this level is *saturated*. The root is at the lowest level (level 0). As we climb up the branches of the tree moving away from the root we move toward *higher* levels (further down in the inverted drawing).

It is very helpful in many analyses to consider an extension of a binary tree. The *extended binary tree* is obtained by adding to each original node, 0, 1, or 2 children of a new distinct type (called *external nodes* or *leaves*) to make the outdegree of all the original nodes (now called *internal*) exactly equal to 2. Figure 1.3 shows the extension of the tree of Figure 1.2, with internal nodes shown as bullets and leaves as squares.

The extended tree has a number of properties that will be of great use in the analysis of sorting algorithms and will be frequently encountered later in this book. These are invariant properties that are true for all trees of a given size regardless of their shape. We discuss these properties in the following definitions and propositions.

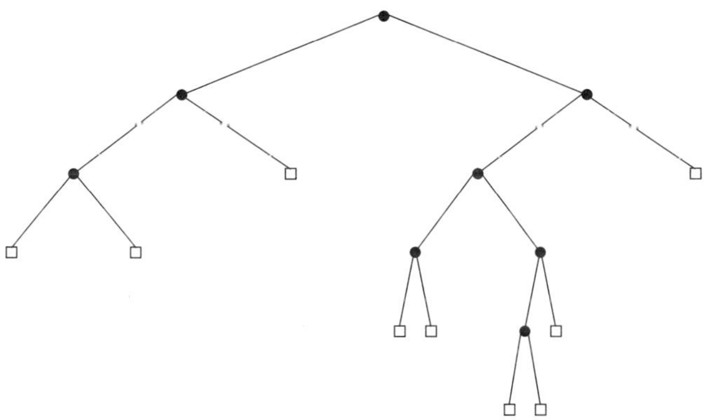

Figure 1.3. An extended binary tree on 8 internal nodes.

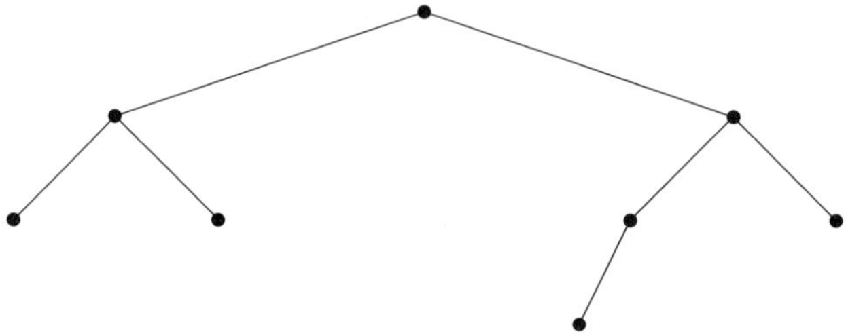

Figure 1.4. A complete binary tree on 8 nodes.

When all the levels of a tree, except possibly the highest level, are saturated, we say the tree is *complete*. The tree of Figure 1.2 is not complete, whereas that in Figure 1.4 is. Furthermore, when all the leaves appear on the same level the tree is said to be *perfect*. The complete tree is so termed because its extension has no "holes" in the levels containing only internal nodes; all these levels are saturated.

The *height* of a binary tree is the length of the longest root-to-leaf path in the tree. So, the height is also the level of the highest leaf. The height of a binary tree T_n on n nodes is denoted by $h(T_n)$ or simply h_n when it is understood which tree is being considered.

Proposition 1.1 *The height h_n of a binary tree on n nodes satisfies:*

$$\lceil \lg(n+1) \rceil \le h_n \le n.$$

Proof. The upper bound is trivial. The given tree is at least as high as a complete tree on n nodes; see Exercise 1.6.3. The complete tree is an instance of the lower bound. Letting h_n^* be the height of the complete tree, we have just argued that $h_n^* \le h_n$. In a complete binary tree of height h_n^*, levels $0, 1, \ldots, h_n^* - 2$ are saturated, and there are internal nodes on level $h_n^* - 1$, and no internal nodes on any higher levels:

$$1 + 2 + \cdots + 2^{h_n^*-2} < n \le 1 + 2 + \cdots + 2^{h_n^*-1},$$

or

$$2^{h_n^*-1} < n + 1 \le 2^{h_n^*}.$$

Taking base-two logarithms:

$$h_n^* - 1 < \lg(n+1) \le h_n^*.$$

However, h_n^* must be an integer. So, $h_n^* = \lceil \lg(n+1) \rceil$. ■

Proposition 1.2 *An extended binary tree on n internal vertices has $n + 1$ leaves.*

Proof. Let L_n be the number of leaves in an extended tree with n internal nodes. The tree has a total of $n + L_n$ nodes. Every internal node has two children (may be both internal, both external, or one of each). Thus the number of nodes (of any type) that are children is $2n$. Except the root, every node in the tree is a child of some parent. Therefore another count of the children is $n + L_n - 1$. Equating the two counts:

$$n + L_n - 1 = 2n,$$

and so

$$L_n = n + 1. \qquad \blacksquare$$

Suppose the internal nodes of a binary tree T_n on n nodes are labeled $1, \ldots, n$ in any manner. Let ℓ_i be the level of node i. The *internal path length* is defined as

$$I(T_n) = \sum_{j=1}^{n} \ell_j.$$

Analogously, suppose that the leaves of the same tree are also labeled $1, \ldots, n+1$ in some arbitrary manner, and that their corresponding depths in the tree are $d_1, \ldots, d_{n+1}$. The *external path length* is defined as

$$X(T_n) = \sum_{j=1}^{n+1} d_j.$$

When it is understood which tree is considered we may use simpler notation for $I(T_n)$ and $X(T_n)$ such as I_n and X_n. The internal and external path length play a major role in modeling the cost of sorting under many algorithms. They are related by the following.

Proposition 1.3 *Let I_n and X_n be the internal and external path lengths of a binary tree of size n. Then*

$$X_n = I_n + 2n.$$

Proof. We prove this proposition by induction on n. At $n = 1$ the statement obviously holds. Suppose the statement is true for some $n \geq 1$, and T_{n+1} is a tree of size $n + 1$, with internal and external path lengths I_{n+1} and X_{n+1}, respectively. There is at least one internal node u whose two children are external. Remove u and the two leaves above it. Replace u with an external node. This operation transforms T_{n+1} into a new binary tree T_n of size n. Let I_n and X_n be the respective internal and external path lengths of T_n. Suppose the level of u in T_{n+1} is ℓ. There are two corresponding leaves in T_{n+1} at level $\ell + 1$ that are lost in the transformation from T_{n+1} into T_n. The loss of these two leaves reduces the external path length by $2(\ell + 1)$, whereas the gain of a new leaf in T_n at level ℓ increases the external path length by ℓ. Thus,

$$X_{n+1} = X_n + 2\ell + 2 - \ell.$$

By the induction hypothesis $X_n = I_n + 2n$. Therefore,

$$X_{n+1} = I_n + 2n + \ell + 2.$$

The transformation from T_{n+1} to T_n also reduces the internal path length by ℓ:

$$I_{n+1} = I_n + \ell.$$

So,

$$X_{n+1} = I_{n+1} + 2(n + 1),$$

completing the induction. ■

A binary tree is called a *binary search tree* when its nodes are labeled in a special way to satisfy a search property. Any data type may be assumed for the labels, with integer, real, and character strings being the most common in application. The labeling set must be a totally ordered set, that is, any two elements x and y can be compared by an ordering or a ranking relation denoted by $\le$. For numeric labels $\le$ has the usual arithmetic meaning of less than or equal to; for strings it means lexicographically less than or equal to, etc. The labels are attached to the nodes so that each node receives one label and all the labels in the left subtree are less than or equal to the root label and all the labels in the right subtree are greater than the root label; this property then propagates to every subtree. In other words, we have a recursive definition of binary search trees: A binary search tree is either empty or

(a) All the labels in the left subtree are less than or equal to the root label.

(b) All the labels in the right subtree are greater than the root label.

(c) The left and right subtrees are binary search trees.

A binary search tree is built from a sequence of data items from an ordered set, the set of numbers say, by allocating a node as the root and attaching the first number as its label. The second number is compared to the root label; if it is less than or equal to (greater than) the root label, it is guided to the left (right); a node is allocated to hold it and linked as a left (right) child of the root. The process repeats on subsequent keys in the same manner. To insert a key a comparison is made at the root; if the key is less than or equal to (greater than) the root label, the key is inserted in the left (right) subtree recursively until termination at an empty tree. That is where the key is adjoined to the tree by allocating a new node to hold the key and linking it as a left (right) child of the node involved in the last comparison before termination according as the key is less than or equal to (greater than) the content of that last

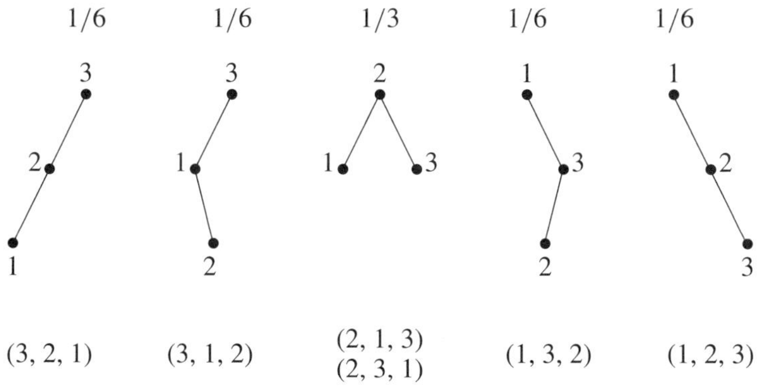

Figure 1.5. Random binary search trees and their probabilities.

node. Figure 1.5 illustrates all binary search trees of order 3 built from a permutation of $\{1, 2, 3\}$. The top row of numbers lists the probabilities, and the bottom row shows the permutations corresponding to each tree.

Several models of randomness may be imposed on binary trees. The model most relevant to the use of binary trees as data structures and as an analysis tool for many sorting algorithms is the *random permutation model*. In this model of randomness, we assume that all $n!$ permutations of $\{1, \ldots, n\}$ are equally likely. These permutations are therefore called random permutations (discussed in some detail in Section 1.10). A tree is then built from a random permutation. Binary search trees are not equally likely under this model of randomness. That is, the uniform model on permutations does not induce a uniform probability distribution on binary search trees. For example, there are six permutations of $\{1, 2, 3\}$ of which four permutations correspond to four linear or zig-zag shaped distinct binary search trees each having probability $\frac{1}{6}$; but the two permutations $(2, 1, 3)$ and $(2, 3, 1)$ give the *same* perfect tree (which has probability $1/3$) as illustrated in Figure 1.5 by the top row of numbers. Unless otherwise explicitly stated, the term *random binary search tree* will refer to a binary search tree grown by the successive insertion of the elements of a random permutation.

The number of descendants of a node in a random binary search tree is instrumental in modeling some sorting algorithms. We shall derive next the distribution of the number of descendants of the node of a given rank in a random binary search tree.

Proposition 1.4 *(Lent and Mahmoud, 1996a). Let $S_j^{(n)}$ be the number of descendants of the node labeled with j in a binary search tree grown from a random permutation of $\{1, \ldots, n\}$. For $k = 0, \ldots, n-2$,*

$$\mathbf{Prob}\{S_j^{(n)} = k\} = \frac{a}{(k+1)(k+2)} + \frac{2b}{(k+1)(k+2)(k+3)},$$

with a and b given by

$$\begin{array}{lll} a = 0, & b = k + 1, & \text{if } k \le \min\{j - 2, n - j - 1\}; \\ a = 1, & b = j - 1, & \text{if } j - 1 \le k \le n - j - 1; \\ a = 1, & b = n - j, & \text{if } n - j \le k \le j - 2; \\ a = 2, & b = n - k - 2, & \text{if } \max\{j - 1, n - j\} \le k \le n - 2. \end{array}$$

Furthermore

$$\mathbf{Prob}\{S_j^{(n)} = n - 1\} = \frac{1}{n}.$$

Proof. The event $\{S_j^{(n)} = n - 1\}$ is the same as the event $\{j$ appears first in the permutation$\}$, which has probability $1/n$.

We find the probabilities for the rest of the feasible values of k, i.e., $k = 0, \ldots, n-2$, by counting the number of permutations giving binary trees in which the node labeled with j has k descendants. We shall construct a permutation Π of $\{1, \ldots, n\}$ that gives rise to a tree with exactly k descendants of j. As in exercise 1.6.7, in such a permutation the subtree rooted at j must contain $k + 1$ consecutive integers, say $x, x + 1, \ldots, x + k$, and j is one of these integers. According to the same exercise, all the descendants must lie after j in Π and whichever of the two numbers $x - 1$, $x + k + 1$ is *feasible* (in the set $\{1, \ldots, n\}$) must occur before j in Π.

Consider first the case when k is small enough and j is not too extremal so that $x - 1$ and $x + k + 1$ are both feasible, that is, $k \le j - 2$ and $k \le n - j - 1$. We can reserve $k + 3$ positions in Π for the integers $x - 1, \ldots, x + k + 1$. These positions can be chosen in $\binom{n}{k+3}$ ways. Place j at the third reserved position, put $x - 1$ and $x + k + 1$ at the first two reserved positions (in two ways), and permute the numbers $x, \ldots, j - 1, j + 1, \ldots, x + k$ over the last k reserved positions in $k!$ ways. To complete the construction of Π, permute the rest of the numbers (not in the set $\{x - 1, \ldots, x + k + 1\}$) over the unreserved positions. So, the number of permutations that give k descendants of j, with a fixed x being the smallest such descendant, is

$$\binom{n}{k+3} \times 1 \times 2 \times k! \times (n - k - 3)! = \frac{2}{(k + 1)(k + 2)(k + 3)} n!. \tag{1.4}$$

The integer x can be any of the numbers $j - k, \ldots, j$; there are $k + 1$ such choices. As we have a total of $n!$ permutations,

$$\mathbf{Prob}\{j \text{ has } k \text{ descendants}\} = \frac{2}{(k + 2)(k + 3)}.$$

We shall consider a similar construction for the case when k exceeds $j - 2$ but is not large enough for $j + k + 1$ to exceed n (that is, $x - 1$ may not be feasible). We consider again a set $\{x, \ldots, x + k + 1\}$ including j to be in the subtree rooted at j. By the condition $j + k + 1 \le n$, the integer $x + k + 1$ is always feasible. The integer $x - 1$ is feasible only if $x \ge 2$. Each of the $j - 1$ choices $x = 2, 3, \ldots, j$ gives us a number of permutations given by (1.4). Here k is large enough for the particular set $\{1, \ldots, k\}$ to be considered in the construction. With $x = 1$, the number $x - 1$ is not feasible (but $x + k + 1$ is). A counting argument similar to that which led to (1.4) gives

$$\binom{n}{k+2} \times 1 \times 1 \times k! \times (n-k-2)! = \frac{1}{(k+1)(k+2)} n!$$

permutations in this case. It follows that

$$\textbf{Prob}\{j \text{ has } k \text{ descendants}\} = \frac{1}{(k+1)(k+2)} + \frac{2(j-1)}{(k+1)(k+2)(k+3)}.$$

The two other cases are when $x - 1$ is feasible but $x + k + 1$ may not be, and the case when both $x - 1$ and $x + k + 1$ may not be feasible. These cases are argued symmetrically. ■

EXERCISES

1.6.1 In terms of levels, where are the leaves of a complete binary tree of size n?

1.6.2 Show that the number of leaves on level $\lfloor \lg(n+1) \rfloor$ in a complete binary tree of size n, is $2(n + 1 - 2^{\lfloor \lg n \rfloor})$.

1.6.3 Show that any binary tree is at least as high as a complete tree of the same size. (Hint: Argue carefully that if the given tree has unsaturated levels, we can transfer nodes from higher levels to fill the holes until it is no longer possible.)

1.6.4 Suppose the $n + 1$ leaves of a binary search tree are numbered $1, \dots, n + 1$ from left to right. Prove that the insertion of a new key into this tree falls at the jth leaf, $j = 1, \dots, n + 1$, if and only if the new key ranks j among all the data items in the new tree.

1.6.5 Let $S_j^{(n)}$ be the number of descendants of the node labeled with j in a random binary search tree constructed from a random permutation of $\{1, \dots, n\}$. Show that

$$\mathbf{E}\left[S_j^{(n)}\right] = H_{n-j+1} + H_j - 2,$$

and

$$\mathbf{E}\left[\left(S_j^{(n)}\right)^2\right] = 2(n+1)H_{n+1} + 2n + (2j - 2n - 7)H_{n-j+2} - (2j+5)H_{j+1} + 10 + \frac{3}{j+1} + \frac{3}{n-j+2}.$$

1.6.6 Let $D_j^{(n)}$ be the depth (number of ancestors) of the node labeled with j in a random binary search tree constructed from a random permutation of

$\{1, \ldots, n\}$. Show that $D_j^{(n)}$ has the same average as the number of descendants of the node labeled j, even though it has a different distribution.

1.6.7 (Devroye, 1991) Let Π_n be a permutation of $\{1, \ldots, n\}$ and let T_n be the tree constructed from it. Prove that the following two statements are equivalent:

(a) the subtree rooted at j consists of nodes labeled with the numbers $d_1 < \cdots < d_{k+1}$.

(b) Let $\mathcal{D}$ be the set of consecutive integers $\{d_1, \ldots, d_{k+1}\} \subseteq \{1, \ldots, n\}$. The set $\mathcal{D}$ includes j and all its other members follow j in Π_n; if $d_1 - 1 \in \{1, \ldots, n\}$, it precedes j in Π_n, and if $d_{k+1} + 1 \in \{1, \ldots, n\}$, it precedes j in Π_n.

1.6.8 Show that the external path length of any binary tree of order n is at least $n \lg n$.

1.6.9 The m-ary tree is a generalization of the binary tree. An m-ary tree is either empty, or has a root and m subtrees distinguished by their position under the root; in turn the m subtrees are m-ary trees. Generalize Propositions 1.1–1.3 to m-ary trees.

1.7 DECISION TREES

A comparison-based sorting algorithm can be modeled by a labeled binary tree, which we shall call the *decision tree*, in the following way. Assume as before that the sorting method is to be applied to the list

$$X_1, X_2, \ldots, X_n.$$

Suppose the sorting method makes the first comparison between X_i and X_j. This first comparison is represented by the root node of a binary tree. The root carries the label $X_i \overset{?}{\le} X_j$. There are only two outcomes for the first comparison: either $X_i \le X_j$ or not. The sorting method may have a different potential routine to follow for each outcome. For example, the algorithm may decide to compare the pair (X_ℓ, X_m), if $X_i \le X_j$, and alternatively check (X_s, X_t), if $X_i > X_j$. The potential continuations are represented by two internal nodes at level 1 in the binary tree (thus, the two nodes are children of the root): a left child labeled $X_\ell \overset{?}{\le} X_m$ (corresponding to the outcome $X_i \le X_j$), and a right child labeled $X_s \overset{?}{\le} X_t$ (corresponding to the alternative outcome $X_i > X_j$). The construction of the tree continues in this fashion—with up to four nodes at level 2, each node represents a possible outcome of the second comparison, then up to eight nodes at level 3, each representing an outcome of the third comparison, and so forth. Each node in this construction represents a comparison of a pair $\{X_i, X_j\}$ and carries this information as its label.

As we climb up a path in the tree, after a node u at level k is reached and the query therein is carried out, $k + 1$ queries have been made with each query of the form of a

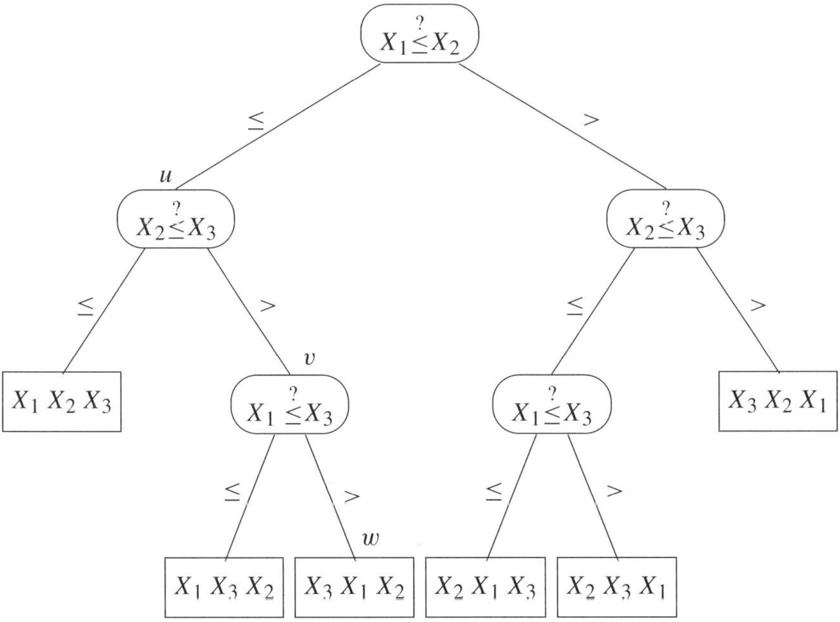

Figure 1.6. The decision tree of an algorithm for sorting three items.

comparison of a particular pair of entries in the list. After these $k + 1$ comparisons, if a certain outcome (say the branch corresponding to less than or equal to (greater than)) provides enough information to sort the list, the sorted list labels an external node attached as a left (right) child of the node u.

As an example, consider the following set of queries to sort a list of three elements X_1, X_2, X_3. We begin with the comparison of X_1 and X_2 (the root node). If $X_1 \leq X_2$, we proceed with comparing X_2 to X_3 (we go left in the tree to level 1 where a node represents the query $X_2 \overset{?}{\leq} X_3$). If the latter query is true, we realize from combining the two queries by the transitivity of the relation $\leq$ that $X_1 \leq X_2 \leq X_3$, completing the sorting task. This is represented by an external node labeled with the ordered list $X_1X_2X_3$; see Figure 1.6. On the other hand, if $X_2 \leq X_3$ turns out to be false, we do not have enough information to make an inference about the relationship between X_1 and X_3; we only know that X_2 is the largest element among the three. One additional question must be asked to determine the relative order of X_1 and X_3; each outcome is sufficient to decide the sorted list; each outcome is represented by a labeled leaf. Mirror-image queries give a symmetric right subtree as shown in Figure 1.6.

An algorithm that works for arbitrary n defines a family of decision trees $D_1, D_2, \ldots$. Each root-to-leaf path in D_n represents the action the algorithm takes on a particular input of size n, that is, the series of queries the algorithm performs on that input. The height of D_n represents the worst-case performance of the algorithm on any input of size n.

Selection algorithms can be modeled by decision trees, too. These trees may be constructed in the same way as sorting decision trees; the only difference is that the leaves of the tree are labeled by the group of order statistics requested from the algorithm. For example, suppose we want to find the maximum among three elements. We may use an algorithm derived from that of Figure 1.6 which sorts three elements. The maximal selection algorithm terminates as soon as enough questions have been asked to identify the maximum and a leaf containing the maximal data element is adjoined to the tree. Figure 1.7 shows the decision tree of this maximal selection algorithm. Note that this is not a worst-case optimal algorithm. By changing the question we ask in the case of a negative answer to our first query, we can obtain the worst-case optimal algorithm of Figure 1.8.

In the decision tree of any complete or partial comparison-based sorting algorithm, every node induces a partial order reflecting how much information has been gathered in climbing up the tree starting at the root. For instance, in any decision tree, the root corresponds to the trivial reflexive partial order $\{X_1 \le X_1, X_2 \le X_2, X_3 \le X_3\}$; no useful information is available prior to climbing up the tree. In the tree of Figure 1.7 the underlying universe is $\{X_1, X_2, X_3\}$. The root's right child corresponds to the partial order $\{X_1 \le X_1, X_2 \le X_2, X_3 \le X_3, X_2 \le X_1\}$, as $X_1 \overset{?}{\le} X_2$ is the only question answered negatively prior to reaching that node; the rightmost leaf labeled with X_1 corresponds to the partial order $\{X_1 \le X_1, X_2 \le X_2, X_3 \le X_3, X_2 \le X_1, X_3 \le X_2, X_3 \le X_1\}$. This partial order is sufficient to claim X_1 as the maximum.

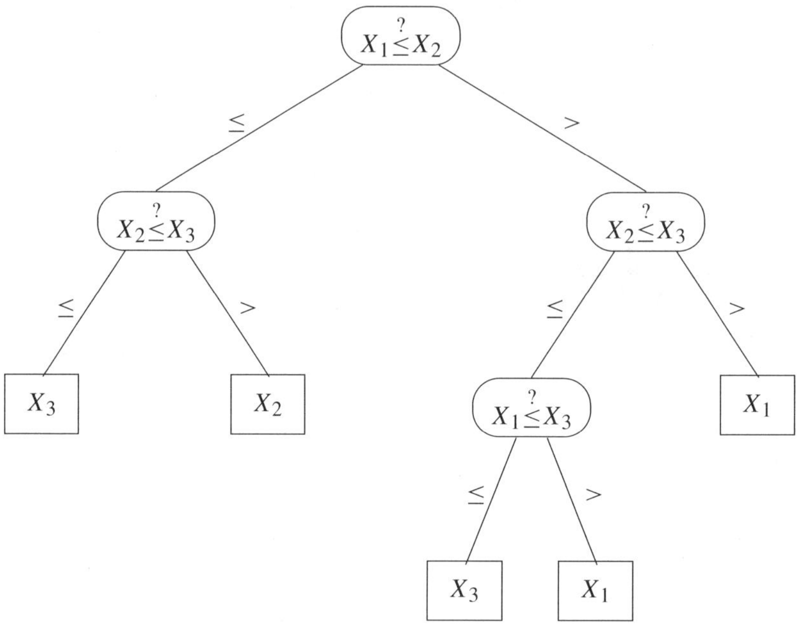

Figure 1.7. The decision tree of an algorithm for finding the maximum of three numbers.

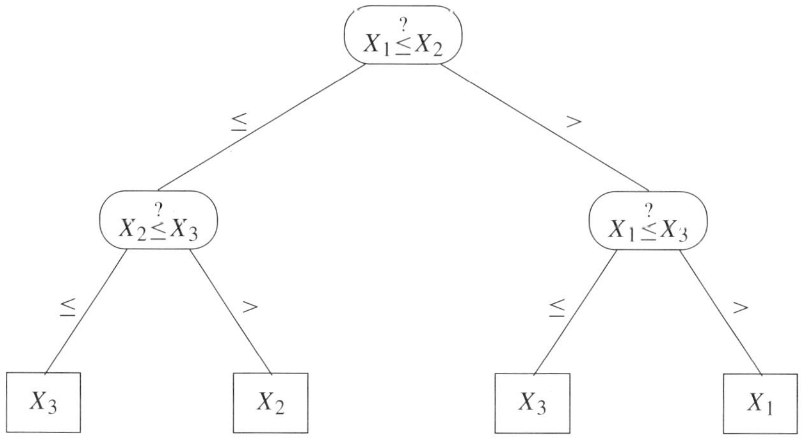

Figure 1.8. The decision tree of an optimal algorithm for finding the maximum of three numbers.

EXERCISES

1.7.1 Devise an iterationless algorithm to find the minimum of four numbers by making only a series of pair comparisons. Illustrate your algorithm by drawing its decision tree.

1.7.2 What partial order corresponds to a leaf of a decision tree of a complete sorting algorithm?

1.7.3 What is the minimum height of the decision tree of any comparison-based sorting algorithm for n keys?

1.7.4 Choose a leaf in the decision tree of Figure 1.6 and determine the partial order induced by every node on the path from the root to that leaf.

1.8 BOUNDS ON SORTING

This book takes a distribution point of view of sorting. It is therefore of interest to identify lower and upper bounds on sorting algorithms so as to put in perspective the range of a random variable associated with a particular sorting algorithm.

In the analysis of algorithms one should be aware of the distinction between the analysis of a specific algorithm for a problem and the complexity of the problem itself, which is a statement about the best algorithm in the whole *class* of possible algorithms for that problem. For example, while we shall consider some $\Theta(n^2)$ comparison-based sorting algorithms, the complexity of sorting by comparisons is $\Theta(n \ln n)$, as we shall demonstrate in the following subsection by the existence of a universal lower bound $\Omega(n \ln n)$, and by the existence of comparison-based sorting algorithms upper bounded by $O(n \ln n)$, with detailed proofs in later chapters.

1.8.1 Lower Bounds on Sorting

Let W_n be the worst-case number of comparisons needed to sort an input sample of size n by some comparison-based sorting algorithm. That is, W_n is the largest number of comparisons the algorithm uses on any input whatever of size n.

We assume our algorithm does not ask any redundant questions, that is, when a node v labeled with the query $a \overset{?}{\le} b$ is reached in the decision tree, this query did not appear before as a label for any node on the path from the root up to v's parent; a and b are not comparable in the partial order induced by v. (Obviously, an algorithm with redundant queries will have more comparisons than one without; the desired lower bound can be established by considering only the class of comparison-based algorithms without redundancy.)

There are $n!$ possible permutations of the ranks of any given sample from an ordered set. The task of sorting is to determine which of these $n!$ permutations of ranks is at hand. Each of the $n!$ permutations is the left-to-right sequence of indexes of the data items contained in a leaf of the decision tree representing the algorithm; there are $n!$ leaves. According to Proposition 1.2, the decision tree has $n! - 1$ internal nodes. The height of the decision tree then represents the number of comparisons of data pairs to determine a permutation corresponding to a leaf at the highest level in the tree; that is, W_n is the height of the decision tree. In view of Proposition 1.1, the height h_k of a tree on k nodes is bounded from below by

$$h_k \ge \lceil \lg(k+1) \rceil.$$

Hence for the decision tree of our sorting algorithm

$$\begin{aligned} W_n &= h_{n!-1} \\ &\ge \lceil \lg n! \rceil \\ &\ge \lg n!. \end{aligned} \tag{1.5}$$

The logarithm can be approximated accurately by Stirling's asymptotic formula:

$$n! = \sqrt{2\pi n}\,\Big(\frac{n}{e}\Big)^n \Big(1 + O\Big(\frac{1}{n}\Big)\Big).$$

Applying the standard calculus inequality

$$\ln(1+x) < x, \qquad \text{for } x > 0,$$

to Stirling's asymptotic form of the factorial, we get

$$\begin{aligned} \lg n! &= n \lg n - \frac{n}{\ln 2} + \frac{1}{2} \lg n + \lg \sqrt{2\pi} + \lg\Big(1 + O\Big(\frac{1}{n}\Big)\Big) \\ &= n \lg n - \frac{n}{\ln 2} + \frac{1}{2} \lg n + \lg \sqrt{2\pi} + O\Big(\frac{1}{n}\Big). \end{aligned} \tag{1.6}$$

Hence, by (1.5),

$$W_n \geq n \lg n - \frac{n}{\ln 2} + \frac{1}{2} \lg n + \lg \sqrt{2\pi} + O\Big(\frac{1}{n}\Big).$$

The algorithm considered in this discussion is an arbitrary comparison-based algorithm. We thus have the following important result.

Theorem 1.1 *Any comparison-based algorithm must make at least* $\lceil \lg n! \rceil = \Omega(n \ln n)$ *comparisons on at least one input instance of size* n.

This lower bound on sorting has been common folklore in the literature of information theory since the 1940s. It was probably first popularized into the literature of sorting in Knuth (1973).

According to Theorem 1.1, each of the many possible comparison-based sorting algorithms will experience $\Omega(n \ln n)$ on some input of size n. There are practical comparison-based algorithms like MERGE SORT that actually make $O(n \ln n)$ comparisons on any input of size n. Thus, we expect that there may exist a worst-case optimal comparison-based algorithm that sorts all inputs of size n in $O(n \ln n)$. According to Theorem 1.1, such an optimal algorithm must also experience $\Omega(n \ln n)$ on some input. A worst-case optimal algorithm achieves the theoretical lower bound and sorts in $\Theta(n \ln n)$ on its worst instance of size n.

It is important to understand that Theorem 1.1 is a statement concerning the complexity of comparison-based algorithms only. It is a statement on the worst behavior of any (including the best possible) comparison-based algorithm, and not on the complexity of sorting at large. Some algorithms like RADIX SORT use some other basic operations on the digits of data to accomplish the task of sorting. Theorem 1.1 *does not* apply to such algorithms. Indeed there are sorting algorithms that are not entirely comparison based and take $o(n \ln n)$ time. For instance, INTERPOLATION SORT has behavior that is asymptotically linear in n.

A lower bound on the average number of comparisons for a comparison-based sorting algorithm is amenable to similar considerations.

Theorem 1.2 *Any comparison-based sorting algorithm requires* $\Omega(n \lg n)$ *comparisons on average.*

Proof. Consider a comparison-based sorting algorithm and its decision tree T_n. Let A_n be the algorithm's average number of comparisons, taken over all equally likely input permutations. As discussed at the beginning of this section, T_n has $n!$ leaves, one corresponding to each input permutation of input ranks. Let the $n!$ leaves of T_n be labeled in any arbitrary manner. The enumeration of leaves is also an enumeration of the permutations, with Π_j the permutation associated with the jth leaf. If the jth leaf is at depth d_j in the decision tree, the permutation Π_j requires d_j comparisons by the algorithm. The algorithm's average number of comparisons is

$$A_n = \sum_{j=1}^{n!} d_j \mathbf{Prob}\{\Pi_j\} = \frac{1}{n!} \sum_{j=1}^{n!} d_j = \frac{1}{n!} X(T_n),$$

where $X(T_n)$ is the external path length of the decision tree. According to Exercise 1.6.8, the decision tree's external path is at least $n! \lg n!$ long. Therefore, $A_n \ge \lg n!$. The proof is completed by the asymptotic development via Stirling's formula as in the transition from (1.5) to (1.6). ∎

1.8.2 Upper Bounds on Sorting

Theorem 1.1 gives a good benchmark for measuring the efficiency of a comparison-based algorithm. If for, example, a comparison-based algorithm always takes $\Theta(n^2)$ on any input of size n, the algorithm is not efficient, and even if the algorithm has other advantages, its use may be recommended only for small or moderate-size files. If a comparison-based algorithm makes $cn \lg n + o(n \ln n)$ on most inputs of size n, for some small constant $c \ge 1$, the algorithm is considered quite good. Fortunately there are several practical comparison-based algorithms that achieve an average behavior of $\Theta(n \lg n)$, with $cn \lg n + o(n \ln n)$ behavior on most inputs, and an occasional worse behavior on some rare inputs. For example, at the expense of space, MERGE SORT *always* makes $n \lg n + O(n)$ comparisons, whereas the in-situ QUICK SORT has an average of $2n \lg n + O(n)$, with $\Theta(n^2)$ behavior on some bad inputs. The existence of algorithms like HEAP SORT and MERGE SORT with $O(n \ln n)$ behavior for all inputs of size n shows that the complexity of comparison-based sorting algorithms is $O(n \ln n)$. This together with the lower bound of $\Omega(n \ln n)$ of Theorem 1.1 shows that the complexity of sorting by comparisons is $\Theta(n \ln n)$. Detailed discussion and proof of the upper bound $O(n \ln n)$ for HEAP SORT and MERGE SORT and other related results will be found in later chapters of this book.

EXERCISES

1.8.1 Is the algorithm discussed in Section 1.7 for sorting three numbers worst-case optimal?

1.8.2 Interpret the following situation in the light of Theorem 1.1. A sorting algorithm first runs a scan on the input to check whether the input of size n is in increasing order or not. The scan obviously takes $n - 1$ comparisons. If the input is in increasing order, the algorithm announces its completion and leaves the input intact claiming it is sorted. If the input is not in increasing order, the algorithm switches to some standard comparison-based algorithm. Does not such an algorithm take $n - 1$ comparisons on some input?

1.8.3 (Demuth, 1956) The lower bound $W_n \geq \lceil \lg n! \rceil$ suggests that there may be a worst-case optimal comparison-based algorithm to sort 5 elements with at least $\lceil \lg 5! \rceil = 7$ comparisons on some input. Design an algorithm that takes at most 7 comparisons for all 5-element inputs. (Hint: After enough comparisons are made to rank any three elements, try to compare the other elements with "middle" data.)

1.8.4 Suppose a sorting algorithm's decision tree contains the query $a \overset{?}{\leq} b$ occurring more than once on some root-to-leaf path. Argue that there is a derived algorithm that runs faster on average.

1.9 BOUNDS ON SELECTION

The problem of selection has its inherent limitations, too. For a long time it was believed that the complexity of the selection of the ith item in a list of size n by comparisons is $\Omega(n \ln n)$. The rationale underlying this belief was based on the view that for general i no good algorithms other than complete sorting then picking the element that lands in position i were known until the 1970s. In the 1970s a number of ingenious designs were considered to show that the selection of the ith item from among n items can take place in $O(n)$ time. The first of these designs was devised by Blum, Floyd, Pratt, Rivest, and Tarjan in 1973. Algorithms that select the ith item were devised to achieve the selection in time bounded from above by $c_i n$, for some constant $c_i > 0$ that depends on i. Since the 1970s research concerning the development of algorithms with lower constants occupied a central position in the attention of computational sciences. For example, the first upper bound of $5.43n$ by Blum, Floyd, Pratt, Rivest, and Tarjan (1973) on the complexity of median finding by comparisons was improved to about $3n$ in Schönhage, Paterson, and Pippenger (1976) and more recently to $2.95n$ in Dor and Zwick (1996a). Though these improved algorithms are very interesting from a theoretical point of view they may not be particularly practical because of the intricacy of their design and their large overhead. We shall discuss a couple of such algorithms in a later chapter.

On a parallel research axis, various techniques of mathematical proof were devised to argue lower bounds and improvements therein. Clearly, partial sorting by comparisons must subject each of the n elements to some comparison, and the complexity of the selection of the ith item from among n keys by a comparison-based algorithm must be at least $\lfloor n/2 \rfloor = \Omega(n)$. After the discovery of $O(n)$ comparison-based selection algorithms for all inputs of size n, the only remaining question in determining the lower bound now concerns the constant in the $\Omega(n)$ lower bound. Such a constant depends on i, the rank of the selected item. Extensive literature on various interesting results and proof techniques concerning specific ranges of i is available. For example, there are exact results for several small fixed values of i stating a lower bound on the exact number of comparisons taken by some algorithm provably known to be worst-case optimal. There are asymptotic results for fixed i, as $n \to \infty$. There are asymptotic results on $i = \lfloor \alpha n \rfloor$, for constant $0 < \alpha < 1$; most of the research along this direction particularly concentrates on lower bounds for the median and quartiles. There are results on uniform bounds for all i. Such uniform bounds may not be the best possible in given ranges of i. There are also algorithms for upper bounds and arguments for lower bounds for the companion problem of finding a subset of keys of given ranks among a list of size n.

It is not our intention to present every known lower bound for every range of i; this may require a separate volume by itself as some of these proofs work by literally considering hundreds of cases. We intend only to present snapshots of some lower bound proof techniques—each snapshot is chosen to illustrate either a simple result or an interesting technique.

1.9.1 Lower Bounds on Selection

Among the proof techniques that were devised for establishing lower bounds on the selection problem are techniques based on the analogy with *tournaments*, *adversary arguments*, and techniques based on decision trees. We shall review each of these techniques in the context of a particular selection problem in this subsection.

The area of bounds on selection continues to thrive as an exciting area of research because many natural questions have not yet been completely answered. The complexity of complete sorting by comparisons, whose leading term is $n \lg n$, is considered to be satisfactorily determined because of the universal lower bound of $\lceil \lg n! \rceil = n \lg n + o(n \ln n)$, and the presence of algorithms (like HEAP SORT) in this class that take $n \lg n + o(n \ln n)$ comparisons on any input of size n. Unlike this state of affairs in complete sorting, in several selection problems there are gaps between the known lower and upper bounds. For example, today the best known lower bound on median selection by comparisons is about $2n$, and the best known upper bound is about $3n$. This gap leaves room for speculation by curious minds. It is possible for the lower bound to come up and for the upper bound to come down. Some believe the correct complexity of median selection by comparisons will turn out to be somewhere in between these two bounds, that is, around $2.5n$.

Selection of the largest number in a set of n numbers is analogous to determining the winner of a tournament of some sporting event (like tennis) with n contenders. The players have inherent strength and the pairings of the tournament (analogous to the comparisons of a ranking algorithm) should decide the best player. Obviously, every player except the winner must lose at least one match against some other player and no two players may lose simultaneously. This forces at least $n-1$ matches. In the language of sorting algorithms, at least $n-1$ comparisons must be made to determine the largest number. The simple-minded algorithm for finding the maximum by a linear scan takes $n - 1$ comparisons, and this cannot be improved as we have just discussed. Hence, the linear scan algorithm is optimal.

Finding the worst case for a comparison-based algorithm for the second best in the tournament has been an enduring challenge in the twentieth century, with history going back to concerns voiced by Lewis Carroll in the late ninteenth century on the organization of tennis tournaments. Today, the standard practice in major tournaments, such as the Wimbledon Tennis Tournament and soccer's World Cup, unequivocally qualifies the best player or team (assuming transitivity of the relation "better than"). However, not enough guarantees are built into the ladder system of these tournaments to guarantee the fairness of the second, third, and lower prizes. For example, if the best tennis player meets the second best at the first match in a single-elimination ladder tournament, the second best is eliminated and the best will march victoriously from match to match all the way to the final to defeat a player ranked lower than second. This other player receives the second prize according to modern organization of tournaments.

The complexity of the number of comparisons (matches) for second number (player) was found in two stages: determining a lower bound (worst number of comparisons for the best algorithm) and then an upper bound (number of comparisons taken by the best algorithm for all inputs). The two bounds coincide giving an exact result!

Finding the second best in a tournament is equivalent to determining a player who can beat all the other players except one, and must lose against that one (the best). In finding the second best, any algorithm must also find the best; as stated in Lemma 1.1. This requires at least $n - 1$ matches. Suppose the algorithm's work for finding the best and second best among n players pairs the eventual champion against $k \le n - 1$ contenders. We shall argue in the next lemma that k is at least $\lceil \lg n \rceil$, providing a key point for the lower bound on the complexity of finding the second largest in a list of n numbers by comparisons.

The proof of the following lemma uses an adversary argument. We shall discuss the essentials of this kind of argument in the next few paragraphs before considering the special adversary argument for the selection of the second largest.

The *adversary* is an opponent of an algorithm who is always determined to challenge the algorithm to do a lot of work. The adversary (also called the *oracle* in some books) does that by contriving a special input that she thinks is particularly bad for this algorithm. For instance, an adversary for a sorting or selection algorithm may try to choose an input that corresponds to the longest path in the decision tree. A perfect adversary is always able to do that for any algorithm. She will then be able to make a correct estimate of the lower bound on all the worst cases of all the algorithms in the class. Adversaries are not always perfect and may not be able to identify the longest possible path in the decision tree of every algorithm in the class. By simpler arguments, less perfect adversaries may still be able to find paths that are nearly the longest in every decision tree given to them. They will thus still be able to identify lower bounds, but perhaps not the sharpest.

It is perhaps best to describe an adversary by an example from pattern recognition games. One popular such pattern recognition game is the two-player game of Master Mind, where a player chooses and hides a pattern consisting of a linear arrangement of four objects. Each player can have one of six colors: Blue (B), Green (G), Purple (P), Red (R), Yellow (Y), or White (W). For example, this player may choose the pattern

$$R \quad Y \quad W \quad W. \tag{1.7}$$

The other player's role is to recognize the given pattern in the least possible number of queries. A query is a guess on the hidden pattern. The first player is obliged by the rules of the game to truthfully answer a query, indicating the number of hits (objects identified by the second player with correct colors and position) as well as the number of near hits (objects identified only by the color, but out of position). For example, if the second player's first guess is the arrangement

$$R \quad B \quad B \quad Y,$$

the first player must answer that the guess hit one position (no obligation to say which position), the leftmost red object in this case, and that there is one near hit (the yellow object). The second player takes the information (one hit and one near hit) and tries to come up with a closer guess, and so on until she eventually gets the hidden pattern (in the official game, there is a limit of 8 tries).

The adversary takes the role of the first player in this game. He actually does not choose any pattern at the beginning but rather keeps a list of possible patterns consistent with his answers up to any point in the game. The adversary is not out there to give his opponent peace of mind. If at some point in the game the second player guesses one of those patterns remaining on the list, the adversary would insist that this was not the pattern, and gives a truthful answer about colors and positions. This answer disqualifies the guessed pattern and possibly a number of other patterns in the remaining list. The adversary, who is truthful in his own way, crosses out all the disqualified patterns. His answer still consistently pertains to any of the patterns remaining on the list. The stubborn adversary admits that a guess is correct only when one pattern consistent with all his answers remains on his list and the second player correctly guesses that last pattern.

Another way to view the adversary's strategy is to think that he actually chooses a pattern, but when guessed correctly by the second player he replaces the pattern by another consistent with all his answers so far. The adversary then claims that the guessed pattern was wrong and gives a correct answer relating to his substituted pattern forcing the unsuspecting second player to go more rounds. (The honest reader is advised not to play Master Mind this way.) At the end of the game, the uninitiated second player has no way of telling that the adversary has been cheating as all the answers given appear consistent with the hidden pattern when finally guessed after sweating.

According to the second view, if the adversary chooses the pattern (1.7) and the second player's first guess is (1.7), the adversary will claim that this pattern is not the correct one, and may give an answer like "no hits or near hits." The adversary then crosses out from the 6^4 possible patterns 3^4 patterns with the colors R, Y, or W appearing in any position, depriving the second player from the pleasure of having a deserved lucky one-shot win. (Of course, this strategy of complete denial right out is not the only way to prolong the game, and may not even be the best choice for the adversary to baffle the second player.)

According to the first view (which has a less cynical outlook toward game playing), if the second player's first guess is (1.7), the adversary has not actually committed to any pattern yet. The adversary sees himself in a perfectly honest position to claim that (1.7) is not the pattern, and may still give an answer like "no hits or near hits." The adversary will remember throughout the rest of the game not to slip into any mistakes like answers forcing Y to appear at some position.

In the context of lower bounds on sorting by comparisons the adversary argument works in the following way. The sorting algorithm is supposed to put in order, via comparisons, a set of data with ranks. An adversary challenges a sorting algorithm by not choosing any particular permutation. The adversary waits to see the progress of the algorithm, always rearranging for himself the input in a way consistent with the partial order observed by the comparisons made up to any stage, and forcing the algorithm to follow longer paths in the tree. In the language of tournaments, the adversary has the capability of choosing the result of any match or controlling who wins in any manner consistent with the results of all previous matches in view of the transitivity of the relation "better than." For example, suppose that in sorting 10 numbers the algorithm reaches a node at depth 13 labeled with the query $X_8 \overset{?}{\le} X_3$ and whose left subtree is a leaf, but the right subtree is not. For an input with $X_8 \le X_3$, the algorithm terminates (in 14 comparisons) on the left leaf, which is labeled by a sorted input with X_8 appearing before X_3 in the permutation. The astute adversary is aware that choosing an input with $X_8 \le X_3$ terminates in 14 steps according to all the comparisons made so far, and also sees the opportunity that choosing an input with $X_8 > X_3$ will force the algorithm to visit the right internal node dictating at least 15 comparisons. The adversary then records for himself that his input is one with $X_8 > X_3$. In future challenges the adversary will remain consistent with this choice and all other true comparisons necessitated by the transitivity of the ordering relation in climbing up the right subtree.

Theorem 1.3 *(Kislitsyn, 1964). Any algorithm for determining the second largest in a list of n numbers by comparisons must exercise at least* $n + \lceil \lg n \rceil - 2$ *comparisons on some input of size n.*

Proof. Consider any algorithm for choosing the second best in a tennis tournament. Lemma 1.1 dictates that such an algorithm finds the best and second best. As simply argued, the algorithm requires at least $n - 1$ comparisons to determine the champion; $n - 1$ players must lose at least once (each against *some* player that may or may not be the champion). Suppose the champion meets and defeats k players on his way to the trophy. Every player who loses to the champion is a candidate for the second position. At least $k - 1$, and possibly all k, players who lost to the champion must lose another match to go out of contention for second position. Thus, the number of matches is at least $(n - 1) + (k - 1) = n + k - 2$.

We shall show that there is an adversary who can always force k to be at least $\lceil \lg n \rceil$ by a specially contrived partial order on the input. In the process of reaching the top, the champion has to prove that he is better than all the other players. The strength of the champion, and any other player is recognized in stages. This process is slowed down if a player's recognized strength grows slowly. Let us define the *strength* of a player at some stage of the game as the number of players known to be at least as bad as himself. At the beginning, in the absence of any prior information every player's recognized strength is 1. The adversary reserves to himself the right of arranging the input (choosing any arbitrary linear total order). His strategy is to choose an order that guarantees slow growth in players' strength. Suppose at some stage the incomparable players A and B, respectively with strength a and b, are about to play a match. The outcome of the match will change the strength: if A wins, A's strength grows to $a + b$, and b's strength remains the same at b with a relative growth of $(a + b)/a$ in A's strength; if B wins, B's strength grows to $a + b$, and a's strength remains the same at a, with a relative growth of $(a + b)/b$ in B's strength. The adversary's strategy then assigns a result to the match according to a and b in whichever manner that minimizes the winner's relative growth in strength. The adversary chooses to let A win if $(a + b)/a \le (a + b)/b$, that is, if $b \le a$. Alternatively, the adversary lets B win if $(a + b)/b \le (a + b)/a$, that is, if $a \le b$. If $a = b$, it make no difference, and an arbitrary choice can be made, say by flipping a coin. The adversary allows the player known to be stronger so far to win again!

If $a \ge b$, then $2a = a + a \ge a + b$. The new strength assigned to A at most doubles; the argument is of course symmetric for player B. After k matches, the champion's k matches will give him strength of at most 2^k. The champion's strength must grow to at least n, for he must prove at least as good as n players (including himself). If k is the minimum number of matches necessary according to this strategy, then

$$2^{k-1} < n \le 2^k.$$

Taking logarithms to base 2, we see that $k = \lceil \lg n \rceil$. ∎

The complexity of determining the median in a list of size n has received tremendous attention in the last quarter of the twentieth century. Intensive research efforts contributed good ideas and proof methods. The problem has a number of variants and we choose the simplest: the problem of selecting the two medians in a list of numbers of size n even.[4] We shall work with even n to simplify the algebra and eliminate the need for ceils and floors. The proof technique involved is based on the decision tree. The proof stands as a good representative for a class of proof methods. Adaptations of the proof can be made to work for other variants of the problem.

We shall show that asymptotically at least $2n$ comparisons are required for finding the two medians among even n numbers. This result means that by the state-of-the-art proof techniques, we can show that for even n, any algorithm for finding the two medians requires at least about $2n$ comparisons on some input of size n. The result was established in the mid-1980s by Bent and John, then was slightly improved in 1996 by Dor and Zwick to $(2 - \varepsilon)n$. The proof given here is that of Dor and Zwick (1996b), who devised a straightforward technique that gives the best known result for

[4]When a sample $\{X_1, \dots, X_n\}$ is of even size, we often define the order statistics $X_{(n/2)}$ and $X_{(n/2+1)}$ as the two medians. Technically speaking, any number in between these two meets the criterion for being a median value. It is customary to define $(X_{(n/2)} + X_{(n/2+1)})/2$ as the sample median for even n.

finding the two medians, after many elaborate and less direct techniques had given weaker results!

For every node in the tree we shall assign a *weight* depending on the partial order (information) determined by the node. In what follows, r will denote the root of the tree.

Suppose we have chosen a weight function $\mathcal{W}$ satisfying the following properties:

P1: $\mathcal{W}(r) \ge 2^{2n-o(n)}$;

P2: $\mathcal{W}(\ell) \le 2^{o(n)}$, for any leaf ℓ;

P3: $\mathcal{W}(u_L) + \mathcal{W}(u_R) \ge \mathcal{W}(u)$, for any internal node u and its left and right children, respectively, u_L and u_R.

Property **P3** implies that each internal node u has a child v such that $\mathcal{W}(v) \ge \frac{1}{2}\mathcal{W}(u)$.

If such a function is at our disposal, then by climbing up the tree along a special path $r = u_0, u_1, \ldots, u_D$, always choosing the child with the larger weight until the leaf u_D (at depth D) is reached, the chosen properties of the function will ensure that

$$\begin{aligned} \frac{\mathcal{W}(u_0)}{\mathcal{W}(u_D)} &= \frac{\mathcal{W}(u_0)}{\mathcal{W}(u_1)} \times \frac{\mathcal{W}(u_1)}{\mathcal{W}(u_2)} \times \cdots \times \frac{\mathcal{W}(u_{D-1})}{\mathcal{W}(u_D)} \\ &\le \underbrace{2 \times 2 \times \cdots \times 2}_{D \text{ times}}, \qquad \text{by Property } \mathbf{P3} \\ &= 2^D. \end{aligned}$$

The point here is that, by the properties of the chosen weight function, there is a leaf in the decision tree at depth

$$D \ge \lg\Big(\frac{\mathcal{W}(u_0)}{\mathcal{W}(u_D)}\Big) \ge \lg\Big(\frac{2^{2n-o(n)}}{2^{o(n)}}\Big) \ge 2n - o(n);$$

there must be a particular arrangement of data ranks that when given to any median pair finding algorithm as input will force the algorithm to perform at least $2n - o(n)$ comparisons.

It only remains to construct a weight function satisfying properties **P1–P3** on the nodes of a decision tree of any median pair finding algorithm. It aids presentation in concise form to introduce additional notation. Denote the partial order associated with a node v by $\mathcal{P}_v$, and denote the set of minima and maxima of P_v in a set A by $\min_v(A)$ and $\max_v(A)$ respectively.

Lemma 1.2 *(Dor and Zwick, 1996b). Let $(S, \preceq)$ be a total order with $|S| = n$ even. Let $A \subseteq S$ with $|A| = n/2$. Let u be a node in the decision tree of an algorithm for determining the two medians of S. Define the node's relative weight function (with respect to A) by*

$$\mathcal{W}_A(u) = \begin{cases} 2^{|\min_v(A)| + |\max_v(A^c)|}, & \text{if } A \text{ is compatible with } P_v; \\ 0, & \text{otherwise.} \end{cases}$$

The absolute weight function associated with a node v is defined as

$$\mathcal{W}(v) = \sum_B \mathcal{W}_B(v),$$

*where the sum is taken over every subset B of size $n/2$. The weight function $\mathcal{W}(v)$ satisfies properties **P1–P3**.*

Proof. At the root r of the decision tree no information is available, except self-reflexivity. So, $P_r = \{(s, s) \ : \ s \in S\}$. Any set B is compatible with this initial partial order, and $\min_r(B) = B$ and $\max_r(B^c) = B^c$. The weight associated with any set B of size $n/2$ at the root is therefore $\mathcal{W}_B(r) = 2^{|B|+|B^c|} = 2^{|S|} = 2^n$. Property **P1** is established by the relation

$$\mathcal{W}(r) = \sum_B \mathcal{W}_B(r) = 2^n \binom{n}{n/2},$$

whose asymptotic equivalent $2^{2n-o(n)}$ follows from Stirling's approximation for binomial coefficients.

When the two medians are determined, the two sets of order statistics $\{X_{(1)}, \dots, X_{(n/2)}\}$ and $\{X_{(n/2+1)}, \dots, X_{(n)}\}$ are determined (Lemma 1.1). Let ℓ be a leaf in the decision tree. Any set compatible with the partial order P_ℓ with an element from $\{X_{(1)}, \dots, X_{(n/2)}\}$ must also contain the two medians and all the higher order statistics. Such a set will be of size larger than $n/2$. The only set of size $n/2$ that is compatible with P_ℓ is the set $B = \{X_{(n/2)+1}, X_{(n/2)+2}, \dots, X_{(n)}\}$. The element $X_{(n/2+1)}$ is known at this time to be smaller than all the elements in B. And so, $|\min_\ell(B)| = 1$. Similarly, $X_{(n/2)}$ is known at this time to be the only maximal element in the complement set $B^c = \{X_{(1)}, \dots, X_{(n/2)}\}$. And so, $|\max_\ell(B^c)| = 1$. Subsequently, the absolute weight $\mathcal{W}(\ell) = 2^{1+1} = 4 \le 2^{o(n)}$; Property **P2** is established.

Let v be a node in the decision tree labeled with the query $a \overset{?}{\le} b$, and let v_L and v_R be its left and right nodes, respectively. Let A be a compatible set of P_v of size $n/2$. We distinguish four cases:

(i) $a, b \in A$.
(ii) $a \in A$, and $b \in A^c$.
(iii) $a \in A^c$, and $b \in A$.
(iv) $a, b \in A^c$.

As assumed in the discussion on lower bounds for sorting, our algorithm does not ask any redundant questions; a query is made only about incomparable elements.

In case (i), the set A is compatible with P_{u_L} and P_{u_R}. At the beginning of the algorithm, no information is available, other than reflexitivity: in the initial partial order associated with the root, every element is in a chain by itself; every element is a minimum and a maximum. All members of A are minimal elements. As we ask questions, more information is revealed and the number of elements in A that are known to be minima will go down. If b is already a member in the set of minima of P_v, a positive answer to the query $a \overset{?}{\leq} b$ removes b from the set of minima in A. If b is not known to be a minimal element, the set of minima does not change. In all cases

$$|\min_{v_L}(A)| \geq |\min_{v}(A)| - 1.$$

By a similar argument, a negative answer may remove a from the minima in A, and

$$|\min_{v_R}(A)| \geq |\min_{v}(A)| - 1.$$

Regardless of the outcome of the query $a \overset{?}{\leq} b$, $|\max_{v_L}(A^c)|$ and $|\max_{v_R}(A^c)|$ will be the same as $|\max_v(A^c)|$ because both a and b are in A (if the query reveals that a or b is a new maximum in A, this new maximum is not in A^c). Thus

$$\begin{aligned}\mathcal{W}_A(v) &= 2^{|\min_v(A)|+|\max_v(A^c)|}\\ &\leq 2^{|\min_{v_L}(A)|+1+|\max_{v_L}(A^c)|}\\ &= 2\mathcal{W}_A(v_L).\end{aligned}$$

By summing the relative weights over every set of size $n/2$, we have the absolute weight $\mathcal{W}(v) \leq 2\mathcal{W}(v_L)$. Similarly, $\mathcal{W}(v) \leq 2\mathcal{W}(v_R)$. Property **P3** is established in case (i).

In case (ii), a set compatible with P_v, for a vertex v, is also compatible with P_{v_R} but not with P_{v_L}. By counting arguments similar to those in case (i), we can quickly find $|\min_{v_R}(A)| = |\min_v(A)|$, and $|\max_{v_R}(A^c)| = |\max_v(A)|$. Thus $\mathcal{W}_A(v_R) = \mathcal{W}_A(v)$, and $\mathcal{W}_A(v_L) = 0$. Summing over all subsets A of size $n/2$ we find $\mathcal{W}(v_L) + \mathcal{W}(v_R) = \mathcal{W}(v)$, establishing Property **P3** in case (ii).

Cases (iii) and (iv) are argued symmetrically. ∎

Order statistics can be found by formulas involving elementary arithmetic functions as in Exercises 1.9.3 and 1.9.4, which constitute other non-comparison-based algorithms. But as one might expect, the number of arithmetic operations involved in finding a particular order statistic grows very fast with the size of the list.

1.9.2 Upper Bounds on Selection

As discussed in some detail in the introductory paragraphs of Section 1.9, the area of bounds on selection is rather vast. Our purpose in this section is to selectively present results indicating that linear upper bounds on selection exist.

To complete the story of choosing the second best, we shall show that Kislitsyn's lower bound (Theorem 1.3) is also an upper bound.

Theorem 1.4 *(Schreier, 1932). There is a comparison-based algorithm that takes at most* $n + \lceil \lg n \rceil - 2$ *comparisons for determining the second largest number in any list of n numbers.*

Proof. Suppose we have n contestants of strengths $a_1, a_2, \ldots, a_n$ in a knock out tournament. (We shall refer to the players by their strength.) Construct a ladder for the best player in the following way: Consider the n contestants as terminal nodes in a tree (of $n-1$ internal nodes and height $\lceil \lg n \rceil$). In the first round match consecutive pairs to come up with $\lfloor n/2 \rfloor$ winners. In the second round match pairs of winners as well as those who moved to this round without playing, and so on till the champion of the tournament is identified. Each internal node represents a match (a comparison) between two players and is labeled with the strength of the winner among the two. One of the internal nodes must represent a match between the best and second best (this is how the second best got knocked out).

Suppose a_j is the champion. Now recompute the labels on the path joining the root to the terminal node a_j. Discard a_j: The player who met a_j in a_j's first match now gets a free ride, she goes up to her parent node (which was previously labeled a_j). This player is paired against the player who competed against a_j in a_j's second match. The winner goes up to a node at distance two from a_j (and was previously labeled a_j), and so forth. At some point a winner along the path will meet the second strongest, and will lose to him. The second strongest will go up, and will continue winning all the way to the root (because the strongest player has been eliminated). At the end of this recalculation along the path, the second strongest contestant will be identified. The number of relabeling operations (comparisons) along that path is at most one less than the height of the tree.

This algorithm finds the champion in $n - 1$ comparisons, then finds the second-best player in at most $\lceil \lg n \rceil - 1$ comparisons. ∎

Figures 1.9 and 1.10 illustrate the proof of Theorem 1.4 on the data set

21 30 16 24 19 17 15

which say are the measures of strength of participants in some scoring scheme. These numbers are shown as the terminal nodes of a binary tree in Figure 1.9. The strongest player's strength is 30, and appears at the root of the tree. The darkened edges show the root to the terminal node containing 30. In the second stage of Schreier's algorithm, the strongest player is removed as in the tree of Figure 1.10 and the labels along that path are recomputed. In this instance the player with strength 21 is not paired in the first round against any one (preferential treatment!) then meets the player with strength 24 (second strongest) in the second round; the second strongest wins the match and all subsequent matches to appear at the root of the tree.

The constructive proof of Theorem 1.4 exhibits an algorithm that selects the second largest in a file of n numbers using no more comparisons than the amount stated in the theorem for whatever input arrangement of these n numbers. The method can be generalized to efficiently find the third best, fourth best, etc. Finding all order

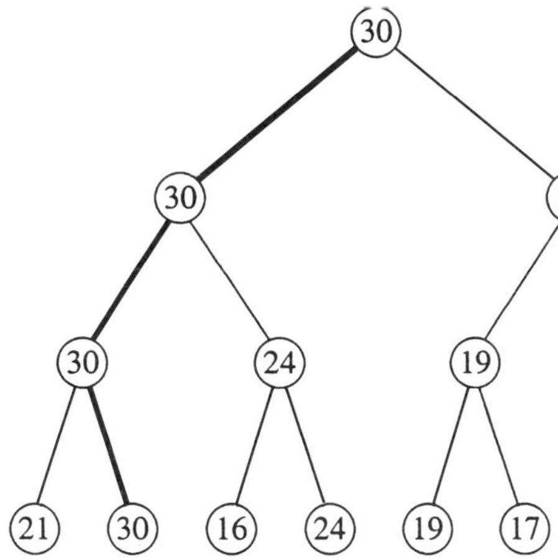

Figure 1.9. The champion selection round in Schreier's algorithm.

statistics by this method is the genesis of a sorting method called HEAP SORT, which will be discussed in Chapter 9.

Kislitsyn's lower bound (Theorem 1.3) together with Schreier's upper bound (Theorem 1.4) tell us that there is no gap between the worst case of some known algorithm and the best known lower bound. The complexity of the selection of second largest is exactly $n + \lceil \lg n \rceil - 2$. This is a selection instance where the chapter has been closed in the sense of worst-case optimality.

Not every selection problem enjoys this status, including instances of prime practical importance like that of median selection where, as discussed in Section 1.9.1, there is a gap between the best algorithm for the worst case and the best known

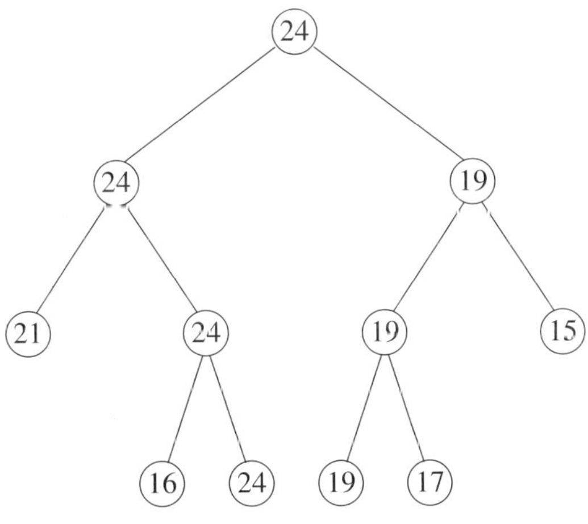

Figure 1.10. The second-best selection round in Schreier's algorithm.

theoretical lower bound. The gap is not in the correct order of the complexity, but is only in the constant associated with it. Of course, the correct order of complexity is an important issue by itself. The linearity of median selection is demonstrated constructively by an algorithm discussed in Chapter 12.

EXERCISES

1.9.1 Let $W_k(n)$ be the worst-case optimal number of comparisons required to find the set of first k order statistics in any given input of n numbers; any algorithm whatever for determining the set of k order statistics in any given input of n numbers will require at least $W_k(n)$ comparisons. Use an adversary argument to show that $W_k(n) \ge W_k(n-1) + 1$, for all $1 \le k \le n$.

1.9.2 Explain why the proof of the lower bound for the selection of two medians does not work for the selection of a single median.

1.9.3 (Meyer, 1969) Given a sample of two numbers X_1 and X_2 prove that

$$\min\{X_1, X_2\} = \frac{X_1 + X_2}{2} - \frac{|X_1 - X_2|}{2}.$$

Derive a similar formula for the maximum.

1.9.4 (Meyer, 1969) Given a sample of three numbers X_1, X_2, and X_3 prove that

$$\min\{X_1, X_2, X_3\} = \frac{1}{4}\big(X_1 + 2X_2 + X_3 - |X_1 - X_2| - |X_2 - X_3| \\ - \big|X_1 - X_3 - |X_1 - X_2| + |X_2 - X_3|\big|\big).$$

Derive a similar formula for the median and the maximum.

1.9.5 Show constructively that it is possible to find the median of five elements using at most six comparisons.

1.10 RANDOM PERMUTATIONS

The majority of sorting algorithms are comparison based. They accomplish the sorting task by making decisions according to the relative smallness of keys. For example, while sorting the given list of numbers $X_1, \ldots, X_n$, a comparison-based algorithm may at some step compare X_i with X_j, and if $X_i \le X_j$, the algorithm will follow a certain path; but if $X_i > X_j$ the algorithm will proceed with an alternate path. We see that the algorithm will, for example, take the same path in the two cases $X_i = 10$ and $X_j = 19$, or $X_i = 30$ and $X_j = 34$. In both cases $X_i \le X_j$ and the algorithm takes the same action in response to the comparison of X_i against X_j. The only data aspect that matters in the outcome of comparing the two entries X_i and

X_j is their relative ranking. For any n distinct keys chosen from an ordered set there corresponds a ranking from smallest to largest, that is, a permutation of $\{1, \ldots, n\}$. This permutation corresponds to some leaf in the decision tree. This permutation is the sequence of indexes (read from left to right) of the data labeling the leaf. This permutation determines exactly the path in the tree from the root to this leaf, or, in other words, determines exactly the sequence of comparisons the algorithm will follow. We may therefore assimilate the actual n keys by that permutation of $\{1, \ldots, n\}$ for which the algorithm will exhibit exactly the same behavior. To unify the analysis we shall always consider permutations of $\{1, \ldots, n\}$. To any actual input of n distinct keys from whatever data set, the corresponding permutation will decide the number of comparisons done on a corresponding arrangement of integers. The timing of comparisons among the actual data keys is only a scaling of the number of comparisons done on the integers of the corresponding permutation.

If the data are "random" according to some probability model, the corresponding permutations will have an induced probability distribution. In practice, many data sets consist of real numbers taken from some continuous distribution. We shall prove that this induces a uniform distribution on permutations. This statement will shortly be made precise by some definitions and propositions. Because of this, we shall take the uniform distribution on permutations, whereby all permutations of $\{1, \ldots, n\}$ are equally likely (each permutation occurs with probability $1/n!$), as the gold standard for the analysis of comparison-based algorithms. The unqualified term *random permutation* will refer to this particular uniform model. Most analyses in this book are conducted under the random permutation model. If any other probability model of randomness on permutations is considered, it will be clearly and explicitly qualified by a description or a special terminology.

When the data contain no duplicates, the speed of a comparison-based sorting algorithm can be made insensitive to the distribution of the data. After all, the data from whatever distribution can be made to appear as if they came from the random permutation model. An example of data with no duplicates is a sample (drawn without replacement) from a discrete distribution. The data can be first subjected to an initial randomizing shuffle so that their ranks become a random permutation. An initial stage to randomize the data may serve as a way of guaranteeing the same uniform average speed of a comparison-based sorting algorithm over all possible distributions of lists with no duplicates. Our forthcoming analyses thus also apply to comparison-based sorting algorithms that perform this randomization prior to sorting, for data with no duplicates even if the ranks do not follow the random permutation probability model.

Let us introduce the working notation. Suppose $X_1, \ldots, X_n$ are distinct real numbers. If X_i is the jth smallest among $X_1, \ldots, X_n$ we say that the *absolute rank* of X_i is j. The absolute rank of X_i is denoted by $Absrank(X_i)$. The range of $Absrank(X_i)$ is $\{1, \ldots, n\}$. The *sequential rank* of X_i, denoted by $Seqrank(X_i)$, is its rank among the keys $X_1, \ldots, X_i$ only. That is, $Seqrank(X_i) = j$, if X_i is the jth smallest among $X_1, \ldots, X_i$. So, the range of $Seqrank(X_i)$ is $\{1, \ldots, i\}$, for each $i \in \{1, \ldots, n\}$.

We assume throughout that the elements of our sample are distinct, as duplicate keys may only occur with probability 0 when the keys are taken from a continuous distribution. That is, the computation of any probability is not altered by ignoring the null set (set with probability zero) of sample space outcomes in which two or more keys are equal.

As we prove in the next proposition, the sequence of sequential ranks is like a "signature" to a permutation. It uniquely characterizes the permutation, and vice versa.

Proposition 1.5 *There is a one-to-one correspondence between permutations of* $\{1, \ldots, n\}$ *and the* n*-long feasible sequences of Sequential ranks.*

Proof. Let n be fixed. Every one of the $n!$ permutations of $\{1, \ldots, n\}$ obviously has exactly one corresponding sequence of sequential ranks. So, we only need to show that there corresponds at most one permutation to any feasible sequence of ranks of length n. We do an induction on n. Suppose that $\Pi = (\pi_1, \ldots, \pi_n)$ and $\Xi = (\xi_1, \ldots, \xi_n)$ are two *different* permutations with the same sequence of sequential ranks $s_1, \ldots, s_n$. Suppose $\pi_i = n = \xi_j$. Thus $s_i = i$ and $s_j = j$. We have three cases:

(a) If $i < j$, then n appears to the left of position j in Π, and π_j is not the largest among $\pi_1, \ldots, \pi_j$, that is, $s_j = \mathit{Seqrank}(\pi_j) < j$, a contradiction.

(b) If $i > j$, this case is symmetric to case (a), only the roles of i and j are interchanged, and Π and Ξ are interchanged.

(c) If $i = j$, drop π_i from Π and ξ_i from Ξ. What is left behind are two permutations of $\{1, \ldots, n-1\}$, each having sequential ranks $s_1, \ldots, s_{i-1}, s_{i+1}, \ldots, s_n$. By the induction hypothesis, these two permutations must be the same, and restoring n at position i to recreate the two original permutations will give two *identical* permutations, a contradiction. ■

Proposition 1.6 *A permutation of* $\{1, \ldots, n\}$ *is a random permutation if and only if*

(a) Each sequential rank is uniformly distributed over its range.

(b) The sequential ranks are mutually independent.

Proof. To prove the "if" part, let Π be a permutation of $\{1, \ldots, n\}$ with corresponding sequential ranks $s_1, \ldots, s_n$, and suppose Π is constructed such that:

(a) For each i,

$$\mathit{Seqrank}(\pi_i) \overset{D}{=} \text{UNIFORM}[1\,..\,i].$$

(b) $\mathit{Seqrank}(\pi_1), \ldots, \mathit{Seqrank}(\pi_n)$ are mutually independent.

A permutation $(\pi_1, \ldots, \pi_n)$ uniquely determines its sequence of sequential ranks $(s_1, \ldots, s_n)$, by Proposition 1.5. By the independence and uniform distribution assumptions (a) and (b):

$$\begin{aligned}
&\mathbf{Prob}\big\{\Pi = (\pi_1, \ldots, \pi_n)\big\}\\
&\quad = \mathbf{Prob}\big\{\Pi(1) = \pi_1, \ldots, \Pi(n) = \pi_n\big\}\\
&\quad = \mathbf{Prob}\big\{\mathit{Seqrank}\big(\Pi(1)\big) = s_1, \ldots, \mathit{Seqrank}\big(\Pi(n)\big) = s_n\big\}\\
&\quad = \mathbf{Prob}\{\mathit{Seqrank}\big(\Pi(1)\big) = s_1\} \ldots \mathbf{Prob}\{\mathit{Seqrank}\big(\Pi(n)\big) = s_n\}\\
&\quad = 1 \times \frac{1}{2} \times \frac{1}{3} \times \cdots \times \frac{1}{n}\\
&\quad = \frac{1}{n!},
\end{aligned}$$

and all permutations are equally likely.

To prove the "only if" part, let us assume that $\Pi = (\pi_1, \ldots, \pi_n)$ is a random permutation. We use the following counting argument to show that $\mathbf{Prob}\{\mathit{Seqrank}(\pi_i) = j\} = 1/i$, for each $1 \le j \le i \in \{1, \ldots, n\}$. Choose and fix $1 \le j \le i \le n$. To have $\mathit{Seqrank}(\pi_i) = j$ when $\pi_i = k$, exactly $j - 1$ numbers less than k must appear among the first $i - 1$ positions of Π. These numbers can be chosen in $\binom{k-1}{j-1}$ ways, then $j-1$ positions are chosen for them from among the first $i-1$ positions (in $\binom{i-1}{j-1}$ ways); these numbers can be permuted in $(j - 1)!$ ways over the chosen positions. So, the number of ways to choose and insert these $j - 1$ numbers to the left of k is $\binom{k-1}{j-1}\binom{i-1}{j-1}(j - 1)!$. To fill the rest of the first $i - 1$ positions, the rest of the numbers must come from the set $\{k + 1, \ldots, n\}$ and then be permuted over the remaining $i - j$ positions. This can be done in $\binom{n-k}{i-j}(i - j)!$. The tail of the permutation can now be constructed by permuting the remaining $n - i$ numbers in the $n - i$ positions following π_i. Hence, the number of permutations with $\mathit{Seqrank}(\pi_i) = j$ and $\pi_i = k$ is

$$\left[\binom{k-1}{j-1}\binom{i-1}{j-1}(j-1)!\right]\left[\binom{n-k}{i-j}(i-j)!\right](n-i)!.$$

The probability of the event $\mathit{Seqrank}(\pi_i) = j$ and $\pi_i = k$ is this number divided by $n!$, as the random permutations are equally likely. So,

$$\begin{aligned}
\mathbf{Prob}\big\{\mathit{Seqrank}(\pi_i) = j\big\} &= \sum_{k=j}^{n} \mathbf{Prob}\big\{\mathit{Seqrank}(\pi_i) = j, \pi_i = k\big\}\\
&= \frac{1}{n!}\binom{i-1}{j-1}(j-1)!\,(i-j)!\,(n-i)!\\
&\quad \times \sum_{k=j}^{n}\binom{k-1}{j-1}\binom{n-k}{i-j}
\end{aligned}$$

$$= \frac{(i-1)!}{(j-1)!\,(i-j)!} \times \frac{(j-1)!\,(i-j)!\,(n-i)!}{n!}$$
$$\times \sum_{k=j}^{n} \binom{k-1}{j-1}\binom{n-k}{i-j}$$
$$= \frac{1}{i\binom{n}{i}} \sum_{k=j}^{n} \binom{k-1}{j-1}\binom{n-k}{i-j}.$$

The last sum is $\binom{n}{i}$, as can be seen by counting the number of ways of choosing i integers from $\{1, \dots, n\}$ by first choosing $k = j, \dots, n$, then forming a subset involving $j - 1$ numbers less than k and $i - j$ numbers greater than k, for some fixed $1 \le j \le i$. After the cancelation of binomial coefficients,

$$\textbf{Prob}\{Seqrank(\pi_i) = j\} = \frac{1}{i},$$

as required for (a).

By Proposition 1.5, a feasible sequence $s_1, \dots, s_n$ of sequential ranks uniquely determines a permutation Ξ. So,

$$\begin{aligned}\textbf{Prob}\Big\{Seqrank\big(\Pi(1)\big) = s_1, \dots, Seqrank\big(\Pi(n)\big) = s_n\Big\} &= \textbf{Prob}\big\{\Pi = \Xi\big\} \\ &= \frac{1}{n!}. \qquad (1.8)\end{aligned}$$

Having proved (a), we can write

$$\begin{aligned}\textbf{Prob}\big\{Seqrank(\pi_1) = s_1\big\} \dots \textbf{Prob}\big\{Seqrank(\pi_n) = s_n\big\} &= 1 \times \frac{1}{2} \times \frac{1}{3} \times \dots \times \frac{1}{n} \\ &= \frac{1}{n!}. \qquad (1.9)\end{aligned}$$

Equating (1.8) and (1.9), we see that for every feasible sequence $s_1, \dots, s_n$ of sequential ranks

$$\begin{aligned}&\textbf{Prob}\big\{Seqrank(\pi_1) = s_1, \dots, Seqrank(\pi_n) = s_n\big\} \\ &\qquad = \textbf{Prob}\big\{Seqrank(\pi_1) = s_1\} \dots \textbf{Prob}\{Seqrank(\pi_n) = s_n\big\},\end{aligned}$$

asserting the independence as required for (b). ■

Suppose X_1, X_2 is a random sample taken from a continuous distribution. By symmetry, the events

$$\{X_1 < X_2\} \qquad \text{and} \qquad \{X_2 < X_1\}$$

are equally likely, and each has probability $\frac{1}{2}$ (the complement event of $\{X_1 < X_2\} \cup \{X_1 > X_2\}$, that is, the event $\{X_1 = X_2\}$, occurs with probability 0). Similarly, for a larger sample $X_1, \ldots, X_n$, from a continuous distribution, the $n!$ events

$$\{X_{\pi_1} < X_{\pi_2} < \ldots < X_{\pi_n}\},$$

for any permutation $(\pi_1, \ldots, \pi_n)$ of $\{1, \ldots, n\}$, are equally likely, each occurring with probability $\frac{1}{n!}$ (and again all the events with ties have probability 0). Whence, *Absrank*$(X_1), \ldots,$ *Absrank*(X_n) is a random permutation and enjoys all the properties of that model of randomness. Notably, the sequential ranks *Seqrank*$(X_1), \ldots,$ *Seqrank*(X_n) are independent and each is uniformly distributed on its range:

$$Seqrank(X_i) \stackrel{D}{=} \text{UNIFORM}[1\,..\,i],$$

as ascertained by Proposition 1.6.

1.10.1 Records

Suppose $X_i, i = 1, \ldots, n$, is a sequence of keys sampled from a continuous distribution. Let π_i be the absolute rank of X_i. With probability 1, the sequence $(\pi_1, \ldots, \pi_n)$ is a random permutation. A value X_j in the sequence is a *record* when $X_j > X_i$ for all $i < j$, or probabilistically equivalently π_j is a record in the random permutation $\Pi_n = (\pi_1, \ldots, \pi_n)$ of $\{1, \ldots, n\}$, when $\pi_j > \pi_i$ for each $i < j$. For example, the permutation

$$\begin{array}{cccccccc} 7 & 6 & 8 & 3 & 1 & 2 & 5 & 4 \\ \uparrow & & \uparrow & & & & & \end{array}$$

has the two records indicated by the upward arrows. The goal of a sorting algorithm is to increase the number of records to the limit.

Let I_j be the indicator of the event that the jth key is a record (I_j assumes the value 1 if the jth key is a record, and assumes the value 0 otherwise). Proposition 1.6 asserts that these indicators are independent and the jth key is a record (the jth indicator is 1) with probability $1/j$. Mean and variance calculations for I_i are immediate. We have

$$\mathbf{E}[I_j] = 0 \times \frac{j-1}{j} + 1 \times \frac{1}{j} = \frac{1}{j},$$

and

$$\mathbf{Var}[I_j] = \mathbf{E}[I_j^2] - \mathbf{E}^2[I_j] = \mathbf{E}[I_j] - \mathbf{E}^2[I_j] = \frac{1}{j} - \frac{1}{j^2}.$$

The probability generating function $g_j(z)$ of the jth indicator is

$$\sum_{k=0}^{\infty} z^k \mathbf{Prob}\{I_j = k\} = \frac{j-1}{j} + \frac{z}{j}.$$

Let R_n be the number of records in the permutation. Then R_n is

$$R_n = I_1 + I_2 + \cdots + I_n.$$

Accordingly we have

$$\begin{aligned}
\mathbf{E}[R_n] &= \mathbf{E}[I_1] + \mathbf{E}[I_2] + \cdots + \mathbf{E}[I_n] \\
&= \sum_{j=1}^{n} \frac{1}{j} \\
&= H_n \qquad (1.10) \\
&\sim \ln n. \qquad (1.11)
\end{aligned}$$

It also follows from the independence of the indicators that[5]

$$\begin{aligned}
\mathbf{Var}[R_n] &= \mathbf{Var}[I_1] + \mathbf{Var}[I_2] + \cdots + \mathbf{Var}[I_n] \\
&= \sum_{j=1}^{n} \Big(\frac{1}{j} - \frac{1}{j^2}\Big) \\
&= H_n - H_n^{(2)} \\
&\sim \ln n. \qquad (1.12)
\end{aligned}$$

Let $r_n(z)$ be the probability generating function of R_n. As R_n has a representation as a sum of independent indicators, its probability generating function is the product of the probability generating function of these indicators:

$$\begin{aligned}
r_n(z) &= g_1(z) \dots g_n(z) \\
&= \prod_{j=1}^{n} \frac{z+j-1}{j} \\
&\stackrel{\text{def}}{=} \frac{\langle z \rangle_n}{n!}. \qquad (1.13)
\end{aligned}$$

As a generating function, $\langle x \rangle_n = \sum_{k=1}^{n} {n \brack k} x^k$ generates the sequence ${n \brack k}$ of signless Stirling numbers of order n. Extracting coefficients gives us the exact probability distribution of the number of records:

$$\mathbf{Prob}\{R_n = k\} = \frac{{n \brack k}}{n!}.$$

[5]See the appendix for the definition of harmonic numbers.

The exact probability structure of R_n is simple enough to admit a transparent passage to a normal limiting distribution by elementary calculation without resorting to a general condition like Lindeberg's, for example. We will show next that a suitably normed version of R_n has a limit distribution by directly taking the limit of the moment generating function of that normed version by appealing only to standard properties of Stirling numbers. We shall use the asymptotic mean (cf. (1.11)) for centering and the asymptotic standard deviation (cf. (1.12)) as a scaling factor. Let us consider

$$\frac{R_n - \ln n}{\sqrt{\ln n}}.$$

In 1944 Goncharov first discovered a result for this normalized number of records in the context of normality of the number of cycles. Bijections between cycles and records were later reported by Foata in 1965. We shall see in our forthcoming discussion on cycles of permutations that the number of cycles is distributed like the number of records. The direct proof below for the asymptotic normality of records runs along the main lines of Mahmoud (1991).

Theorem 1.5 *(Goncharov, 1944). The number of records R_n in a random permutation of the set $\{1, \dots, n\}$ has the Gaussian tendency:*

$$\frac{R_n - \ln n}{\sqrt{\ln n}} \xrightarrow{\mathcal{D}} \mathcal{N}(0, 1).$$

Proof. Let $\phi_n(t)$ be the moment generating function of $(R_n - \ln n)/\sqrt{\ln n}$. Condition on R_n to obtain

$$\begin{aligned}
\phi_n(t) &= \mathbf{E}[e^{(R_n - \ln n)t/\sqrt{\ln n}}] \\
&= \sum_{k=1}^{n} e^{(k-\ln n)t/\sqrt{\ln n}} \, \mathbf{Prob}\{R_n = k\} \\
&= e^{-\sqrt{\ln n}\, t} \sum_{k=1}^{n} \left(e^{t/\sqrt{\ln n}}\right)^k \frac{\left[{n \atop k}\right]}{n!} \\
&= e^{-\sqrt{\ln n}\, t} \frac{e^{t/\sqrt{\ln n}}(e^{t/\sqrt{\ln n}} + 1) \dots (e^{t/\sqrt{\ln n}} + n - 1)}{n!} \\
&= e^{-\sqrt{\ln n}\, t} \frac{\Gamma(e^{t/\sqrt{\ln n}} + n)}{\Gamma(n + 1)\, \Gamma(e^{t/\sqrt{\ln n}})}.
\end{aligned}$$

Using Stirling' approximation of the gamma function (as $n \to \infty$)

$$\begin{aligned}\phi_n(t) &\sim \frac{e^{-\sqrt{\ln n}\,t}}{\Gamma(e^{t/\sqrt{\ln n}})} n^{\exp(t/\sqrt{\ln n})-1} \\ &= \frac{e^{-\sqrt{\ln n}\,t}}{\Gamma(e^{t/\sqrt{\ln n}})} \exp\bigl\{(e^{t/\sqrt{\ln n}} - 1)\ln n\bigr\} \\ &= \frac{e^{-\sqrt{\ln n}\,t}}{\Gamma(e^{t/\sqrt{\ln n}})} \exp\biggl\{\biggl[\biggl(1 + \frac{t}{\sqrt{\ln n}} + \frac{t^2}{2\ln n} \\ &\qquad + O\Bigl(\frac{t^3}{\ln^{3/2} n}\Bigr)\biggr) - 1\biggr]\ln n\biggr\}.\end{aligned}$$

Simplifying and then taking the limit as $n \to \infty$ for fixed t yields the simple expression $e^{t^2/2}$, which is the moment generating function of the standard normal distribution. Convergence of moment generating functions implies convergence in distribution and the proposition follows. ■

1.10.2 Inversions

In a permutation $\Pi_n = (\pi_1, \dots, \pi_n)$ of $\{1, \dots, n\}$, a pair (π_i, π_j) is an *inversion* if $\pi_i > \pi_j$ with $i < j$, that is, when a bigger element precedes a smaller one in the permutation. The permutation

$$7 \quad 6 \quad 8 \quad 3 \quad 1 \quad 2 \quad 5 \quad 4$$

has 19 inversions.

Clearly, an algorithm to sort Π_n (or equivalently a data set with ranks $\pi_1, \dots, \pi_n$) must remove all inversions therein and it is evident that analysis of inversions in a permutation will play a role in the analysis of some sorting algorithms.

Let V_k be the number of inversions caused by π_k, that is, the number of items greater than π_k and appearing in positions $1, \dots, k-1$ in the permutation. So, $V_k = j$ if and only if π_k is less than j of the first $k-1$ elements of the permutation. In other words, $V_k = k - Seqrank(\pi_k)$. Proposition 1.6 then states that

$$V_k \overset{D}{=} \text{UNIFORM}[0\,..\,k-1],$$

and $V_1, \dots, V_n$ are independent. The uniform distribution of V_k admits simple calculation of its mean and variance:

$$\mathbf{E}[V_k] = \sum_{j=0}^{k-1} j\,\mathbf{Prob}\{V_k = j\} = \sum_{j=0}^{k-1} \frac{j}{k} = \frac{k(k-1)}{2k} = \frac{k-1}{2},$$

and

$$\begin{aligned}\mathbf{Var}[V_k] &= \Bigl(\sum_{j=0}^{k-1} \frac{j^2}{k}\Bigr) - \Bigl(\frac{k-1}{2}\Bigr)^2\\ &= \frac{k(k-1)(2k-1)}{6k} - \frac{(k-1)^2}{4}\\ &= \frac{k^2-1}{12}.\end{aligned}$$

Let Y_n be the total number of inversions in a random permutation. Then

$$Y_n = V_1 + \cdots + V_n.$$

Taking expectations,

$$\begin{aligned}\mathbf{E}[Y_n] &= \mathbf{E}[V_1] + \cdots + \mathbf{E}[V_k]\\ &= \sum_{k=1}^{n} \frac{k-1}{2}\\ &= \frac{n(n-1)}{4} \qquad (1.14)\\ &\sim \frac{n^2}{4}.\end{aligned}$$

Taking the variance and using the independence of the V_k's we obtain

$$\begin{aligned}\mathbf{Var}[Y_n] &= \mathbf{Var}[V_1] + \cdots + \mathbf{Var}[V_n]\\ &= \sum_{k=1}^{n} \frac{k^2-1}{12}\\ &= \frac{1}{72} n(n-1)(2n+5) \qquad (1.15)\\ &\sim \frac{n^3}{36}.\end{aligned}$$

Proposition 1.7 *The number of inversions, Y_n, in a random permutation of $\{1, \ldots, n\}$ has the Gaussian tendency*

$$\frac{Y_n - n^2/4}{n^{3/2}} \xrightarrow{\mathcal{D}} \mathcal{N}\Bigl(0, \frac{1}{36}\Bigr).$$

Proof. A standard verification of Lindeberg's condition yields the result as follows. Consider the centered random variables $\tilde{V}_k = V_k - \mathbf{E}[V_k]$, where $V_k = \text{UNIFORM}[0\,..\,k-1]$ is the number of inversions caused by the kth item. We have $\sum_{k=1}^{n} \tilde{V}_k = Y_n - \mathbf{E}[Y_n]$. Let $s_n^2 = \sum_{k=1}^{n} \mathbf{Var}[\tilde{V}_k] = \sum_{k=1}^{n} \mathbf{Var}[V_k]$, which is given by (1.15). For any fixed $\varepsilon > 0$ we want to verify

$$\lim_{n\to\infty} \frac{1}{s_n^2} \sum_{k=1}^{n} \int_{\{|\tilde{V}_k| > \varepsilon s_n\}} \tilde{V}_k^2 \, dP = 0,$$

where P is the uniform probability measure on the sample space of permutations. This Lebesgue–Stieltjes integration in question is simply a summation because our case is discrete. The last limit is equal to

$$\lim_{n\to\infty} \frac{1}{s_n^2} \sum_{k=1}^{n} \sum_{|j - \mathbf{E}[V_k]| > \varepsilon s_n} \bigl(j - \mathbf{E}[V_k]\bigr)^2 \mathbf{Prob}\{V_k = j\}.$$

However, for large n, the set $\{V_k = j : |j - \mathbf{E}[V_k]| > \varepsilon s_n\}$ is empty because s_n^2 grows cubically, whereas $\tilde{V}_k < k \le n$. For large n, the inner summation is zero. ■

1.10.3 Cycles

Let $\Pi_n = (\pi_1, \pi_2, \dots, \pi_n)$ be a permutation of $\{1, 2, \dots, n\}$. We call a chain of indexes $i_1, i_2, \dots, i_k$ such that $\pi_{i_1} = i_2, \pi_{i_2} = i_3, \dots, \pi_{i_{k-1}} = i_k$, and $\pi_{i_k} = i_1$, a *cycle* of the permutation. We denote such a cycle by the parenthesized list $(i_1, i_2, i_3, \dots, i_k)$. Put together, all the cycles of a permutation give an alternative representation to permutations. For example, (4 3 8) is a cycle of the permutation

$$\begin{pmatrix} 1 & 2 & 3 & 4 & 5 & 6 & 7 & 8 \\ 7 & 6 & 8 & 3 & 1 & 2 & 5 & 4 \end{pmatrix} \tag{1.16}$$

and a cycle representation of that permutation is

$$(1 \quad 7 \quad 5)\ (2 \quad 6)\ (4 \quad 3 \quad 8). \tag{1.17}$$

Obviously, by its cyclic nature, a cycle can be represented in many equivalent ways by cyclically shuffling its members. For example, the cycle (4 3 8) can also be represented as (3 8 4) or (8 4 3). By starting at different positions in the chain of indexes of a cycle we get different representations—a cycle of length k can be cyclically shuffled in k different ways. Among the various cycle representations of a permutation Π_n, we consider the canonical cycle representation as the one that cyclically shuffles the members of each cycle so that the first element of each cycle is the largest in the cycle and permutes the cycles in a such a way as to put their first elements in increasing order.

For sorting purposes off-line algorithms can be designed to detect the cycles in the ranks of the input and fix them by exchanges. Analyzing the number of cycles in a permutation does not require any new techniques. In fact there is a one-to-one correspondence between n-long permutations with k cycles and n-long permutations with k records. We shall prove this correspondence in the next two propositions. It will then follow that cycles have the same probabilistic qualities as records.

Proposition 1.8 *(Foata, 1965). To a permutation of* $\{1, \ldots, n\}$ *with* k *cycles there corresponds a unique permutation of* $\{1, \ldots, n\}$ *with* k *records.*

Proof. Remove the parentheses from the canonical representation of a permutation of $\{1, \ldots, n\}$. In a left-to-right scan of the resulting permutation, the first element of the first cycle is a record, and remains so till the first element of the second cycle appears to dictate a new record and so on. The largest elements of the k cycles now provide k records. ∎

As an illustration of the proof of Proposition 1.8, consider the permutation (1.16), with the cycle structure (1.17). Massage a cycle representation (1.17) in two steps: First cyclically shuffle each cycle to bring its largest element to the beginning of the cycle:

$$(7\quad 5\quad 1)\ (6\quad 2)\ (8\quad 4\quad 3);$$

second, obtain the canonical representation by rearranging the three cycles according to an increasing order of their first elements:

$$(6\quad 2)\ (7\quad 5\quad 1)\ (8\quad 4\quad 3).$$

If we drop the parentheses, we obtain

$$\begin{array}{cccccccc} 6 & 2 & 7 & 5 & 1 & 8 & 4 & 3\ ; \\ \uparrow & & \uparrow & & & \uparrow & & \end{array}$$

the first element in each of the three cycles provides a record in the last permutation (indicated by an upward arrow under the record).

Reversing the construction is easier. Starting with a permutation of $\{1, \ldots, n\}$ with k records, we can put parentheses around stretches that begin with a record and end with an element preceding the next record to obtain k cycles.

Proposition 1.9 *(Foata, 1965). To a permutation of* $\{1, \ldots, n\}$ *with* k *records, there corresponds a unique permutation of* $\{1, \ldots, n\}$ *with* k *cycles.*

Proof. Let

$$\begin{array}{ccccccccccc} \pi_{i_1,1} & \pi_{i_2,1} \ldots \pi_{i_{\ell_1},1} & \pi_{i_1,2} & \pi_{i_2,2} \ldots \pi_{i_{\ell_2},2} \ldots & \pi_{i_1,k} & \pi_{i_2,k} \ldots \pi_{i_{\ell_k},k} \\ \uparrow & & \uparrow & & \uparrow & \end{array}$$

be a permutation of $\{1, \ldots, n\}$ with the k records $\pi_{i_1,1}, \pi_{i_2,2}, \ldots, \pi_{i_k,k}$ as indicated by the upward arrows. Consider the group of elements with the same second index as the successive members of one cycle to get the cycle representation

$$(\pi_{i_1,1}\pi_{i_2,1} \ldots \pi_{i_{\ell_1},1})\ \ (\pi_{i_1,2}\pi_{i_2,2} \ldots \pi_{i_{\ell_2},2}) \ldots (\pi_{i_1,k}\pi_{i_2,k} \ldots \pi_{i_{\ell_k},k}),$$

with k cycles. ∎

The constructive proofs of Propositions 1.8 and 1.9 illustrate the concept of equality in distribution. The permutation with k records that corresponds to Π_n, a permutation of k cycles, is not necessarily Π_n. If we think of the permutations of $\{1, \ldots, n\}$ as points of a sample space, the number of cycles of a permutation may not be the same as its number of records. Yet, the *number* of permutations with k cycles is the same as the *number* of permutations with k records. Specifically, the number of permutations of $\{1, \ldots, n\}$ with k cycles is $\left[{n \atop k}\right]$, Stirling's number of the first kind of order n and degree k. For equiprobable permutations, the probability that a random permutation Π_n of $\{1, \ldots, n\}$ has k cycles is

$$\mathbf{Prob}\{\Pi_n \text{ has } k \text{ cycles}\} = \left[{n \atop k}\right] \frac{1}{n!}.$$

Let C_n be the number of cycles of a random permutation of $\{1, \ldots, n\}$, and let R_n be the number of records of the same permutation. While generally, $R_n \neq C_n$, their distribution functions are the same, and

$$C_n \overset{D}{=} R_n.$$

The number of cycles in a permutation has the exact same probabilistic character as the number of records. In particular, it has the same mean, variance, and exact and limit distributions. In random permutations of $\{1, \ldots, n\}$, there are $H_n \sim \ln n$ cycles on average, and the variance of the number of cycles is $H_n - H_n^{(2)} \sim \ln n$. The number of cycles satisfies Goncharov's theorem (Theorem 1.5).

1.10.4 Runs

An ascending trend in a permutation (a block of ascending elements that is not a sub-block of a longer block of ascending elements) is called a run. Runs represent natural trends in the data that may be favorable to some sorting algorithms.

Formally, in a permutation $(\pi_1, \ldots, \pi_n)$ of $\{1, \ldots, n\}$, the subarrangement $\pi_i, \ldots, \pi_j$ of consecutive elements is a *run*, if $\pi_{i-1} > \pi_i$, $\pi_i < \pi_{i+1} < \ldots < \pi_j$, and $\pi_j > \pi_{j+1}$, with the interpretation of boundaries $p_0 = n + 1$, $p_{n+1} = 0$. For example, the permutation

$$7 \quad 6 \quad 8 \quad 3 \quad 1 \quad 2 \quad 5 \quad 4$$

can be broken up into the five runs

$$\boxed{7} \quad \boxed{6 \quad 8} \quad \boxed{3} \quad \boxed{1 \quad 2 \quad 5} \quad \boxed{4}\,.$$

Let $\mathcal{R}_n$ be the number of runs in a random permutation of size n. A random permutation of size n "grows" from a random permutation of size $n - 1$ by inserting n at any of the n gaps. If n falls at the end of a run it will lengthen that run by one place, but the number of runs remains the same (this can happen in $\mathcal{R}_{n-1}$ ways). Alternately, if n falls within a run, it breaks up that run further into two runs, increasing the total number of runs by 1. Therefore,

$$\mathcal{R}_n \mid \mathcal{R}_{n-1} = \begin{cases} \mathcal{R}_{n-1}, & \text{with probability } \frac{\mathcal{R}_{n-1}}{n}; \\ \mathcal{R}_{n-1} + 1, & \text{with probability } 1 - \frac{\mathcal{R}_{n-1}}{n}. \end{cases} \tag{1.18}$$

So, given the sigma field $\mathcal{F}_{n-1}$ (the entire history of the growth process up to and including stage $n - 1$),

$$\mathbf{E}\big[\mathcal{R}_n \mid \mathcal{F}_{n-1}\big] = \mathcal{R}_{n-1} + \Big(1 - \frac{\mathcal{R}_{n-1}}{n}\Big) = \frac{n-1}{n}\mathcal{R}_{n-1} + 1. \tag{1.19}$$

One sees that

$$\mathbf{E}\big[n\mathcal{R}_n \mid \mathcal{F}_{n-1}\big] = (n-1)\mathcal{R}_{n-1} + n,$$

and $n\mathcal{R}_n$ is "almost" a martingale, but not quite because of the presence of the shift factor n on the right-hand side. The variable $n\mathcal{R}_n$ can easily be turned into a martingale. A linear transformation accomplishes this martingalization. Let $M_n = n\mathcal{R}_n + b_n$ for an appropriately chosen factor b_n. We want M_n to be a martingale; so, we want

$$\mathbf{E}\big[M_n \mid \mathcal{F}_{n-1}\big] = M_{n-1}.$$

We let this happen by setting

$$\mathbf{E}\big[n\mathcal{R}_n + b_n \mid \mathcal{F}_{n-1}\big] = \mathbf{E}\big[n\mathcal{R}_n \mid \mathcal{F}_{n-1}\big] + b_n = (n-1)\mathcal{R}_{n-1} + n + b_n$$

equal to $M_{n-1} = (n-1)\mathcal{R}_{n-1} + b_{n-1}$, for all $n \ge 2$. This gives a recurrence for b_n. We must have

$$b_n = b_{n-1} - n = b_1 - (2 + 3 + \cdots + n).$$

We want the mean of our martingale to be 0, dictating that

$$\mathbf{E}[M_1] = \mathcal{R}_1 + b_1 = 1 + b_1 = 0$$

Hence,

$$M_n = n\mathcal{R}_n - \frac{1}{2}n(n+1)$$

is a martingale. This immediately yields

$$\mathbf{E}[M_n] = n\mathbf{E}[\mathcal{R}_n] - \frac{1}{2}n(n+1) = 0,$$

giving

$$\mathbf{E}[\mathcal{R}_n] = \frac{n+1}{2}.$$

The martingalization process just discussed applies to the more general situation where a random variable X_n satisfies $\mathbf{E}[X_n \mid \mathcal{F}_{n-1}] = c_n X_{n-1} + d_n$. One sets up the linear martingale transform $Y_n = a_n X_{n-1} + b_n$, then solves the recurrences for a_n and b_n in terms of c_n and d_n. Chao (1997) gives a powerful martingale central limit theorem and applies it to several characteristics of permutations.

To finish off with the martingale central limit theorem, we should get the variance of $\mathcal{R}_n$. From the square of (1.18), we have

$$\begin{aligned}\mathbf{E}\big[\mathcal{R}_n^2 \mid \mathcal{F}_{n-1}\big] &= \mathcal{R}_{n-1}^2 \frac{\mathcal{R}_{n-1}}{n} + (\mathcal{R}_{n-1}+1)^2\Big(1 - \frac{\mathcal{R}_{n-1}}{n}\Big)\\ &= \Big(1 - \frac{2}{n}\Big)\mathcal{R}_{n-1}^2 + \frac{2n-1}{n}\mathcal{R}_{n-1} + 1. \qquad (1.20)\end{aligned}$$

Now we take expectations of both sides to get a recurrence for second moments:

$$\mathbf{E}\big[\mathcal{R}_n^2\big] = \Big(1 - \frac{2}{n}\Big)\mathbf{E}\big[\mathcal{R}_{n-1}^2\big] + n + \frac{1}{2}.$$

Let

$$a_n = 1 - \frac{2}{n}, \qquad g_n = n + \frac{1}{2}.$$

Our recurrence becomes

$$\mathbf{E}[\mathcal{R}_n^2] = a_n \mathbf{E}[\mathcal{R}_{n-1}^2] + g_n,$$

which can be iterated:

$$\begin{aligned}\mathbf{E}[\mathcal{R}_n^2] &= a_n a_{n-1}\mathbf{E}[\mathcal{R}_{n-2}^2] + a_n g_{n-1} + g_n\\ &\ \ \vdots\\ &= a_n a_{n-1}\dots a_2 \mathbf{E}[\mathcal{R}_1^2] + \sum_{j=2}^{n} a_n a_{n-1}\dots a_{j+1} g_j.\end{aligned}$$

The product $a_n a_{n-1}\dots a_{j+1}$ is telescopic and simplifies to

$$a_n a_{n-1}\dots a_{j+1} = \frac{n-2}{n}\times\frac{n-3}{n-1}\times\frac{n-4}{n-2}\times\frac{n-5}{n-3}\times\dots\times\frac{j}{j+2}\times\frac{j-1}{j+1} = \frac{j(j-1)}{n(n-1)}.$$

So, $a_n a_{n-1}\dots a_2 = 0$ and

$$\mathbf{E}[\mathcal{R}_n^2] = \frac{1}{n(n-1)}\sum_{j=2}^{n} j(j-1)\Big(j + \frac{1}{2}\Big) = \frac{1}{12}(3n+4)(n+1).$$

This gives the variance

$$\mathbf{Var}[\mathcal{R}_n] = \mathbf{E}[\mathcal{R}_n^2] - \mathbf{E}^2[\mathcal{R}_n] = \frac{n+1}{12}.$$

This narrow concentration admits a law of large numbers. By Chebyshev's inequality,

$$\mathbf{Prob}\left\{\left|\frac{\mathcal{R}_n}{n+1} - \frac{1}{2}\right| > \varepsilon\right\} \le \frac{\mathbf{Var}[\mathcal{R}_n/(n+1)]}{\varepsilon^2} = \frac{\mathbf{Var}[\mathcal{R}_n]}{\varepsilon^2(n+1)^2} = \frac{1}{12\varepsilon^2(n+1)} \to 0,$$

for any fixed $\varepsilon > 0$. That is, $\mathcal{R}_n/(n+1) \xrightarrow{P} \frac{1}{2}$. Multiply by the deterministic convergence $(n+1)/n \to 1$, to get $\mathcal{R}_n/n \xrightarrow{P} \frac{1}{2}$. It will be useful for the work below to write this relation as

$$\mathcal{R}_n = \frac{1}{2}n + o_P(n), \tag{1.21}$$

where $o_P(n)$ is a quantity that is negligible in comparison to n *in probability*. The proof of asymptotic normality for runs appeals to the martingale central limit theorem (see the appendix). The backward difference $\triangledown M_k = M_k - M_{k-1}$ is a martingale difference as it satisfies $\mathbf{E}[\triangledown M_k \mid \mathcal{F}_{k-1}] = 0$, and so is any scaling $t(n) \triangledown M_k$.

Proposition 1.10 *In a random permutation of* $\{1, \ldots, n\}$, *the number of runs*, $\mathcal{R}_n$, *satisfies a central limit theorem:*

$$\frac{\mathcal{R}_n - \frac{1}{2}n}{\sqrt{n}} \xrightarrow{\mathcal{D}} \mathcal{N}\left(0, \frac{1}{12}\right).$$

Proof. The variables $n^{-3/2} \triangledown M_k$, for $k = 1, \ldots, n$, are a martingale difference sequence. We shall show that it satisfies the conditional Lindeberg's condition and the $\frac{1}{12}$-conditional variance condition. From its definition

$$\begin{aligned}\triangledown M_k &= \left[k\mathcal{R}_k - \frac{1}{2}k(k+1)\right] - \left[(k-1)\mathcal{R}_{k-1} - \frac{1}{2}(k-1)k\right] \\ &= k \triangledown R_k + R_{k-1} - k.\end{aligned}$$

By the growth rules (1.18), $\triangledown R_k \in \{0, 1\}$, and so $| \triangledown M_k| < k$. It then follows that for any fixed $\varepsilon > 0$, the sets $\{| \triangledown M_k| > \varepsilon n^{3/2}\}$ are empty for large enough n; the conditional Lindeberg's condition is verified.

Let

$$V_n = \sum_{k=1}^{n} \mathbf{E}\left[\left(\frac{\triangledown M_k}{n^{3/2}}\right)^2 \middle| \mathcal{F}_{k-1}\right].$$

We have

$$V_n = \frac{1}{n^3}\sum_{k=1}^{n} \mathbf{E}\Big[\big(kR_k - (k-1)R_{k-1} - k\big)^2 \Big| \mathcal{F}_{k-1}\Big]$$

$$= \frac{1}{n^3}\bigg(\sum_{k=1}^{n}\Big\{(k-1)^2R_{k-1}^2 + 2k(k-1)R_{k-1} + k^2\mathbf{E}\big[R_k^2 \mid \mathcal{F}_{k-1}\big]$$

$$- 2k^2\mathbf{E}\big[R_k \mid \mathcal{F}_{k-1}\big] + k^2 - 2k(k-1)R_{k-1}\mathbf{E}\big[R_k \mid \mathcal{F}_{k-1}\big]\Big\}\bigg).$$

Substituting the conditional expectation of the first moment with (1.19), and that of the second moment with (1.20), we get

$$V_n = \frac{1}{n^3}\sum_{k=1}^{n} R_{k-1}(k - R_{k-1}).$$

We use the representation (1.21) for R_{k-1}, and we further have

$$\begin{aligned} V_n &= \frac{1}{n^3}\sum_{k=1}^{n}\Big(\frac{1}{4}k^2 + o_P(k^2)\Big) \\ &= \frac{1}{12} + o_P(1) \\ &\xrightarrow{P} \frac{1}{12}; \end{aligned}$$

the $\frac{1}{12}$-conditional variance condition is verified.

The sum $\sum_{k=1}^{n} n^{-3/2} \bigtriangledown M_k = n^{-3/2}M_n$ converges in distribution to a random variable with characteristic function $\mathbf{E}[\exp(-\frac{1}{24}t^2)]$, which is that of $\mathcal{N}(0, \frac{1}{12})$. And so

$$\frac{\mathcal{R}_n - \frac{1}{2}n - \frac{1}{2}}{\sqrt{n}} \xrightarrow{\mathcal{D}} \mathcal{N}\Big(0, \frac{1}{12}\Big).$$

The numerator's nonessential factor $-\frac{1}{2}$ is adjusted by Slutsky's theorem: Add the almost-sure convergence $\frac{1}{2}/\sqrt{n} \to 0$ to get the result as stated. ■

EXERCISES

1.10.1 Let $\Pi_n = (\pi_1, \dots, \pi_n)$ be a random permutation of $\{1, \dots, n\}$. Let a new permutation Π_n' be obtained by a circular shuffle of the first element. That is, $\Pi_n' = (\pi_2, \pi_3, \dots, \pi_n, \pi_1)$. Prove that Π_n' is a random permutation.

1.10.2 Suppose the nonuniform probability distribution $p_1, \ldots, p_{n!}$ is imposed on $\Pi_n^{(1)}, \ldots, \Pi_n^{(n!)}$, the $n!$ permutations of $\{1, \ldots, n\}$; with $\sum_{i=1}^{n!} p_i = 1$. Let $\Xi_n = (\xi_1, \ldots, \xi_n)$ be picked from this distribution (with probability p_i). Assume now that Ξ_n is *randomized* into $\Xi_n' = (\xi_1', \ldots, \xi_n')$ by picking ξ_1' *uniformly at random* from the set $\{\xi_1, \ldots, \xi_n\}$, (with probability $1/n$), then picking ξ_2' from among the remaining elements with probability $1/(n-1)$, and so on. Prove that all randomized permutations are equally likely.

1.10.3 Assuming the binary search tree is constructed by progressively inserting the elements of a random permutation of $\{1, 2, \ldots, n\}$, what is the probability of inserting the nth element at the jth leaf (in a left-to-right numbering of leaves of the tree after $n - 1$ insertions)?

1.11 AN ANALYTIC TOOLKIT

Most of the probability theory used in this book is standard and can be found in many classic textbooks. We shall not develop probability theory. However, for quick referencing, we provide in the appendix statements of classic probability theory results used throughout the book.

A certain analytic toolkit has emerged in the analysis of the algorithms. By contrast to probability theory, this analytic toolkit cannot be found in many textbooks and remains somewhat obscure and limited to research outlets. To disseminate these techniques we shall outline the salient features of this analytic toolkit. We shall present the setup in its general form and leave the mechanics to the various chapters as the analytic kit will be encountered a few times in the book.

Many algorithms give rise naturally to recurrence equations on averages, variances, moments, and various forms of moment generating functions. In many cases one is able to express the probabilistic behavior of a sorting algorithm in terms of a recurrence for its moment generating function. A certain property of the algorithm, like its number of comparisons or exchanges, is typically a nonnegative integer random variable (X_n say), when the algorithm sorts an input of n random keys (according to some well-specified model of randomness). When it converges within a disc including the origin, the moment generating function

$$\phi_n(t) = \mathbf{E}\big[e^{tX_n}\big] = \sum_{k=0}^{\infty} \mathbf{E}\big[X_n^k\big]\frac{t^k}{k!}$$

captures the distribution of X_n. Sometimes, to avoid convergence issues, one deals with the characteristic function $\phi_n(it)$, which always exists because

$$|\phi_n(it)| = \big|\mathbf{E}\big[e^{itX_n}\big]\big| \le \mathbf{E}\big[\big|e^{itX_n}\big|\big] = 1.$$

Other times it is more convenient to deal with the probability generating function

$$\phi_n(\ln t) = \mathbf{E}\big[t^{X_n}\big].$$

These approaches are only minor technical variations on the main theme to handle a convergence issue or get what works.

It is often helpful to set up a bivariate generating function

$$\Phi(t, z) = \sum_{j=0}^{\infty} a_j \phi_j(t) z^j.$$

Sometimes such a bivariate generating function is called a super moment generating function, as it generates moment generating functions. One then gets a representation for $\phi_j(t)$ from the basic mechanics of the algorithm; subsequently a representation for $\Phi(t, z)$ becomes available.

The usual routine is to formulate an equation for $\Phi(t, z)$ by multiplying both sides of a representation of $\phi_n(t)$ (a recurrence for example) by z^n and summing over n in the range of validity of the recurrence. One often gets an equation for $\Phi(t, z)$. For recursive algorithms such an equation is usually a functional equation.

Derivatives of this equation give us equations (or functional equations) for the moments. For instance, the kth derivative with respect to t, at $t = 1$, gives

$$\begin{aligned} \frac{\partial^k}{\partial t^k} \Phi(0, z) &= \sum_{j=0}^{\infty} a_j \frac{d^k \phi_j(t)}{dt^k} z^j \bigg|_{t=0} \\ &= \sum_{j=0}^{\infty} a_j \mathbf{E}[X_j^k] z^j, \end{aligned}$$

a generating function for the kth moment of the cost. When the representation of $\Phi(t, z)$ is a functional equation (both sides involve $\Phi(t, z)$), taking the first and second derivatives of the functional equation gives functional equations on generating functions of the first and second moments.

1.11.1 The Saddle Point Method

What can one do with an equation or a functional equation on $\Phi(t, z)$, a super generating function of $\phi_n(t)$? The equations arising from algorithms vary in complexity and the techniques for solving them accordingly vary widely. In many cases an exact solution is not readily available, whereas an asymptotic one is. One obvious thing to try is to recover $\phi_n(t)$ from $\Phi(t, z)$. The function $\phi_n(t)$ is the nth coefficient of $\Phi(t, z)$ and the process is that of extraction of coefficients.

We use the notation $[z^n]$ for the operator that extracts the coefficient of z^n from a function (possibly bivariate) of z. One form of functional equations that we encounter sometimes in the analysis of algorithms leads to $\phi_n(t)$, or a component of it, given as a coefficient of some bivariate function

$$[z^n] \Psi^k(t, z),$$

with k large. In this case, We may utilize Cauchy's classical formula:

$$[z^n]\Psi^k(t,z) = \frac{1}{2\pi i}\oint_\Gamma \frac{\Psi^k(t,z)}{z^{n+1}}\,dz, \tag{1.22}$$

where Γ is a closed contour enclosing the origin. The need arises for asymptotic evaluation of integrals, as the parameter k gets large. This situation can be handled by the saddle point method. We shall say a quick word about the principle involved and the interested reader who may wish to learn more about the method can refer to a chapter in a specialized book on asymptotic integration, like Wong's (1989).

The *saddle point method* is a heuristic based on working with expansions of functions and is used often to asymptotically evaluate integrals of the form

$$I(\lambda) = \frac{1}{2\pi i}\int_\Gamma g(z)e^{\lambda h(z)}\,dz,$$

over a contour Γ (not necessarily closed), for large values of the *real* parameter λ. The saddle point method will thus be helpful for extracting the nth coefficient from the representation (1.22) particularly when k is large, which requires a minor extension for the cases where both g and h become bivariate, and the value of the integral itself depends on t.

The functions g and h will both be assumed to behave reasonably well. For our particular applications these two functions will be analytic, thus differentiable within a domain containing Γ. The basic principle is that the exponential function (of a complex variable) grows very fast with the growth in its real part. The contour Γ is deformed to another contour on which the integral has the same leading asymptotic term as $I(\lambda)$, only introducing a small error in lower-order terms. Having an exponential factor in its integrand, the integral picks up the most contribution from around a neighborhood of a maximal value of the exponential factor.

Of the family of possible such contours, we choose Γ' which passes through z_0, a point that maximizes the real part of the function $h(z)$, and $h(z)$ drops off as sharply as possible as we move away from z_0 on Γ'. Such a curve with fastest rate of decrease in the real part of $h(z)$ is called the *path of steepest descent*. Integrating over the path of steepest descent admits good approximation to the original integral from only a very small neighborhood around z_0. It turns out from careful examination of Cauchy-Riemann conditions that on the path of steepest descent the imaginary part of $h(z)$ remains constant (the steepest descent path is sometimes called the *constant phase path*). Constant phase prevents large oscillations that otherwise, on some other contour, may interfere with and even supersede the main contribution at $h(z_0)$ as we move away from z_0, because we are letting λ be very large. The situation needs closer attention when there are several local maxima on Γ'. For all our purposes we shall be dealing with simple cases with only one essential maximal point.

When all the conditions for an essential saddle point are checked and Γ' is indeed the steepest descent path, the integral picks up the main contribution from only a small neighborhood around z_0. The function $h(z)$ attains a maximum at z_0, that is,

$h'(z_0) = 0$. Hence, the function $h(z)$ has an expansion near z_0 given by:

$$h(z) = h(z_0) + \frac{1}{2}h''(z_0)(z - z_0)^2 + O\big((z - z_0)^3\big);$$

so does the function $g(z)$:

$$g(z) = g(z_0) + O(z - z_0).$$

Because the function $h(z)$ is in the exponential factor, the second term in its expansion is critical, whereas for $g(z)$, which is only a multiplicative factor, it is sufficient to work with its leading term $g(z_0)$, as $z \to z_0$. Working with the expansions,

$$\begin{aligned} I(\lambda) &= \frac{1}{2\pi i}\int_{\Gamma'} \big[g(z_0) + O(z - z_0)\big] \exp\Big\{\lambda\big[h(z_0) + \frac{1}{2}h''(z_0)(z - z_0)^2 \\ &\qquad + O\big((z - z_0)^3\big)\big]\Big\}\, dz \\ &\sim \frac{g(z_0)}{2\pi i} e^{\lambda h(z_0)} \int_{\Gamma'} \exp\Big\{\frac{\lambda}{2}h''(z_0)(z - z_0)^2\Big\}\, dz. \end{aligned}$$

The change of variable $v^2 = -\lambda h''(z_0)(z - z_0)^2$ yields

$$I(\lambda) \sim \frac{g(z_0)e^{\lambda h(z_0)}}{2\pi\sqrt{\lambda h''(z_0)}} \int_{v_1}^{v_2} e^{-v^2/2}\, dv;$$

the beginning and ending points v_1 and v_2 usually form a real interval containing 0 (the image of z_0 under the v transformation). Again because the integrand is an exponential function, we can extend the range of integration to the entire real line, only introducing an asymptotically negligible error. So,

$$I(\lambda) \sim \frac{g(z_0)e^{\lambda h(z_0)}}{\sqrt{2\pi\lambda h''(z_0)}} \int_{-\infty}^{\infty} \frac{e^{-v^2/2}}{\sqrt{2\pi}}\, dv.$$

The remaining integral is that of the standard normal density, and is therefore 1. When all the conditions check out, this procedure gives a working asymptotic formula, as $\lambda \to \infty$,

$$I(\lambda) \sim \frac{g(z_0)e^{\lambda h(z_0)}}{\sqrt{2\pi\lambda h''(z_0)}}.$$

1.11.2 The Mellin Transform

The *Mellin transform* of a function $f(x)$ is

$$\int_0^{\infty} f(x)x^{s-1}\, dx.$$

The resulting transform depends on s and will be denoted by

$$\mathcal{M}\{f(x); s\}.$$

The Mellin transform usually exists in fundamental domains in the complex s plain, often taking the form of a vertical strip

$$a < \Re s < b,$$

for $a < b$ real values. The function $f(x)$ can be recovered from its transform $\mathcal{M}\{f(x); s\}$ by the inversion integral

$$f(x) = \frac{1}{2\pi i} \int_{c-i\infty}^{c+i\infty} \mathcal{M}\{f(x); s\}\, x^{-s} ds.$$

This inversion formula is typical of a general class of integral transform inversion formulas that include the Mellin, Laplace, and Fourier transforms where the inversion is given by a line integral in the complex plane.

The Mellin transform and its inversion play a prominent role in simplifying harmonic sums of the general form

$$\sum_{k=0}^{\infty} \lambda_k f(a_k x).$$

The function $f(x)$ is called the *base function* and is involved in the harmonic sum with different scales for its argument. These harmonic sums appear often in algorithms for digital data. A function

$$g(x) = \sum_{k=0}^{\infty} \lambda_k f(a_k x)$$

(well defined for $x > 0$) has a simple Mellin transform. The formula for the transform $g^*(s) = \mathcal{M}\{g(x); s\}$ is obtained easily from

$$g^*(s) = \int_0^\infty \sum_{k=0}^{\infty} \lambda_k f(a_k x) x^{s-1}\, dx = \sum_{k=0}^{\infty} \lambda_k \int_0^\infty f(a_k x) x^{s-1}\, dx.$$

The change of the order of summation and integration is permissible in view of the absolute convergence of the series at every $x > 0$, to $g(x)$. In the kth integral, the change of the integration variable $a_k x = v_k$ gives

$$g^*(s) = \sum_{k=0}^{\infty} \lambda_k a_k^{-s} \int_0^\infty f(v_k) v_k^{s-1}\, dv_k = f^*(s) \sum_{k=0}^{\infty} \lambda_k a_k^{-s}.$$

One then resorts to contour integration to recover $g(x)$ from its Mellin transform via the inversion integral

$$\frac{1}{2\pi i}\int_{c-i\infty}^{c+i\infty}\Bigl(f^*(s)\sum_{k=0}^{\infty}\lambda_k a_k^{-s}\Bigr)x^{-s}\,ds,$$

which is taken over an infinite vertical line $\Re s = c$ in the fundamental strip of existence of $g^*(s)$, say $a < \Re s < b$.

The line integral involved in the inversion formula is usually estimated asymptotically (for large x) by the method of "closing the box." One links the inversion integral to a contour integral, which is to be evaluated by residues of the singularities of the integrand. Typically, these singularities are poles of low order aligned on the real axis, and equispaced singularities aligned (symmetrically about the horizontal line) on vertical lines outside the domain of existence.

It is common folklore that the poles on vertical lines give small oscillations in the solution, and that the dominant behavior in inversion formulas comes from singularities of higher order, as we shall soon see in an example. Consider the contour integral

$$\oint_\Lambda = -\int_{c-iM}^{c+iM} - \int_{c+iM}^{d+iM} - \int_{d+iM}^{d-iM} - \int_{d-iM}^{c-iM},$$

where Λ is a box connecting the four corners given in the limits of the four integrals; the minus signs account for traversing the contour in the clockwise direction. The contour encloses a number of the poles to the right of the line $\Re s = c$, and we choose the altitude M so that the contour does not pass through any of the poles of the integrand. The common types of Mellin transforms are functions that decrease exponentially fast as one moves away from the horizontal axis, such as the Gamma function. By a careful limit process, letting $M \to \infty$ without letting the top or bottom sides of the box pass through any poles, the integrals on the top and bottom sides of the box vanish. Another consequence of the exponential decay of the Mellin transform in the vertical direction is that the right integral $\int_{d+iM}^{d-iM}$ is bounded by $O(x^{-d})$ as we shall see shortly. As d is arbitrary, we can take it large to minimize the error. In the limit, the contour Λ grows to Λ_∞, a contour encompassing the right boundary of the infinite strip $c < \Re s < d$, and

$$\frac{1}{2\pi i}\int_{c-i\infty}^{c+i\infty} = -\frac{1}{2\pi i}\oint_{\Lambda_\infty} + O(x^{-d}).$$

A large value of d will allow Λ_∞ to encompass all the the poles lying to the right of the fundamental strip. The evaluation of the line integral has been reduced to a computational task. By Cauchy's classical residue theorem, the equivalent integral on the closed contour Λ_∞ is $2\pi i$ times the sum of the residues of the poles of the integrand inside Λ_∞. It follows that

$$\frac{1}{2\pi i}\int_{c-i\infty}^{c+i\infty} = -\sum\{\text{residues of singularities to the right of the fundamental strip}\} + O(x^{-d}).$$

Example 1.1 To see the mechanical procedure of evaluating a harmonic sum in action, consider

$$A(x) = \sum_{k=0}^{\infty}\left(1 - e^{-x/2^k}\right).$$

The Mellin transform of the harmonic sum $A(x)$ is

$$\begin{aligned} A^*(s) &= \int_0^\infty A(x)x^{s-1}\,dx \\ &= \mathcal{M}\{1 - e^{-x}; s\}\sum_{k=0}^{\infty} 2^{sk}. \end{aligned}$$

The sum converges to $1/(1 - 2^s)$ in the domain

$$|2^s| < 1,$$

that is, $\Re s < 0$. On the other hand, the Mellin transform of $1 - e^{-x}$ is what is sometimes referred to as the Cauchy–Saalschütz analytic continuation of the Gamma function. This transform is $-\Gamma(s)$ in the domain

$$-1 < \Re s < 0.$$

Both $\mathcal{M}\{1 - e^{-x}; s\}$ and the infinite series exist in the vertical strip $-1 < \Re s < 0$, and

$$A^*(s) = -\frac{\Gamma(s)}{1 - 2^s}$$

is well defined for $\Re s \in (-1, 0)$. The inverse Mellin transform is

$$A(x) = \frac{1}{2\pi i}\int_{c-\infty}^{c+i\infty} A^*(s)x^{-s}\,ds,$$

for any $c \in (-1, 0)$. Shifting the line of integration to a vertical line at $d > 0$ requires compensation by the residues of the poles of $A^*(s)$:

$$A(x) = -\sum\{\text{Residues of poles to the right of the strip } -1 < \Re s < 0\} + O(x^{-d}).$$

The transform $A^*(s)$ has singularities at the roots of the equation

$$2^s = 1 = e^{2\pi ik}, \qquad k = 0, \pm 1, \pm 2, \ldots .$$

There are singularities at $s_k = 2\pi ik/\ln 2$, for $k = 0, \pm 1, \pm 2, \ldots$. The Gamma function contributes additional singularities at $0, -1, -2, -3, \ldots$. Each of the integrand's singularities is therefore a simple pole, except the pole at 0, which is a double pole (see Figure 1.11). We shall see that this double pole gives the asymptotically dominant term.

Proceeding with the inversion,

$$A(x) = -\operatorname*{Res}_{s=0} A^*(s)x^{-s} - \sum_{\substack{k=-\infty \\ k\neq 0}}^{\infty} \operatorname*{Res}_{s=s_k} A^*(s)x^{-s} + O\big(x^{-d}\big).$$

The residue at 0 is

$$\lim_{s\to 0} \frac{d}{ds}\Big(-\frac{s^2\Gamma(s)x^{-s}}{1-2^s}\Big) = -\lg x - \frac{\gamma}{\ln 2} - \frac{1}{2},$$

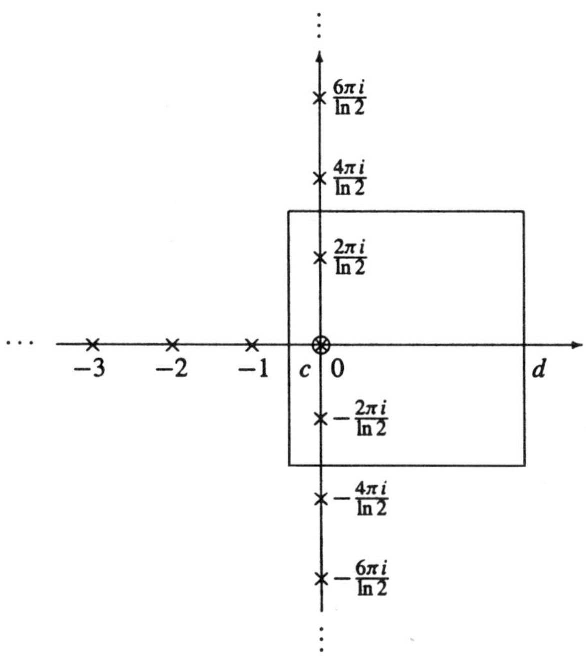

Figure 1.11. The poles of the Mellin transform $A^*(s)$; $\times$ indicates a simple pole and $\otimes$ indicates a double pole.

where $\gamma = 0.5772\ldots$ is Euler's constant. The residue at s_k, with $k \neq 0$, is

$$\lim_{s \to s_k} -\frac{(s - s_k)\Gamma(s)x^{-s}}{1 - 2^s} = \frac{1}{\ln 2}\Gamma\Big(\frac{2\pi ik}{\ln 2}\Big)e^{-2\pi ik \lg x}.$$

For any arbitrary $d > 0$, the sum is

$$A(x) = \lg x + \frac{\gamma}{\ln 2} + \frac{1}{2} - Q(\lg x) + O(x^{-d}),$$

where Q is the function

$$Q(u) = \frac{1}{\ln 2} \sum_{\substack{k=-\infty \\ k \neq 0}}^{\infty} \Gamma\Big(\frac{2\pi ik}{\ln 2}\Big)e^{-2\pi iku}.$$

A few technical points in the example are worthy of some interpretation. The error $O(x^{-d})$ introduced in inverting the Mellin transform is valid for any fixed d. What this really means is that the error is less than the reciprocal of any polynomial in x; it might for instance be exponentially small.

The function $Q(u)$ is an oscillating function of a truly ignorable magnitude. For all u, the magnitude of the function is bounded by $0.1573158421 \times 10^{-5}$. Note that these oscillations occur in the lower-order term, anyway, and do not make the harmonic sum behave in an unstable manner. As a function of x, the sum is a "wave" of a very small amplitude flickering around a "steady" component.

1.11.3 Poissonization

A combinatorial problem like sorting n random keys is sometimes hard to approach for fixed n. Occasionally, the same problem is more tractable if, instead of fixed n, one assumes that the size of the combinatorial problem (number of keys to be sorted in the case of sorting problems) is random according to some probability distribution. For example, one may consider sorting $N(z)$ keys where $N(z)$ follows a probability distribution with parameter z. Which probability models are helpful? The Poisson distribution has proved to be an instrumental tool leading to *Poissonization*. The general idea in Poissonization is that the behavior of a fixed-population problem should be close to that of the same problem under a Poisson model having the fixed-population problem size as its mean. To a lesser extent, geometric random variables have been used and indeed there is work involving *geometrization* like Grabner's (1993) work on incomplete digital trees.

One of the earliest applications of Poissonization can be traced back to the work of Kac (1949). Over the past few decades the idea has resurfaced in various forms and guises. The difficult step has always been in relating results from the Poisson model back to the fixed-population model, the model of prime interest. Interpreting the Poissonization results for the fixed-population model is called *de-Poissonization*.

A variety of ad hoc de-Poissonization solutions have been proposed. Perhaps the earliest Poissonization in the analysis of algorithms is due to Jacquet, and Régnier (1986). Poissonization and de-Poissonization have also been considered by some as a mathematical transform followed by an inversion operation. The first to treat it as such were Gonnet and Munro (1984) [Poblete (1987) follows up on the inversion of the transform]. Other useful related results may be found in Holst (1986), Aldous (1989), Rais, Jacquet, and Szpankowski (1993), Arratia and Tavaré (1994), and Jacquet and Szpankowski (1995).

For a fixed-population sorting problem, analysis via Poissonization is carried out in two main steps: Poissonization, followed by de-Poissonization. Instead of having a population of fixed size, we first determine the number of keys to be sorted by a draw from a Poisson distribution with parameter z. A recurrence or what have you of techniques is then set up and the problem is completely solved under the Poisson model.

Of course, z, being the parameter of a Poisson distribution, must be a real, positive number. However, all calculations can be extended to complex numbers. For example, for any complex z, we can define

$$A_k(z) = \frac{z^k}{k!}e^{-z}.$$

as the analytic continuation of $\mathbf{Prob}\{N(z) = k\}$; that is to say $A_k(z)$ is a function whose value coincides with $\mathbf{Prob}\{N(z) = k\}$ for every integer k and real z, or $A_k(z)$ is the *analytic continuation* of $\mathbf{Prob}\{N(z) = k\}$.

For subsequent de-Poissonization, we *formally* allow z to be a complex number in order to manipulate the resulting generating function by considering its analytic continuation to the z complex plane. This allows us to use the full power of complex analysis. De-Poissonization is then achieved by a combination of Mellin transform methods and asymptotic integration over a circle of radius n in the z complex plane.

Suppose X_n is the number of operations a sorting algorithm performs on a set of n random keys. Let the exponential generating function of the means sequence be

$$M(z) = \sum_{j=0}^{\infty} \mathbf{E}[X_j]\frac{z^j}{j!}.$$

The generating function $e^{-z}M(z)$ has a Poissonization interpretation. Had we started with $N(z) = \text{POISSON}(z)$ random number of keys instead of a fixed n, the average cost of sorting would have been

$$\begin{aligned}\mathbf{E}\big[X_{N(z)}\big] &= \sum_{j=0}^{\infty} \mathbf{E}\big[X_{N(z)} \mid N(z) = j\big]\,\mathbf{Prob}\{N(z) = j\} \\ &= \sum_{j=0}^{\infty} \mathbf{E}\big[X_j\big]\,\frac{z^j e^{-z}}{j!} \\ &= e^{-z}M(z).\end{aligned}$$

The intermediate step

$$\mathbf{E}\big[X_{N(z)}\big] = \sum_{j=0}^{\infty} \mathbf{E}[X_j]\, \frac{z^j e^{-z}}{j!}$$

entices one to think of $\mathbf{E}\big[X_{N(z)}\big]$ as a mathematical transform on the sequence $\mathbf{E}[X_j]$, $j = 0, 1, 2, \ldots$; Gonnet and Munro (1984) were the first to recognize it as such. A sequence $\{a_n\}_{n=0}^{\infty}$ will be called *Poissonized* when it is under the Poisson transform. For example, $\mathbf{E}[X_{N(z)}] = e^{-z}M(z)$ is the Poissonized average of X_j. We shall speak of Poissonized averages, variances, and probability distributions.

One can analogously define the Poisson transform for any sequence of numbers a_k, $k = 0, 1, \ldots$. For such a sequence, the *Poisson transform* is formally

$$\mathcal{P}\big[\{a_k\}_{k=0}^{\infty}; z\big] = \sum_{j=0}^{\infty} a_j \frac{z^j e^{-z}}{j!}.$$

A sequence to transform may be an arbitrary deterministic sequence, a sequence of means, variances, moment generating functions, or probabilities. The Poisson transform with respect to z, is a function of z and can be denoted simply as a function like $\mathcal{P}(z)$. Consider a cone $S_\theta = \{z : |\arg z| \le \theta\}$; certain rates of growth of $\mathcal{P}(z)$ along a large circle (as in Figure 1.12) inside and outside S_θ provide good asymptotic approximations of sequences by their Poisson transform.

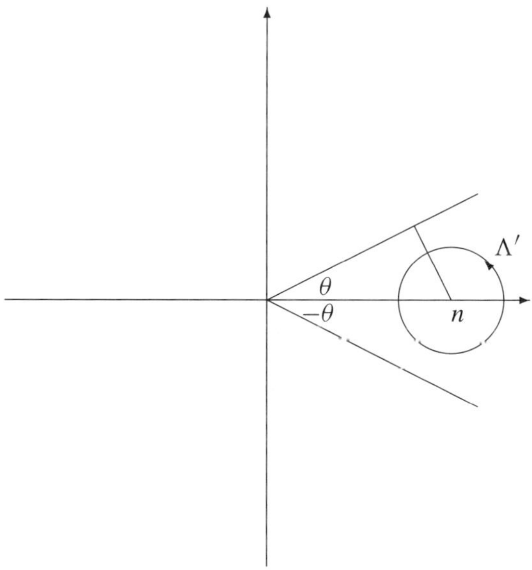

Figure 1.12. The cone of de-Poissonization.

Lemma 1.3 *(Jacquet and Szpankowski, 1998). Let $\{a_n\}_{n=0}^{\infty}$ be a sequence of real numbers. Suppose that the Poisson transform $\mathcal{P}(z)$ exists and can be analytically continued as an entire function of complex z. Fix $\theta \in (0, \pi/2)$ and let S_θ be the cone $\{z : |\arg z| \le \theta\}$. Suppose that there exist positive constants $\alpha < 1$, β_1, β_2, c, and z_0 such that the following conditions hold:*

(i) *For all $z \in S_\theta$ with $|z| \ge z_0$,*

$$|\mathcal{P}(z)| \le \beta_1 |z|^c;$$

(ii) *for all $z \notin S_\theta$ with $|z| \ge z_0$,*

$$|\mathcal{P}(z)e^z| \le \beta_2 |z|^c e^{\alpha|z|}.$$

Then for large n,

$$a_n = \mathcal{P}(n) + O\big(n^{c-\frac{1}{2}} \ln n\big).$$

Proof. Extract a_n from its Poisson transform by Cauchy's formula:

$$a_n = \frac{n!}{2\pi i} \oint_\Lambda \frac{e^z \mathcal{P}(z)}{z^{n+1}}\, dz,$$

where Λ is any closed contour enclosing the origin. To make use of saddle point techniques, choose Λ to be the circle $z = ne^{i\phi}$, for $0 \le \phi \le 2\pi$, with radius $n > z_0$. Break up the line integral into two parts

$$\frac{n!}{2\pi i} \oint_\Lambda \frac{e^z \mathcal{P}(z)}{z^{n+1}}\, dz = \frac{n!}{2\pi} \int_{-\theta}^{\theta} \frac{\mathcal{P}\big(ne^{i\phi}\big) \exp\big(ne^{i\phi}\big)}{(ne^{i\phi})^n}\, d\phi$$
$$+ \frac{n!}{2\pi} \int_{\theta}^{-\theta} \frac{\mathcal{P}\big(ne^{i\phi}\big) \exp\big(ne^{i\phi}\big)}{(ne^{i\phi})^n}\, d\phi.$$

The Stirling approximation of $n!$ by $n^n e^{-n} \sqrt{2\pi n}\,(1 + O(1/n))$ cancels out several factors. It is seen that the second integral involves the behavior of the Poisson transform outside the cone (condition (ii)). This second integral introduces a term that is only $O(n^{c+\frac{1}{2}} e^{-(1-\alpha)n})$. So,

$$a_n = \frac{n!}{2\pi n^n e^{-n}} \int_{-\theta}^{\theta} \mathcal{P}\big(ne^{i\phi}\big) \exp\big\{n(e^{i\phi} - i\phi - 1)\big\}\, d\phi + O\big(n^{c+\frac{1}{2}} e^{-(1-\alpha)n}\big).$$

In the integral, set $x = \phi\sqrt{n}$ to get

$$a_n = \frac{1 + O(1/n)}{\sqrt{2\pi}} \int_{-\theta\sqrt{n}}^{\theta\sqrt{n}} \mathcal{P}\big(ne^{ix/\sqrt{n}}\big) \exp\Big\{n\Big(e^{ix/\sqrt{n}} - \frac{ix}{\sqrt{n}} - 1\Big)\Big\}\, dx$$
$$+ O\big(n^{c+\frac{1}{2}} e^{-(1-\alpha)n}\big).$$

To proceed with further approximations, expand both functions in the integrand for fixed x. However, these expansion are only valid for small x; near the limits of the integral, x is not small enough to warrant use of the expansions. So, further split the integral inside the cone as

$$\int_{-\theta\sqrt{n}}^{\theta\sqrt{n}} = \int_{-\theta\sqrt{n}}^{-\ln n} + \int_{-\ln n}^{\ln n} + \int_{\ln n}^{\theta\sqrt{n}}.$$

As $\ln n/\sqrt{n} \to 0$, for n large enough, the expansions are valid over the entire range of the middle integral. The first and third (symmetric) integrals contribute a negligible error because

$$\begin{aligned}
&\left|\int_{\ln n}^{\theta\sqrt{n}} \mathcal{P}\Big(ne^{ix/\sqrt{n}}\Big) \exp\Big\{n\Big(e^{ix/\sqrt{n}} - \frac{ix}{\sqrt{n}} - 1\Big)\Big\}\, dx\right| \\
&\quad \le \int_{\ln n}^{\theta\sqrt{n}} \left|\mathcal{P}\big(ne^{ix/\sqrt{n}}\big)\right| \left|\exp\Big\{n\Big(e^{ix/\sqrt{n}} - \frac{ix}{\sqrt{n}} - 1\Big)\Big\}\right| dx.
\end{aligned}$$

But then by condition (i):

$$\begin{aligned}
\left|\int_{\ln n}^{\theta\sqrt{n}}\right| &\le \int_{\ln n}^{\theta\sqrt{n}} \beta_1 n^c \exp\Big\{n\Big(\cos\Big(\frac{x}{\sqrt{n}}\Big) - 1\Big)\Big\}\, dx \\
&\le \beta_1 n^c \exp\Big\{n\Big(\cos\Big(\frac{\ln n}{\sqrt{n}}\Big) - 1\Big)\Big\} \int_{\ln n}^{\theta\sqrt{n}} dx \\
&= \beta_1 n^c \exp\Big\{n\Big(1 - \frac{\ln^2 n}{2n} + O\Big(\frac{\ln^4 n}{n^2}\Big)\Big) - n\Big)\Big\}\, [\theta\sqrt{n} - \ln n] \\
&\le \beta_1 \theta n^c e^{-\xi \ln^2 n} \sqrt{n},
\end{aligned}$$

for some positive constant ξ; the latter bound decays exponentially fast and is $O(n^{c-\frac{1}{2}} \ln n)$. Of course, $\left|\int_{-\theta\sqrt{n}}^{-\ln n}\right|$ gives the same asymptotic error.

We now return to the range where expansions are valid; for any fixed x the exponential function gives

$$\begin{aligned}
\exp\Big\{n\Big(e^{ix/\sqrt{n}} - \frac{ix}{\sqrt{n}} - 1\Big)\Big\} &= \exp\Big\{n\Big(\Big[1 + \frac{ix}{\sqrt{n}} - \frac{x^2}{2n} - \frac{ix^3}{6n^{3/2}} + O\Big(\frac{1}{n^2}\Big)\Big] \\
&\qquad - \frac{ix}{\sqrt{n}} - 1\Big)\Big\} \\
&= \exp\Big\{-\frac{x^2}{2} - \frac{ix^3}{6\sqrt{n}} + O\Big(\frac{1}{n}\Big)\Big\} \\
&= e^{-x^2/2}\Big\{1 - \frac{ix^3}{6\sqrt{n}} + O\Big(\frac{1}{n}\Big)\Big\}.
\end{aligned}$$

Being analytic, the Poisson transform can be expanded in series form around n, giving

$$\mathcal{P}(z) = \mathcal{P}(n) + \mathcal{P}'(n)(z-n) + \frac{1}{2!}\mathcal{P}''(n)(z-n)^2 + \cdots = \mathcal{P}(n) + O\big(\mathcal{P}'(n)(z-n)\big),$$

and again $\mathcal{P}'$ can be extracted by Cauchy's formula:

$$\mathcal{P}'(n) = \frac{1}{2\pi i}\oint_{\Lambda'} \frac{\mathcal{P}(z)}{(z-n)^2}\, dz,$$

where Λ' is a circle centered at n and of any radius sufficiently small for the entire circle to lie inside the de-Poissonization cone, that is, Λ''s radius is less than $n \sin\theta$ (see Figure 1.12); let us take it to be $R_n = \frac{1}{2} n \sin\theta$. As Λ' lies entirely inside S_θ, condition (i) now comes again into the picture to show that

$$\big|\mathcal{P}'(n)\big| = \frac{1}{2\pi}\oint_{\Lambda'} \frac{|\mathcal{P}(z)|}{|z-n|^2}\,|dz| \le \frac{1}{2\pi}\oint_{\Lambda'} \frac{\beta_1 n^c}{R_n^2}\,|dz|.$$

The integral $\oint_{\Lambda'} |dz|$ is the perimeter of the Λ', and so

$$\big|\mathcal{P}'(n)\big| \le \frac{\beta_1 n^c}{R_n}.$$

In the integral $\int_{-\ln n}^{\ln n}$, z comes from a restricted cone of (infinitesimal as $n \to \infty$) angle $-\ln n/\sqrt{n} \le \phi \le \ln n/\sqrt{n}$. Each z on the arc of Λ lying inside the restricted cone is close enough to n, with $|z-n| < n \times (\ln n/\sqrt{n}) = \sqrt{n}\ln n$. The error term $O\big(\mathcal{P}'(n)(z-n)\big)$ is $O(n^{c+\frac{1}{2}} \ln n/R_n)$. Within the restricted cone, the Poisson transform has the expansion

$$\mathcal{P}(z) = \mathcal{P}(n) + O\big(n^{c-\frac{1}{2}} \ln n\big).$$

So, finally,

$$a_n = O\big(n^{c+\frac{1}{2}} e^{-(1-\alpha)\theta}\big) + \frac{1+O(1/n)}{\sqrt{2\pi}} \int_{-\ln n}^{\ln n} e^{-x^2/2}\Big\{1 - \frac{ix^3}{6\sqrt{n}} + O\Big(\frac{1}{n}\Big)\Big\}\Big(\mathcal{P}(n) + O\big(n^{c-\frac{1}{2}}\ln n\big)\Big)\,dx + O\big(n^{c-\frac{1}{2}}\ln n\big).$$

The dominant component in this integrand is $\mathcal{P}(n)e^{-x^2/2}$; upon extending the limits of the integral from $-\infty$ to $+\infty$, we only introduce another $O(1/n)$ error; the dominant remaining integral is $\mathcal{P}(n)$ times 1 (the area under the standard normal density). ■

If a_n is a sequence of average values, the De-Poissonization Lemma states that under suitable growth conditions on its Poisson transform, the fixed-population averages are nearly the same as the Poissonized average with n replacing z.

The rough idea behind de-Poissonization is the following. For any fixed n, the coefficient of z^n in the power series expansion of $e^z\mathcal{P}(z)$ about the origin is $a_n/n!$. Finding a_n is then only a matter of extracting a coefficient from a generating function. This is routinely done by considering a contour integral around the origin in the z complex plane. In particular, when we choose the contour to be the circle $|z| = n$, we get

$$a_n = \frac{n!}{2\pi i} \oint \frac{e^z \mathcal{P}(z)}{z^{n+1}}\, dz.$$

Such integrals typically have most of their value contributed by a small arc at the intersection of the circle with the positive real line (i.e., at the saddle point $z = n$); the rest of the contour adds only an ignorable correction. Over this small arc, the value of the function $\mathcal{P}(z)$ is well approximated by $\mathcal{P}(n)$. Taking this now-constant factor outside the integral, performing the remaining elementary integration, and applying Stirling's approximation to $n!$, we see that all factors cancel out, except $\mathcal{P}(n)$. The point is that a_n can be accurately approximated by $\mathcal{P}(n)$, with a diminishing error as $n \to \infty$. The De-Poissonization Lemma gives sufficient conditions to make this approximation valid.

1.11.4 The Dirichlet Transform

Another transform that appears in solving recurrences is the Dirichlet transform which, like the Poisson transform, transforms sequences of real numbers. For a sequence $\{a_n\}_{n=1}^{\infty}$, we define the *Dirichlet transform* as the generating function

$$D(s) \overset{\text{def}}{=} \sum_{n=1}^{\infty} \frac{a_n}{n^s},$$

for some domain of convergence in the s complex plane.

An inversion formula, proved next, recovers certain finite sums of the members of the sequence. This inversion formula is called the Mellin–Perron inversion, and is typical of the general class of integral transform inversion formulas that include the Mellin, Laplace, and Fourier transforms where the inversion is given by a line integral in the complex plane. By methods similar to the closing-the-box method, such line integrals are actually estimated within small errors by residues of poles within the box.

Lemma 1.4 *(Mellin-Perron). Let $a_1, a_2, \ldots$ be a sequence of real numbers with an absolutely convergent Dirichlet generating function $D(s)$, for $\Re s > \alpha > 0$. Then, for $\beta > \alpha$,*

$$\sum_{k=1}^{n-1} (n-k)a_k = \frac{n}{2\pi i} \int_{\beta - i\infty}^{\beta + i\infty} \frac{D(s)\, n^s}{s(s+1)}\, ds.$$

Proof. Let $\beta > \alpha$ and consider the line integral

$$\int_{\beta-i\infty}^{\beta+i\infty} \frac{D(s)\, n^s}{s(s+1)}\, ds = \int_{\beta-i\infty}^{\beta+i\infty} \sum_{k=1}^{\infty} \Big(\frac{a_k}{k^s}\Big) \frac{n^s}{s(s+1)}\, ds.$$

As the series $\sum_{k=1}^{\infty} a_k k^{-s}$ converges absolutely for $\Re s > \alpha$, so does the series $\sum_{k=1}^{\infty} a_k n^s/[s(s+1)k^s]$; it is permissible to interchange the sum and integral signs. So,

$$\frac{n}{2\pi i} \int_{\beta-i\infty}^{\beta+i\infty} \frac{D(s)\, n^s}{s(s+1)}\, ds = \frac{n}{2\pi i} \sum_{k=1}^{\infty} a_k \int_{\beta-i\infty}^{\beta+i\infty} \frac{(n/k)^s}{s(s+1)}\, ds. \qquad (1.23)$$

Introduce the function

$$J(x) = \frac{1}{2\pi i} \int_{\beta-i\infty}^{\beta+i\infty} \frac{x^s}{s(s+1)}\, ds.$$

To evaluate the line integral in $J(x)$, for $x > 1$, consider the circle centered at the origin and having radius $R > \beta$. Let its two intercepts with the vertical line $\Re s = \beta$ be the points $s_1 = \beta + ia$ and $s_2 = \beta - ia$, with $a = \sqrt{R^2 - \beta^2}$. Consider the closed contour Λ consisting of the vertical line segment connecting s_1 to s_2 and the closing arc Λ_1, the arc of the circle connecting the points s_1 and s_2 to the left of the line $\Re s = \beta$; see Figure 1.13. The contour integral

$$\oint_{\Lambda} \frac{x^s}{s(s+1)}\, ds = \int_{\beta-ia}^{\beta+ia} \frac{x^s}{s(s+1)}\, ds + \int_{\Lambda_1} \frac{x^s}{s(s+1)}\, ds,$$

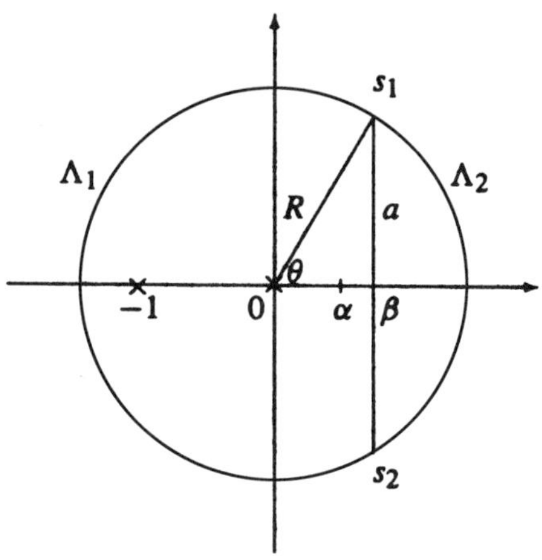

Figure 1.13. Evaluating $J(x)$ via residues.

is to be evaluated via the residues of the poles of its integrand by Cauchy's classical formula:

$$\int_{\beta-ia}^{\beta+ia} \frac{x^s}{s(s+1)}\,ds = 2\pi i\Big[\operatorname*{Res}_{s=0} \frac{x^s}{s(s+1)} + \operatorname*{Res}_{s=-1} \frac{x^s}{s(s+1)}\Big] - \int_{\Lambda_1} \frac{x^s}{s(s+1)}\,ds.$$

Letting $R \to \infty$, $a \to \infty$; the integral on the straight line segment becomes $2\pi i J(x)$. Thus,

$$J(x) = 1 - \frac{1}{x} - \frac{1}{2\pi i}\lim_{R\to\infty} \int_{\Lambda_1} \frac{x^s}{s(s+1)}\,ds.$$

As is customary in complex analysis, the limit in question is 0 because

$$\begin{aligned}\left|\int_{\Lambda_1} \frac{x^s}{s(s+1)}\,ds\right| &\le \int_{\Lambda_1} \left|\frac{x^s}{s(s+1)}\right| |ds| \\ &\le \int_{\Lambda_1} \frac{x^{\Re s}}{|s|(|s|-1)}\,|ds| \\ &\le \int_{\arg s_1}^{\arg s_2} \frac{x^\beta}{R(R-1)}\,R\,d\theta \\ &\le \frac{2\pi x^\beta}{R-1} \\ &\to 0, \qquad \text{as } R\to\infty.\end{aligned}$$

Thus $J(x) = 1 - 1/x$, if $x > 1$.

By a similar contour integral, closing the contour with Λ_2, the arc of the circle extending to the right between s_1 and s_2, one sees that $J(x) \equiv 0$, for all $x \le 1$, as the contour encloses no poles of the integrand. The limiting process by which one shows that, as $R \to \infty$, $\int_{\Lambda_2} \to 0$ is basically the same where one uses the condition $x < 1$.

Going back to (1.23),

$$\begin{aligned}\frac{n}{2\pi i}\int_{\beta-i\infty}^{\beta+i\infty} \frac{D(s)\,n^s}{s(s+1)}\,ds &= n\sum_{k=1}^{\infty} a_k J\Big(\frac{n}{k}\Big) \\ &= n\sum_{k=1}^{n-1}\Big(1-\frac{k}{n}\Big)a_k. \qquad \blacksquare\end{aligned}$$

The Mellin-Perron theorem is very useful for handling recurrence equations that are commonplace in divide-and-conquer algorithms. The sorting algorithm MERGE SORT falls in this class. Algorithms that divide a given input into two parts are forced to deal with ceils and floors. When the input is of size n, the two parts are not equal for odd n. Suppose a_n is some efficiency measure for such a divide-and-conquer algorithm. The algorithm creates two parts of sizes $\lceil n/2 \rceil$ and $\lfloor n/2 \rfloor$, on which it recursively repeats its operations. The result of a divide-and-conquer algorithm is constructed from the recursive operations; combining the results of the two recursions may add a toll b_n to the efficiency measure. A recurrence equation of the general form

$$a_n = a_{\lceil n/2 \rceil} + a_{\lfloor n/2 \rfloor} + b_n \tag{1.24}$$

is associated with such an algorithm. MERGE SORT will provide a perfectly natural example in this class of algorithms.

A technique for handling these recurrences and finding asymptotic solutions by integral transform methods is due to Flajolet and Golin (1994). This technique is discussed next.

The method consists in first getting rid of all ceils and floors by taking forward and backward differences of the recurrence. The Dirichlet generating function for the sequence of differences is then constructed and inverted.

The starting point in solving the general form (1.24) is to get rid of the ceils and floors involved by differencing operations. The forward and backward difference operators are denoted by $\triangle$ and $\triangledown$: Operating on a sequence q_n,

$$\triangle q_n = q_{n+1} - q_n,$$

$$\triangledown q_n = q_n - q_{n-1}.$$

Taking the backward difference of (1.24) we have

$$\triangledown a_n = \left(a_{\lceil n/2 \rceil} + a_{\lfloor n/2 \rfloor} + b_n\right) - \left(a_{\lceil (n-1)/2 \rceil} + a_{\lfloor (n-1)/2 \rfloor} + b_{n-1}\right).$$

For even indexes $n = 2m$, this simplifies to:

$$\begin{aligned} \triangledown a_{2m} &= (a_m + a_m + b_{2m}) - (a_m + a_{m-1} + b_{2m-1}) \\ &= (a_m - a_{m-1}) + (b_{2m} - b_{2m-1}) \\ &= \triangledown a_m + \triangledown b_{2m}, \end{aligned}$$

a recurrence without ceils and floors. For odd indexes $n = 2m + 1$, the procedure is exactly the same; one gets

$$\triangledown a_{2m+1} = \triangledown a_{m+1} + \triangledown b_{2m+1},$$

also a recurrence without ceils and floors. If we let m run its course, from 1 to ∞ on the recurrence for $\triangledown a_{2m}$, and from 0 to ∞ on the recurrence for $\triangledown a_{2m+1}$, the left-hand sides of the two recurrences will produce the entire sequence $\triangledown a_n$, $n = 1, 2, \ldots$, term by term, and so will the right-hand sides generate the entire sequence $\triangledown b_n$, $n = 1, 2, \ldots$. However, the right-hand side of each recurrence will generate $\triangledown a_1, \triangledown a_2, \triangledown a_3, \ldots$. This overlap gives rise to a clumsy Dirichlet transform for the sequence of backward differences on a_n, because convergence issues clutter the formal definitions. For example, the Dirichlet transform for the recurrence on even indexes gives the term $\nabla a_3/6^s$, and the transform for the recurrence on odd indexes gives the term $\nabla a_3/5^s$. It is not easy to combine such terms in one convergent Dirichlet generating function. This minor hurdle is circumvented by applying a second layer of differencing operators to remove the overlap. Applying the forward difference operator to the recurrence on backward differences gives

$$\begin{aligned}\triangle \nabla a_{2m} &= \nabla a_{2m+1} - \nabla a_{2m} \\ &= (\nabla a_{m+1} + \nabla b_{2m+1}) - (\nabla a_m + \nabla b_{2m}) \\ &= \triangle \nabla a_m + \triangle \nabla b_{2m}, \end{aligned} \tag{1.25}$$

and for odd indexes

$$\triangle \nabla a_{2m+1} = \triangle \nabla b_{2m+1}. \tag{1.26}$$

We next develop the Dirichlet generating function $D(s)$ of the sequence $\{\triangle \nabla a_n\}_{n=1}^{\infty}$:

$$D(s) \overset{\text{def}}{=} \sum_{n=1}^{\infty} \frac{\triangle \nabla a_n}{n^s}.$$

Because of the parity consideration, we break up the infinite sum into odd and even indexes and apply (1.25) and (1.26) to obtain

$$\begin{aligned} D(s) &= \sum_{m=1}^{\infty} \frac{\triangle \nabla a_{2m}}{(2m)^s} + \sum_{m=0}^{\infty} \frac{\triangle \nabla a_{2m+1}}{(2m+1)^s} \\ &= \sum_{m=1}^{\infty} \frac{\triangle \nabla a_m + \triangle \nabla b_{2m}}{(2m)^s} + \sum_{m=0}^{\infty} \frac{\triangle \nabla b_{2m+1}}{(2m+1)^s} \\ &= \frac{1}{2^s} \sum_{m=1}^{\infty} \frac{\triangle \nabla a_m}{m^s} + \sum_{k=1}^{\infty} \frac{\triangle \nabla b_k}{k^s} \\ &= \frac{D(s)}{2^s} + \eta(s), \end{aligned}$$

where $\eta(s)$ is the Dirichlet transform of the *known* sequence $\{\triangle \nabla b_k\}$. Reorganizing the last equality,

$$D(s) = \frac{\eta(s)}{1 - 2^{-s}}.$$

Forward and backward operators telescope nicely to unfold the final term of a summation (the nth term) via the formula in the next lemma.

Lemma 1.5

$$\sum_{k=1}^{n-1} (n-k) \bigtriangleup \bigtriangledown a_k = a_n - na_1.$$

Proof. By definition we take $a_0 = 0$. Break up the finite sum as follows:

$$\sum_{k=1}^{n-1} (n-k) \bigtriangleup \bigtriangledown a_k = \sum_{k=1}^{n-1} n \bigtriangleup \bigtriangledown a_k - \sum_{k=1}^{n-1} k \bigtriangleup \bigtriangledown a_k.$$

The first sum on the right unwinds as

$$\begin{aligned}\sum_{k=1}^{n-1} n \bigtriangleup \bigtriangledown a_k &= n(\bigtriangleup \bigtriangledown a_1 + \bigtriangleup \bigtriangledown a_2 + \cdots + \bigtriangleup \bigtriangledown a_{n-1}) \\ &= n\big[(\bigtriangledown a_2 - \bigtriangledown a_1) + (\bigtriangledown a_3 - \bigtriangledown a_2) \\ &\quad + (\bigtriangledown a_4 - \bigtriangledown a_3) + \cdots + (\bigtriangledown a_n - \bigtriangledown a_{n-1})\big] \\ &= n(\bigtriangledown a_n - \bigtriangledown a_1).\end{aligned}$$

In a like manner, the second sum unwinds as

$$\begin{aligned}\sum_{k=1}^{n-1} k \bigtriangleup \bigtriangledown a_k &= (\bigtriangledown a_2 - \bigtriangledown a_1) + 2(\bigtriangledown a_3 - \bigtriangledown a_2) + \cdots + (n-1)(\bigtriangledown a_n - \bigtriangledown a_{n-1}) \\ &= -(\bigtriangledown a_1 + \bigtriangledown a_2 + \cdots + \bigtriangledown a_{n-1}) + (n-1) \bigtriangledown a_n.\end{aligned}$$

Combining the two sums we obtain

$$\begin{aligned}\sum_{k=1}^{n-1} (n-k) \bigtriangleup \bigtriangledown a_k &= \bigtriangledown a_n - n \bigtriangledown a_1 + (\bigtriangledown a_1 + \cdots + \bigtriangledown a_{n-1}) \\ &= (a_n - a_{n-1}) - n(a_1 - a_0) \\ &\quad + \big[(a_1 - a_0) + \cdots + (a_{n-1} - a_{n-2})\big] \\ &= a_n - na_1.\end{aligned}$$

■

In the recurrence (1.24) we shall assume bounds on the growth of the sequence $\{b_n\}_{n=1}^{\infty}$. This is done to ensure the existence of the Dirichlet transform of $\bigtriangleup \bigtriangledown b_n$ in some manageable domain in the complex plane. In almost all recurrences of the general form (1.24) arising naturally in an algorithmic application reasonable growth bounds like $b_n = O(n)$ exist. The growth rate $b_n = O(n)$, as $n \to \infty$, means that there is a constant K and an integer n_0, such that $|b_n| < Kn$, for all $n \ge n_0$. So, the second-order differences $|\bigtriangleup \bigtriangledown b_n| = |(b_{n+1} - b_n) - (b_n - b_{n-1})| \le |b_{n+1}| + 2|b_n| + |b_{n-1}| < K(n+1) + 2Kn + K(n-1) = 4Kn$, for all $n > n_0$. The formal

Dirichlet transform

$$\Big|\sum_{n=1}^{\infty} \frac{\triangle \nabla b_n}{n^s}\Big| = \Big|\sum_{n=1}^{n_0} \frac{\triangle \nabla b_n}{n^s}\Big| + \Big|\sum_{n_0+1}^{\infty} \frac{\triangle \nabla b_n}{n^s}\Big|$$
$$\le O(1) + \sum_{n=n_0+1}^{\infty} \Big|\frac{\triangle \nabla b_n}{n^s}\Big|$$
$$\le O(1) + \sum_{n=n_0+1}^{\infty} \frac{4Kn}{|n^s|}$$
$$= O(1) + 4K \sum_{n=1}^{\infty} \frac{1}{|n^{s-1}|},$$

which finitely exists for $\Re s > 2$.

A combination of Lemmas 1.4 and 1.5 culminates into an exact solution to a recurrence equation with ceils and floors. We can take the line integrals involved in the inversion at any $\beta > 2$. We shall take $\beta = 3$.

Theorem 1.6 *(Flajolet and Golin, 1994). Let $b_1, b_2, \ldots$ be a sequence of numbers with growth $b_n = O(n)$. The recurrence equation*

$$a_n = a_{\lceil n/2 \rceil} + a_{\lfloor n/2 \rfloor} + b_n$$

has the solution

$$a_n = a_1 n + \frac{n}{2\pi i} \int_{3-i\infty}^{3+i\infty} \frac{\eta(s) n^s}{s(s+1)(1-2^{-s})}\, ds,$$

where $\eta(s)$ is the Dirichlet transform of the sequence of second-order differences $\triangle \nabla b_n$.

Applying Theorem 1.6 requires evaluation of a line integral. It is typical in this Mellin-like analysis to resort to evaluating the integral by relating it to the singularities of its integrand. Typically, these singularities are poles of low order lying on the real axis and equispaced singularities aligned on vertical lines symmetrically about the horizontal line.

It is common folklore that the poles on vertical lines give small oscillations in the solution, and that the dominant behavior in inversion formulas comes from singularities of higher order, as will be verified when we follow this route. To evaluate line integrals like those in the statement of Theorem 1.6 we close a contour—starting with the line integral $\int_{3-iM}^{3+iM}$, for some large M, we relate it to a line integral on the closed contour Λ consisting of the line segment joining $3-iM$ to $3+iM$, and Λ_1, an arc of a large circle centered at the origin and with radius $\sqrt{M^2+9}$. The number M is suitably chosen so that the circle does not pass through any poles; see Figure 1.14. We have

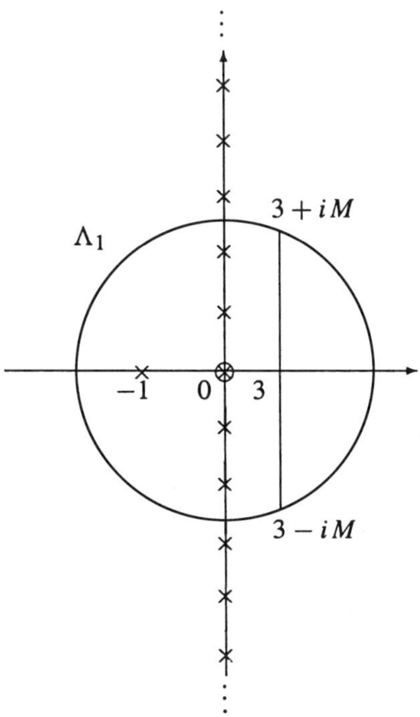

Figure 1.14. Evaluating the line integral via residues.

$$\int_{3-iM}^{3+iM} + \int_{\Lambda_1} = \oint_{\Lambda} .$$

The contour encloses a number of the poles to the left of the line $\Re s = 3$. By a careful limit process, letting $M \to \infty$ without letting the circle pass through any poles, the integral on Λ_1 vanishes, by considerations, not repeated here, like those used in the proof of Lemma 1.4 for determining the limit of a line integral. In the limit, the contour grows to the infinite semicircle Λ_∞, with the vertical line $\Re s = 3$ as its diameter, and

$$\int_{3-i\infty}^{3+i\infty} = \oint_{\Lambda_\infty} .$$

The evaluation of the line integral has been reduced to a computational task. By Cauchy's classical residue theorem, the equivalent integral on the closed contour Λ_∞ is $2\pi i$ times the sum of the residues of the poles of the integrand inside Λ_∞.

1.11.5 Rice's Method

Upon extracting coefficients from closed-form generating functions, certain types of summations whose summands alternate signs and contain binomial coefficients often appear. The general form of these sums is

$$\sum_{k=a}^{n} (-1)^k f(k) \binom{n}{k},$$

starting at some small integer a. This type of sum will be called an *alternating sum.*

It is not always easy to get a closed-form expression for an alternating sum. Even numerical evaluation is often fraught with difficulty because binomial coefficients tend to be rather large and the alternation of signs then demands accurate subtraction of very large numbers, which is a challenging arithmetic exercise on computers.

Accurate approximations can be worked out for alternating sums via *Rice's method*, a method based in complex variable theory. The idea in the method is to extend $f(k)$ to its analytic continuation $f(z)$ in the complex plain. An alternating sum then has an interpretation as the residue calculation for the poles *inside* some closed contour. By Cauchy's residue theorems we can enlarge the contour and work with the poles *outside* the old contour and compensate by the new line integral. When the new contour is properly chosen, the new line integral gives an asymptotically ignorable error.

This outline of Rice's method hinges on the connection between a sum and a line integral, established next.

Lemma 1.6 *Let $f(k)$ be a function defined at all nonnegative integers, $k = 0, 1, \ldots$. Suppose $f(z)$ is the analytic continuation of f. Then*

$$\sum_{k=a}^{n} (-1)^k f(k) \binom{n}{k} = -\frac{1}{2\pi i} \oint_{\Lambda} f(z) \beta(-z, n+1)\, dz,$$

where β is the classical Beta function, and Λ is a closed contour enclosing the integers $a, a+1, \ldots, n$, and no other integers.

Proof. The Beta function has the representation:

$$\begin{aligned} \beta(-z, n+1) &= \frac{\Gamma(-z)\Gamma(n+1)}{\Gamma(-z+n+1)} \\ &= \frac{n!\,\Gamma(-z)}{(n-z)\Gamma(n-z)} \\ &\;\;\vdots \\ &= \frac{n!\,\Gamma(-z)}{(n-z)(n-z-1)\ldots(-z)\Gamma(-z)} \\ &= \frac{n!}{(n-z)(n-1-z)\ldots(-z)}. \end{aligned} \tag{1.27}$$

The last representation manifests clearly that the Beta function $\beta(-z, n+1)$ has simple poles at the integers $k = 0, 1, \ldots, n$, of which only $a, \ldots, n$ are inside Λ.

Evaluate the contour integral via residues of poles inside Λ:

$$-\frac{1}{2\pi i}\oint_{\Lambda} f(z)\beta(-z, n+1)\,dz = -\frac{1}{2\pi i} \times 2\pi i \sum_{k=a}^{n} \underset{z=k}{Res}\, \frac{n!\, f(z)}{(n-z)\ldots(-z)}$$

$$= \sum_{k=a}^{n} \lim_{z\to k} \frac{n!\, f(z)(z-k)}{\big[(n-z)\ldots(k+1-z)\big](z-k)}$$

$$\times \frac{1}{\big[(-1)^k (z-(k-1))\ldots z\big]}$$

$$= \sum_{k=a}^{n} (-1)^k f(k)$$

$$\times \frac{n!}{\big[(n-k)\times\ldots\times 2\times 1\big]\big[1\times 2\times\ldots\times k\big]}.$$

■

Lemma 1.6 translates an alternating sum into a line integral on a closed contour Λ surrounding the points $a, a+1, \ldots, n$. The correspondence of the sum to a contour integral gives us an alternative for computing residues. We can work with the residues of poles outside the contour, but inside a new enlarged one.

To asymptotically estimate this contour integral, enlarge the contour Λ into a new contour Λ' to enclose poles of the integrand outside Λ. By Cauchy's residue theorem

$$-\frac{1}{2\pi i}\oint_{\Lambda} f(z)\beta(-z, n+1)\,dz = -\frac{1}{2\pi i}\oint_{\Lambda'} f(z)\beta(-z, n+1)\,dz$$

$$+ \sum \big\{\text{Residues of } f(z)\beta(-z, n+1) \text{ outside } \Lambda \text{ and inside } \Lambda'\big\}.$$

The larger the new contour is, the more poles outside Λ it will catch, and as is common in complex analysis, the compensating line integral $\oint_{\Lambda'}$ gives a smaller error. The situation is a trade-off between how complicated a residue computation we are willing to tackle, versus how small an error we wish to control. Often, a large contour catching only a few dominant poles outside Λ is sufficient to bring out the main character of the sum with only ignorable error. This heuristic is illustrated next by an example.

Example 1.2 Let us evaluate

$$S_n = \sum_{k=2}^{n} \frac{(-1)^k}{1-2^{-k}} \binom{n}{k}.$$

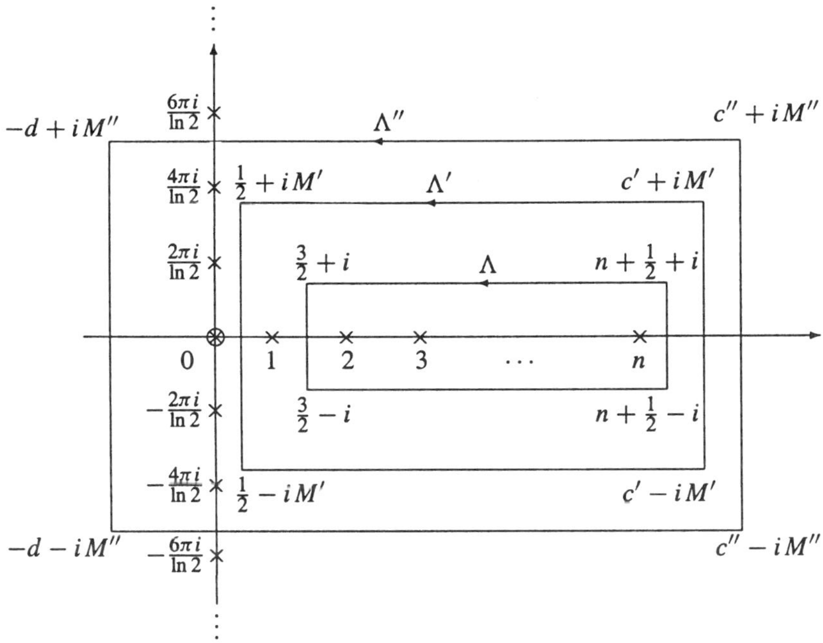

Figure 1.15. Changing the integration contour by Rice's method; $\times$ indicates a simple pole and $\otimes$ indicates a double pole.

The analytic continuation of $1/(1-2^{-k})$ to the z complex plane is $1/(1-2^{-z})$. Take the contour of integration Λ to be the rectangle connecting the four corners $3/2+i$, $n+\frac{1}{2}+i$, $n+\frac{1}{2}-i$, and $3/2-i$; see Figure 1.15.

By Rice's method, if we enlarge Λ into Λ', the rectangular contour that connects the four points $1/2+iM'$, $c'+iM'$, $c'-iM'$, and $1/2-iM'$, for $c' > n+1$, the given alternating sum is given by

$$S_n = \sum\left\{\text{Residues of } \frac{\beta(-z, n+1)}{1-2^{-z}} \text{ outside } \Lambda \text{ and inside } \Lambda'\right\} - \frac{1}{2\pi i}\oint_{\Lambda'} \frac{\beta(-z, n+1)}{1-2^{-z}}\, dz;$$

the easiest way to compute these residues is via the working formula of rational functions given in (1.27). The only pole of the integrand between the two contours Λ and Λ' is at 1. This pole gives the residue

$$\underset{z=1}{Res} \frac{\beta(-z, n+1)}{1-2^{-z}} = 2n.$$

The error term requires some work. For any fixed x, the Gamma function $\Gamma(x+iy)$ is $O(|y|^{x-1/2}e^{-\pi|y|/2})$, as $|y| \to \infty$. So, $|\Gamma(x+iy)| \le K_1|y|^{x-1/2}e^{-\pi|y|/2}$, for some positive constant K_1 and all $|y|$ larger than some y_0. This is helpful in estimating the error introduced by shifting the line integral to a larger contour with $M' > y_0$. On the top side of Λ', $z = x + iM'$, for $1/2 \le x \le c'$. As we let $M' \to \infty$, the integral on the top side of the rectangle diminishes to 0, which can be seen from:

$$\begin{aligned}\Bigl|\int_{z=1/2+iM'}^{c'+iM'} \frac{\beta(-z, n+1)}{1-2^{-z}}\,dz\Bigr| &\le \int_{x=1/2}^{c'} \Bigl|\frac{1}{1-2^{-x-iM'}}\Bigr| \\ &\qquad \times \Bigl|\frac{\Gamma(n+1)\,\Gamma(-x-iM')}{\Gamma(n+1-x-iM')}\Bigr|\,dx \\ &\le \int_{1/2}^{c'} \frac{K_1}{1-2^{-x}}(M')^{-x-1/2}e^{-\pi M'/2} \\ &\qquad \times \Bigl|\frac{\Gamma(n+1)}{\Gamma(n+1-x-iM')}\Bigr|\,dx.\end{aligned}$$

By the Stirling approximation of Gamma functions, the ratio $|\Gamma(n+1)/\Gamma(n+1-x-iM')|$ is bounded by $K_2 n^x$, for some constant K_2 and for all n exceeding some n_0. Hence, for any large enough fixed n,

$$\begin{aligned}\Bigl|\int_{z=1/2+iM'}^{c'+iM'} \frac{\beta(-z, n+1)}{1-2^{-z}}\,dz\Bigr| &\le \frac{\sqrt{2}\,K_1K_2e^{-\pi M'/2}}{(\sqrt{2}-1)\sqrt{M'}} \int_{1/2}^{c'} \Bigl(\frac{n}{M'}\Bigr)^x dx \\ &\le \frac{\sqrt{2}\,K_1K_2e^{-\pi M'/2}\,(n/M')^{1/2}}{(\sqrt{2}-1)\sqrt{M'}\,\ln(M'/n)},\end{aligned}$$

valid for any fixed $n \ge n_0$, with $M' \ge \max\{y_0, n\}$. Taking the limit,

$$\lim_{M'\to\infty} \Bigl|\int_{z=1/2+iM'}^{n+1+iM'} \frac{\beta(-z, n+1)}{1-2^{-z}}\,dz\Bigr| = 0.$$

The argument for the line integral on the bottom side is the same. On the right side, for very large values of c' the Gamma function in the denominator grows very fast in magnitude and annihilates the integral; on the line $z = c' + iy$, for $-M' \le y \le M'$, we have

$$\Bigl|\int_{z=c'+iM'}^{c'-iM'} \frac{\beta(-z, n+1)}{1-2^{-z}}\,dz\Bigr| \le \int_{y=-M'}^{M'} \frac{|\Gamma(n+1)|}{|1-2^{-c'-iy}|}\Bigl|\frac{\Gamma(-c'-iy)}{\Gamma(-c'+n+1-iy)}\Bigr|\,dy,$$

and again if c' is very large, for any fixed n, we have $|\Gamma(-c'-iy)/\Gamma(-c'+n+1-iy)| < K_2(c')^{-(n+1)}$, that is, the line integral on the right side of Λ' is bounded by

$$\int_{-M'}^{M'} \frac{n!}{1-2^{-c'}} \times K_2(c')^{-(n+1)}\,dy \le \frac{2K_2M'}{(1-2^{-c'})(c')^{n+1}}\,n!;$$

as $c' \to \infty$ at the right rate, the bound on the integral tends to 0. For example, for any fixed n we can take $c' = M'$, then let $M' \to \infty$.

The only nonzero error comes from the left side of the rectangle. On the line $z = 1/2 + iy$, for $-M' \le y \le M'$, we have for all $n \ge n_0$,

$$\begin{aligned}\left|\int_{z=1/2-iM'}^{1/2+iM'} \frac{\beta(-z, n+1)}{1-2^{-z}}\, dz\right| &\le \int_{y=-M'}^{M'} \frac{|\Gamma(-1/2-iy)|}{|1-2^{-1/2-iy}|} \\ &\quad \times \left|\frac{\Gamma(n+1)}{\Gamma(1/2+n-iy)}\right| dy \\ &\le \int_{-M'}^{M'} \frac{|\Gamma(-1/2-iy)|\, K_2 n^{1/2}}{1-1/\sqrt{2}}\, dy \\ &\le \frac{K_2\sqrt{2n}}{\sqrt{2}-1} \int_{-\infty}^{\infty} |\Gamma(-1/2-iy)|\, dy.\end{aligned}$$

The integral in the bound contributes only a constant; the magnitude of the Gamma function decays exponentially fast. The integral on the left side of Λ' contributes $O(\sqrt{n}\,)$. As $n \to \infty$, we have an asymptotic estimate for our sum

$$S_n = 2n + O(\sqrt{n}\,).$$

The $O(\sqrt{n}\,)$ term is only an artifact of assuming the right side of the rectangle at $\Re z = 1/2$. Had we taken that right side at any $\Re z = c \in (0, 1)$, we would have gotten an error bounded by $O(n^c)$. For example, we may take the right side at .01 and claim the error to be $O(n^{.01})$.

The error cannot be improved below an exponent of n unless we shift the left side of the rectangle to the left of the vertical axis. For example, if we consider the contour Λ'', a rectangle connecting the four corners $-d \pm iM''$ and $c'' \pm iM''$ instead as our enlargement of Λ (with $d > 0$, $c'' > c'$, and $M'' > M'$), several of the hidden terms in our $O(\sqrt{n}\,)$ calculation with Λ' will come out. Of course, more poles enter the calculation, but that is the expense one pays for increased accuracy; see Figure 1.15.

Working with Λ'', the arguments are exactly the same for the top, bottom, and right sides: For any fixed n, however large, they all still tend to 0 as we let $M'' \to \infty$. The line integral on the left side of the rectangle Λ'' introduces an error that is $O(n^{-d})$, for arbitrary $d > 0$. Such an error cannot take the form of the reciprocal of a polynomial of n; this error may be exponentially small in n. As we let M'' approach infinity the rectangle Λ'' grows to encompass all the poles aligned on the vertical axis; the enlarged contour grows to swallow all the poles outside the original contour Λ. For any arbitrarily large $d > 0$, our summation can be approximated by

$$O(n^{-d}) + \sum\{\text{Residues of poles outside } \Lambda''\}.$$

The function $1/(1 - 2^{-z})$ has poles at solutions of the equation

$$2^{-z} = 1 = e^{2\pi ik}, \qquad k = 0, \pm 1, \pm 2, \ldots .$$

There is a simple pole of $1 - 2^{-z}$ at every $z_k = 2\pi ik/\ln 2$, for $k = \pm 1, \pm 2, \ldots$. Recall that the Beta function has a simple pole at 0. Thus $\beta(-z, n+1)/(1 - 2^{-z})$ has simple poles at each z_k, $k = \pm 1, \pm 2, \ldots$, and a double pole at 0 (refer to Figure 1.15). We have already computed the residue at $z = 1$. The double pole at 0 contributes

$$\operatorname*{Res}_{z=0} \frac{\beta(-z, n+1)}{1 - 2^{-z}} = -\frac{H_n}{\ln 2}.$$

Each of the poles z_k, $k \neq 0$, contributes

$$-\frac{1}{\ln 2}\beta(-z_k, n+1).$$

Collectively the simple poles on the vertical axis contribute

$$\delta(n) = -\frac{1}{\ln 2} \sum_{\substack{k=-\infty \\ k\neq 0}}^{\infty} \beta(-z_k, n+1).$$

A better approximation to the given summation is

$$S_n = 2n - \frac{H_n}{\ln 2} + \delta(n) + O(n^{-d}),$$

for arbitrarily large positive d (but must be held fixed, as we let $n \to \infty$). The major correction in working with Λ'', instead of Λ', is the harmonic number H_n, which grows as the natural logarithm. It is small in comparison with the linear term. The function $\delta(n)$ is bounded in magnitude for all n and contributes a term of oscillatory nature. For large n, one can use the Stirling approximation for $\beta(-z_k, n + 1) = \Gamma(-z_k)\Gamma(n+1)/\Gamma(n+1-z_k)$ to represent these summand terms in the form

$$\beta(-z_k, n+1) = \Gamma(-z_k)n^{z_k}\Big(1 + O\Big(\frac{1}{n}\Big)\Big).$$

Thus

$$|\delta(n)| \le \frac{1}{\ln 2} \sum_{\substack{k=-\infty \\ k\neq 0}}^{\infty} \Big|\Gamma\Big(\frac{2\pi ik}{\ln 2}\Big)\Big|\Big(1 + O\Big(\frac{1}{n}\Big)\Big).$$

The term $1 + O(1/n)$ can be taken out of the summation as the O is uniform in k. The remaining sum divided by $\ln 2$ is $0.1573158421 \times 10^{-5}$, the same constant that appeared in our bound on the oscillations of Example 1.1 on Mellin transform. Of course the O term can be made arbitrarily small. For example, the $1 + O(1/n)$ term is absolutely bounded by $1 + 10^{-6}$ and the entire oscillating term has amplitude not exceeding $0.1573159994 \times 10^{-5}$, for large enough n. These very small oscillations can be neglected in comparison with the logarithmic correction to the linear term.

EXERCISES

1.11.1 (Rais, Jacquet, and Szpankowski, 1993) Suppose $\{X_n\}$ is a sequence of nonnegative integer random variables. Let $\phi_n(u)$ be the probability generating function of X_n. Show that if the bivariate generating function $\sum_{n=0} \phi_n(u)z^n/n!$ is $O(|e^z|)$, the fixed-population probabilities $\mathbf{Prob}\{X_n = k\}$ can be approximated by Poissonized probabilities (the Poisson transform of the probabilities). Specifically, if $N(z) = \text{POISSON}(z)$, show that

$$\mathbf{Prob}\{X_n = k\} = \mathbf{Prob}\big\{X_{N(n)} = k\big\} + \mathcal{E}(n),$$

for a small error function $\mathcal{E}(n)$, uniformly in k. Quantify the error by O bounds.

1.11.2 Suppose X_n is a sequence of random variables. Is the Poissonized variance the same as the Poisson generating function of the sequence of variances $\mathbf{Var}[X_n]$? In other words, if $N(z)$ is a POISSON(z) random variable, is the function $\mathbf{Var}[X_{N(z)}]$ the same as $V(z) = \sum_{n=0}^{\infty} \mathbf{Var}[X_n]z^n e^{-z}/n!$? How then would one go about obtaining $\mathbf{Var}[X_n]$ via de-Poissonization?

1.11.3 (Delange, 1975) Let $\nu(n)$ be the number of 1's in the binary expansion of n. For example, $\nu(14) = \nu((1110)_2) = 3$. Let $Z(n) = \sum_{j=1}^{n-1} \nu(j)$. Show that

$$Z(n) = Z\Big(\Big\lceil \frac{n}{2} \Big\rceil\Big) + Z\Big(\Big\lfloor \frac{n}{2} \Big\rfloor\Big) + \Big\lfloor \frac{n}{2} \Big\rfloor.$$

(Hint: Only "half" of the numbers $1, 2, \dots, n$, are odd.)

1.11.4 Asymptotically solve the Delange recurrence of Exercise 1.11.3.

1.11.5 Solve the functional equation

$$f(x) = 2f\Big(\frac{x}{2}\Big) + e^{-x} - 1 + x,$$

asymptotically, as $x \to \infty$.

1.11.6 (Hwang, 1993) For positive x near 0, approximate the function

$$\sum_{k=1}^{\infty} \frac{1}{e^{kx} - 1}.$$

1.11.7 (Hwang, 1993) Estimate each of the following constants:

(a) $\sum_{k=1}^{\infty} \frac{1}{2^k - 1}$.

(b) $\sum_{k=1}^{\infty} \frac{1}{2^k + 1}$.

(c) $\sum_{k=1}^{\infty} \ln\big(1 + \frac{1}{2^k}\big)$.

1.11.8 Asymptotically approximate the sum

$$\sum_{k=2}^{n} \frac{(-1)^k}{1 - 2^k} \binom{n}{k},$$

as $n \to \infty$.

2

Insertion Sort

In its standard forms INSERTION SORT is a very simple algorithm that belongs to the class of naive sorting algorithms. Although INSERTION SORT is not very efficient it is sometimes a method of choice for small and medium-size files because of its simplicity and capability of handling on-line data.

Straightforward transparent code can be quickly written for this algorithm. It is sometimes favored by practitioners when the situation calls for a rapid production of a working, easy-to-maintain, and easy-to-share computer program when efficiency is not a major concern—when the size of the file to be sorted is relatively small any reasonable sorting method will do the job in a fraction of a second on a modern computer. A nonstandard implementation of INSERTION SORT, suggested by Melville and Gries (1980), optimizes it into a parsimonious algorithm. We shall only hint in passing at this complex and nonstandard implementation in this introduction. Later in the chapter we shall focus on the standard algorithms for sorting by insertion.

INSERTION SORT works by adding data in stages to a sorted file. At the ith stage, the algorithm operates to insert the ith key, say K, into a sorted data file containing $i - 1$ elements. Searching (successfully or unsuccessfully) a sorted file is a relatively easy matter and can be accomplished by many alternative methods. For example, at the ith stage we can conduct a linear search either going forward starting from the top or going backward starting from the bottom of the data file (let us assume the data are kept as an array $A[1 \mathinner{..} i-1]$). Once an insertion location is found (a gap between two elements, say $A[j] \le K < A[j+1]$), we must then make room for the new key by moving the block of data $A[j+1 \mathinner{..} i-1]$ down to occupy positions $j+2, j+3, \ldots, i$. Then we can insert K at its correct position $j+1$; the block of data $A[1 \mathinner{..} i]$ is now sorted.

By using an efficient search method like BINARY SEARCH, INSERTION SORT can be made competitively efficient in a situation where only data comparisons are involved with no or few data movements. Consider, for example, the way Bridge players collect a dealt hand of 13 cards. Most players like to keep the hand sorted by suits, and within one suit, the cards are arranged in decreasing order from left to right. The actual algorithm that many players use is a hybrid of BUCKET SORT and INSERTION SORT. Bridge players usually pick the dealt hand one card at a time. The usual arrangement puts the spades as the leftmost group in the hand, then the hearts, and the diamonds followed by the clubs. After picking $i - 1$ cards, whatever cards are picked are already kept in groups categorized by suit and sorted in decreasing order within a suit. The player finds out the category of the ith card (a simple matter of visual recognition of the color and shape of the suit). This determines the suit or *bucket* in which the card falls. Then an insertion sort using linear search is used to determine the insertion position. The insertion itself in a hand of cards does not require any card movements (usually the player just slides the newly picked card into its position).

Melville and Gries (1980) suggested introducing "holes" in the array—empty positions to receive new arrivals. For example, to store n keys from a sample $X_1, \dots, X_n$, we can use $2n + 1$ slots to create a sorted array of the ordered sample of the form

$$hole \quad X_{(1)} \quad hole \quad X_{(2)} \quad hole \ \dots \ hole \quad X_{(n)} \quad hole. \tag{2.1}$$

A new key falls in one of the holes and can be inserted without moving any data. After some insertions, contiguous clusters of data will begin to form. A new item falling within a cluster will force all the greater elements of the cluster to move down one slot until the first gap below the cluster is filled (as in the so-called hashing with linear probing method). Sooner or later, the array will become too dense and finding the next hole may take a long time. At this stage, the algorithm suspends its routine insertion and enters a *reconfiguration phase*. This phase redistributes the new collection of data over a larger portion of the host array and restores the form (2.1). Suitable optimization criteria can be chosen for the number of gaps, and the need to reconfigure. With these criteria the average number of date moves is $O(n)$ to insert n elements. With an appropriate encoding of an efficient searching method like BINARY SEARCH to search in the presence of gaps, the total search time is $O(n \ln n)$. The average total running time of the optimized algorithm is therefore kept as low as $O(n \ln n)$ as in parsimonious sorting algorithms. This optimization makes the algorithm efficient, but not simple and implementation is fraught with difficulty. Practitioners prefer other standard parsimonious algorithms.

For the rest of the chapter we shall discuss standard implementations of INSERTION SORT.

2.1 A GENERAL FRAMEWORK

A *search strategy* is a collection of algorithms used at the different stages of insertion. Each algorithm in the sequence can be represented by a deterministic insertion extended binary search tree, henceforth to be called the *insertion tree*. For example, a strategy may use the following algorithm for its 10th insertion. At the 10th stage we wish to insert a new key K in a sorted array of 9 elements. The search algorithm may probe position 3 first and if $K < A[3]$, the algorithm probes position 1. If $K < A[1]$, the new key is smaller than all the elements in $A[1 \,..\, 9]$. But if $K > A[1]$, the algorithm will have to probe position 2 next, and it will be immediately determined whether K is the second or third smallest according to whether $K < A[2]$ or not.

TABLE 2.1. The Probe Sequences of a Search Algorithm

Rank(K)	Probe Sequence
1	3 1
2	3 1 2
3	3 1 2
4	3 8 7 6 5 4
5	3 8 7 6 5 4
6	3 8 7 6 5
7	3 8 7 6
8	3 8 7
9	3 8 9
10	3 8 9

Similarly, a deterministic probe sequence corresponds to each possible ranking of K. Our algorithm may be completely specified by its probe sequences as in Table 2.1. This particular search algorithm may be of no practical interest. It is only given for the sake of illustrating the general framework.

We can construct an insertion tree as an extended binary search tree that reflects the essence of this algorithm as follows. The first probe (position 3) will be the root label. Depending on whether $K < A[3]$ or not, the next two potential probes are 1 and 8 (see the second entry of all the probe sequences in Table 2.1). We make these the left and right children of the root. Depending on the outcome of the first two comparisons, the process may just be terminated or continued with one of the potential third probes 2, 7, 9 (see the third entry of the probe sequences in Table 2.1); we make these possibilities the four descendants at level 2 (with a leaf accounting for the possibility of termination). The insertion tree construction proceeds in this fashion until every gap between two keys is represented by a leaf. Upon completing the construction we obtain the tree of Figure 2.1. Now, the different probe sequences

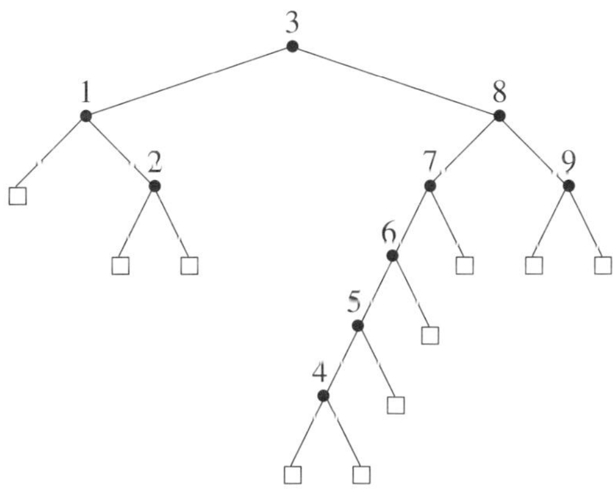

Figure 2.1. The insertion tree of the search algorithm of Table 2.1.

are represented by the labels of the internal nodes on the root-to-leaf paths of the extended binary tree—assuming the 10 leaves are labeled 1 to 10 from left to right, the labels on the path from the root to the jth leaf, $j = 1, \dots, 10$, represent the sequence of probes used while inserting the 10th key if its rank is j.

We emphasize at this point that the insertion tree is completely deterministic. If, for example, the rank of the 10th key is 3, the algorithm will always probe positions 3, 1, then 2. It is the randomness of the rank of the 10th key that introduces a stochastic effect in the insertion process. Under the random permutation model the rank of the 10th key is distributed like a UNIFORM[1 .. 10] random variable. We have seen in Exercise 1.10.3 that the insertion of a 10th key in a random binary search tree of 9 internal nodes is equally likely to fall at any of the 10 leaves of the tree—all probe sequences are equally probable candidates to appear during the insertion of the 10th key.

Many reasonable search strategies can be used in conjunction with INSERTION SORT. To name a few we mention FORWARD LINEAR SEARCH, BACKWARD LINEAR SEARCH, BINARY SEARCH, FIBONACCI's SEARCH, and the many variations available therein. In fact, the different stages of insertion are independent and may even use different insertion algorithms. For example, an INSERTION SORT algorithm may use FORWARD LINEAR SEARCH for the 10th insertion, then BINARY SEARCH for the 11th, and FIBONACCI's SEARCH for the 12th. No one will actually try to do that because in doing so INSERTION SORT will lose its edge of simplicity; such an insertion strategy is far more complicated than necessary. In practice, programmers prefer to use the same simple algorithm (like BACKWARD LINEAR SEARCH) throughout all the stages of INSERTION SORT. Technically speaking, a search algorithm like BACKWARD LINEAR SEARCH for n elements is different from BACKWARD LINEAR SEARCH for $n + 1$ elements, as is evident from their different insertion trees or probe sequences. Nevertheless, in a programming language like PASCAL, for all $n \geq 1$ the different BACKWARD LINEAR SEARCH algorithms can be coded as one procedure that may be invoked at the different stages of INSERTION SORT with different parameters.

In this chapter we shall first present a general distribution theory that encompasses most reasonable search strategies. We then focus on two specific implementations of INSERTION SORT—one that uses BACKWARD LINEAR SEARCH (to be called plainly LINEAR INSERTION SORT) and the other assumes the insertion is done using BINARY SEARCH (to be called BINARY INSERTION SORT). Many different variations on these basic themes will be left for the exercises. Our purpose for presenting LINEAR INSERTION SORT implementation is mostly pedagogic. LINEAR INSERTION SORT introduces the insertion sorting method in one of its simplest forms, and indeed all calculations in the analysis are quite straightforward. We shall establish a Gaussian law for a suitably normalized version of the number of comparisons. The simplicity of LINEAR INSERTION SORT will admit easy calculations for the rate of convergence to the normal limit distribution, too, a question of practical importance. The extreme simplicity of LINEAR INSERTION SORT allows computation of the exact probability distribution.

On the other hand, BINARY INSERTION SORT must be considered since binary search is one of the most efficient known searching algorithms. An insertion sort based on BINARY SEARCH is bound to be efficient (as far as the number of comparisons is concerned). If data movements are not a concern (such as the case in Bridge hand collection), or if Melville and Gries' adaptation is implemented, BINARY INSERTION SORT becomes as efficient as parsimonious sorting algorithms. We shall also find a Gaussian law for BINARY INSERTION SORT. Calculation of moments is more involved for this method. We shall see, for example, that the variance of the total number of comparisons of BINARY INSERTION exhibits periodic fluctuations of small magnitude.

2.2 A SUFFICIENT CONDITION FOR NORMALITY

The sequence of n insertion algorithms used by a search strategy corresponds to a sequence of n insertion trees. Let h_i be the height of the ith insertion tree, $i = 1, \dots, n$. Let the number of data comparisons required by the ith algorithm be X_i. Then clearly, C_n, the total number of data comparisons is

$$C_n = X_1 + X_2 + \cdots + X_n. \tag{2.2}$$

Taking expectations, we get

$$\mathbf{E}[C_n] = \mathbf{E}[X_1] + \mathbf{E}[X_2] + \cdots + \mathbf{E}[X_n]. \tag{2.3}$$

It underlies our basic random permutation probability model that the random variables X_i are independent (see Proposition 1.6). Thus taking the variance of C_n is only a matter of adding up the variances $\mathbf{Var}[X_i]$. Introduce

$$s_n^2 = \mathbf{Var}[C_n] = \sum_{i=1}^{n} \mathbf{Var}[X_i].$$

The next theorem characterizes a sufficient condition that guarantees a limiting normal behavior. The condition relates the rate of growth of the variance of C_n to the height h_n. This condition is satisfied by almost all practical search strategies.

Theorem 2.1 *(Lent and Mahmoud, 1996b). Suppose INSERTION SORT uses a search strategy that corresponds to a deterministic sequence of insertion trees of heights h_n, $n = 1, 2, \dots$. Let C_n be the total number of comparisons to sort n keys, and s_n^2 be its variance. If the sequence of heights is nondecreasing with*

$$h_n = o(s_n), \qquad \text{as } n \to \infty,$$

then

$$\frac{C_n - \mathbf{E}[C_n]}{s_n} \xrightarrow{\mathcal{D}} \mathcal{N}(0, 1).$$

Proof. Suppose the ith stage of insertion makes X_i data comparisons. Let $Y_i = X_i - \mathbf{E}[X_i]$, for $i = 1, \dots, n$. The event $\{|Y_i| > h_n\}$ implies $\{|Y_i| > h_i\}$, as the heights of the sequence of trees considered are non-decreasing. That is,

$$\mathbf{Prob}\{|Y_i| > h_n\} \le \mathbf{Prob}\{|Y_i| > h_i\}.$$

However,

$$\begin{aligned}\{|Y_i| > h_i\} &= \{X_i - \mathbf{E}[X_i] > h_i\} \cup \{\mathbf{E}[X_i] - X_i > h_i\} \\ &\subseteq \{X_i > h_i\} \cup \{\mathbf{E}[X_i] > h_i\}.\end{aligned}$$

The two sets in the latter union are empty because $X_i \le h_i$; consequently the average $\mathbf{E}[X_i]$ is also less than h_i. So

$$\mathbf{Prob}\{|Y_i| > h_i\} \le \mathbf{Prob}\big\{\{X_i > h_i\} \cup \{\mathbf{E}[X_i] > h_i\}\big\} = 0;$$

consequently, uniformly for $i = 0, \dots, n$,

$$\mathbf{Prob}\{|Y_i| > h_n\} \le \mathbf{Prob}\big\{|Y_i| > h_i\} = 0.$$

It follows from our assumption $h_n = o(s_n)$ that for any $\varepsilon > 0$ there is an integer N_ε so that $h_n < \varepsilon s_n$, for all $n \ge N_\varepsilon$. Then uniformly for $i = 0, \dots, n$,

$$\sum_{|k| \ge \varepsilon s_n} k^2\, \mathbf{Prob}\{Y_i = k\} \le \sum_{|k| > h_n} k^2\, \mathbf{Prob}\{Y_i = k\} = 0.$$

Hence for $n \ge N_\varepsilon$,

$$\sum_{i=1}^{n} \sum_{|k| \ge \varepsilon s_n} k^2\, \mathbf{Prob}\{Y_i = k\} = 0,$$

or

$$\lim_{n\to\infty} \frac{1}{s_n^2} \sum_{i=1}^{n} \sum_{|k| \ge \varepsilon s_n} k^2\, \mathbf{Prob}\{Y_i \le k\} = 0,$$

verifying Lindeberg's condition; convergence in distribution to normality follows from Lindeberg's central limit theorem. ∎

2.3 LINEAR INSERTION SORT

This is perhaps the simplest implementation of INSERTION SORT as it integrates the search, data movement, and insertion all in a single algorithm. We assume the data come on-line as an input stream and are to be sorted in an array $A[1\,..\,n]$. (An off-line version is discussed in Exercise 2.2.) The first key is placed in $A[1]$. The second key is compared with $A[1]$. If the second key is less than $A[1]$, it is placed "above" it, by first moving the content of $A[1]$ down to position 2 then replacing $A[1]$ with this second key. If the second key is larger than the first, it is placed in $A[2]$. The algorithm continues in this fashion, and right before the ith insertion, the block of data in $A[1 \mathinner{..} i-1]$ is in sorted order. To insert a new key K, its value is first obtained by performing an input operation. We then start a search from the bottom of the sorted block $A[1 \mathinner{..} i-1]$, that is, we first probe position $i-1$ using a backward-going working index (j say, initialized to $i-1$). If $K > A[j]$ we insert K at the bottom of the block at position j. However, if $K < A[j]$, we copy the key in $A[j]$ at the position underneath it (thus there are now two copies of this key at positions j and $j+1$). We decrement j and continue our probing at position j. So long as we find elements of A that are less than K we keep pulling them down. When for the first time a stopper key (if any) less than K is found (at position j), we insert K right under it. In the preceding step the key at $A[j+1]$ was copied at position $A[j+2]$, and it is all right to proceed with a destructive assignment of K to overwrite the leftover copy at position $j+1$. Figure 2.2 illustrates the insertion tree for (BACKWARD) LINEAR INSERTION SORT to search for a position for the 6th key. In case K is smaller than all the keys inserted so far, it should go to the top of A. To avoid a problem of having the array index "go through the roof" (that is, having $j = 0$), we must check every step of the way up that j is still greater than 0, before probing position j, or else an attempt to access the nonexistent $A[0]$ will be made, giving rise to an error. Thus we may write the backbone loop of LINEAR INSERTION SORT in the form

while $(j > 0)$ **and** $(K < A[j])$ **do**

which modern programming languages handle with the so-called short-circuit option that does not evaluate the second condition $(K < A[j])$ unless the first condition $(j > 0)$ is true.

Index comparisons can add a substantial number of comparisons, one index comparison (comparison of integers) for every data comparison. This may be inconsequential in case the data are of some complex nature (long character strings, say,

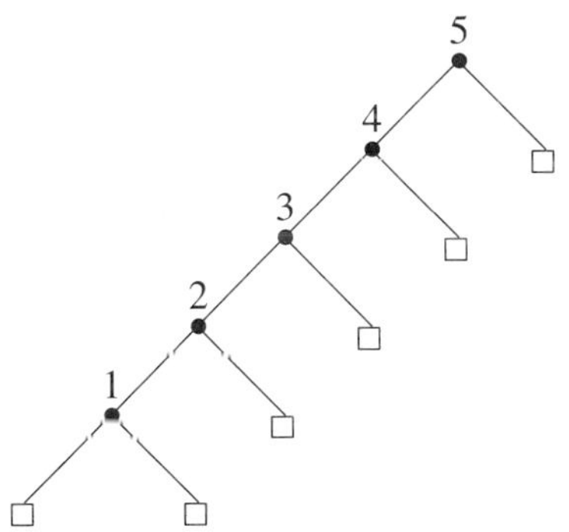

Figure 2.2. The insertion tree of LINEAR INSERTION SORT for the 6th key.

like DNA sequences). However, if a comparison between two data items takes an amount of time comparable to index comparisons, the issue may not be ignored. For example, if the keys themselves are integers, the comparison $K < A[j]$ takes an amount of time comparable to the comparison $j > 0$.

Index comparisons can be eliminated altogether by introducing a sentinel value as a stopper—a dummy key value to occupy the top array position and is less than all the other possible values for this type of data. This suggests that we extend the range of our subscripts to include 0, that is, set up the array $A[0 \,..\, n]$. In the case of numeric data, we let $A[0] = -\infty$. This ensures that some key in the array is less than K and if no other stopper is found, $A[0]$ will act as a one; there is no need ever to check whether the index is within bounds. In practice $-\infty$ is replaced by the smallest admissible value in the data type we are sorting. For example, if we are sorting integers, PASCAL provides $-MAXINT$ as the smallest integer it can deal with and this is the "practical" negative infinity for this type. Supposing instead we are sorting ASCII characters; the nonprintable null character, the first in the collating sequence, may be thought of as the practical $-\infty$. The required loop can then be simplified to the plain

$$\textbf{while}\ K < A[j]\ \textbf{do}$$

We shall assume the presence of this sentinel in the algorithm presented in the following text and in subsequent analysis. As mentioned before, this eliminates any index comparisons. On the other hand, the probability distribution of data comparisons is very slightly altered, an effect that whithers away anyhow in the asymptotics. The complete (BACKWARD) LINEAR INSERTION SORT algorithm is presented in Figure 2.3.

Let X_i be the number of comparisons made to insert the ith key and let C_n be the overall number of comparisons to sort the n data items. For LINEAR INSERTION SEARCH (with a sentinel) the number of comparisons for the ith insertion is UNIFORM$[1 \,..\, i]$.

```
A[0] ← −∞;
for i ← 1 to n do
   begin
        j ← i − 1;
        read(K);
        while A[j] > K do
             begin
                  A[j + 1] ← A[j];
                  j ← j − 1;
             end;
        A[j + 1] ← K;
   end;
```

Figure 2.3. The LINEAR INSERTION SORT algorithm.

When we insert the ith key, it will have to cross over every datum in $A[1 \mathinner{..} i-1]$ that is larger than it, that is, we have to fix every inversion caused by the ith key. If the sequential rank $X_i = j$, the new key must be moved to position j from the top of the array. Regardless of what order the larger keys that preceded X_i appeared in the input, they are arranged in order and now occupy the bottom $i - j$ positions of $A[1 \mathinner{..} i-1]$ (the key X_i causes $i - j$ inversions). To cross over these $i - j$ data, X_i must be compared with these $i - j$ keys plus one more key (the stopper at position $j - 1$, the first key from the bottom that does not exceed X_i). Therefore, the number of comparisons needed to insert X_i is $1 + V_i$, where V_i is the number of inversions caused by X_i. Summing over all the stages of insertion

$$C_n = \sum_{i=1}^{n} (1 + V_i) = n + Y_n,$$

where Y_n is the total number of inversions in the input stream (review Section 1.10.2). The mean and variance calculations for the number of inversions (1.14) and (1.15) provide us with direct answers for LINEAR INSERTION SORT. For the mean number of comparisons, from (1.14) we have

$$\mathbf{E}[C_n] = n + \mathbf{E}[Y_n] = \frac{n(n+3)}{4} \sim \frac{n^2}{4}, \tag{2.4}$$

and for its variance, from (1.15) we have

$$s_n^2 = \mathbf{Var}[C_n] = \mathbf{Var}[n + Y_n] = \mathbf{Var}[Y_n] = \frac{n(n-1)(2n+5)}{72} \sim \frac{n^3}{36}. \tag{2.5}$$

The asymptotic normality of the number of comparisons can be demonstrated either from Proposition 1.7 or from Theorem 2.1. The normality of Theorem 2.1 holds for LINEAR INSERTION SORT: The height of the nth insertion tree is $h_n = n$, whereas $s_n \sim \frac{1}{6} n^{3/2}$. Hence, according to Theorem 2.1, LINEAR INSERTION SORT has a normal limit behavior,

$$\frac{C_n - \mathbf{E}[C_n]}{s_n} \xrightarrow{\mathcal{D}} \mathcal{N}(0, 1),$$

which, according to (2.4), is

$$\frac{C_n - \frac{1}{4} n(n+3)}{s_n} \xrightarrow{\mathcal{D}} \mathcal{N}(0, 1). \tag{2.6}$$

The asymptotic formula (2.5) simplifies the last convergence in distribution as follows. First noting that

$$\frac{s_n}{n^{\frac{3}{2}}} \to \frac{1}{6},$$

we can use the multiplicative form of Slutzky's theorem to get (from (2.6) and the last formula):

$$\frac{C_n - \frac{1}{4}n(n+3)}{n^{\frac{3}{2}}} \xrightarrow{\mathcal{D}} \frac{1}{6}\mathcal{N}(0, 1) = \mathcal{N}\Big(0, \frac{1}{36}\Big).$$

From the latter normal limit behavior and the convergence $\frac{3}{4}n/n^{3/2} \to 0$, we have

$$\frac{C_n - \frac{1}{4}n^2}{n^{3/2}} \xrightarrow{\mathcal{D}} \mathcal{N}\Big(0, \frac{1}{36}\Big).$$

by an application of the additive form of Slutzky's theorem.

A practitioner may now ask how large n should be before the tables of the normal distribution can give a reliable approximation to the actual distribution or what the error is for a given n. We can settle this sort of question by estimating the rate of convergence of the distribution function of

$$C_n^* \stackrel{\text{def}}{=} \frac{C_n - \mathbf{E}[C_n]}{s_n}$$

to the distribution function of the normal random variable $\mathcal{N}(0, 1)$. Let X_i^* be the centered version of X_i:

$$X_i^* = X_i - \mathbf{E}[X_i] = X_i - \frac{1}{2}(i+1).$$

Then C_n^* has a representation as a sum of independent random variables (compare with (2.2) and (2.3)); namely

$$\begin{aligned} C_n^* &= \frac{1}{s_n}\Big(\sum_{i=1}^{n} X_i - \sum_{i=1}^{n} \mathbf{E}[X_i]\Big) \\ &= \frac{1}{s_n}\sum_{i=1}^{n} X_i^*, \end{aligned}$$

with $C_n^* \xrightarrow{\mathcal{D}} \mathcal{N}(0, 1)$ as shown in Theorem 2.1.

The Berry–Ésseen theorem bounds the absolute value of the error committed in using the distribution function of the normal distribution instead of the actual distribution function of C_n^* by the relation

$$\sup_x \Big|\mathbf{Prob}\{C_n^* \le x\} - \mathbf{Prob}\big\{\mathcal{N}(0, 1) \le x\big\}\Big| \le \frac{A}{s_n^3}\sum_{i=1}^{n} \mathbf{E}|X_i^*|^3, \qquad (2.7)$$

where A is the absolute constant $\frac{2}{\sqrt{\pi}}(\frac{7}{3} + \frac{30\sqrt{2}}{\pi}) = 17.87135734\ldots$. The Berry–Ésseen theorem requires a calculation of absolute third moments of X_i^* to get a uniform bound on the error. For this computation we have

$$\begin{aligned}\mathbf{E}|X_i^*|^3 &= \sum_{k=1}^{i} \left|k - \frac{i+1}{2}\right|^3 \mathbf{Prob}\{X_i = k\} \\ &= \frac{1}{i}\left\{\sum_{k=1}^{\lfloor\frac{i+1}{2}\rfloor}\left(\frac{i+1}{2} - k\right)^3 + \sum_{k=\lfloor\frac{i+1}{2}\rfloor+1}^{i}\left(k - \frac{i+1}{2}\right)^3\right\}.\end{aligned}$$

A lengthy calculation (left as an exercise) shows that

$$f(n) \stackrel{\text{def}}{=} \sum_{i=1}^{n} \mathbf{E}|X_i^*|^3 = \frac{n^4}{128} + \frac{n^3}{64} - \frac{3n^2}{128} - \frac{n}{32} + g(n),$$

where $g(1) = 0$, and for $n \geq 2$, $g(n)$ is defined in terms of H_n, the nth harmonic number, as

$$g(n) = \begin{cases} \dfrac{1}{32}H_{n-1} - \dfrac{1}{32n(n-1)} - \dfrac{1}{32}H_{\frac{1}{2}(n-3)}, & \text{if } n \text{ is odd;} \\ \dfrac{1}{32}H_{n-1} - \dfrac{1}{64}H_{\frac{1}{2}n-1}, & \text{if } n \text{ is even.} \end{cases}$$

It is clear that uniformly in $n \geq 1$,

$$f(n) \leq \frac{n^4}{128} + \frac{n^3}{64} + \frac{1}{32}H_{n-1}.$$

All lower-order terms on the right-hand side of this inequality are $o(n^4)$. If n is large enough, all the lower order terms can be made less than Mn^4, for any arbitrary constant M. For example, if $n \geq 20$, $\frac{1}{64}n^3 \leq 0.1\frac{n^4}{128}$, and $\frac{1}{32}H_{n-1} \leq 0.000088694\frac{n^4}{128}$. So, for all $n \geq 20$,

$$f(n) \leq 1.10000887\frac{n^4}{128}. \tag{2.8}$$

From the asymptotic representation (2.5), we immediately see that

$$\frac{1}{s_n^3}\sum_{i=1}^{n} \mathbf{E}|X_i^*|^3 \sim \frac{216 f(n)}{n^{9/2}} \sim \frac{216n^4}{128n^{\frac{9}{2}}} = \frac{27}{16\sqrt{n}}.$$

Hence the order of the rate of convergence is the slow $1/\sqrt{n}$.

We can actually answer a specific question like "What is the error in the probability calculation $\mathbf{Prob}\{C_{40000} > 400070003\}$, if the approximating asymptotic distribution is used instead?" The rate of convergence is slow as already observed, and we are choosing such a large number of items ($n = 40000$) to have meaningful approximations with reasonably small errors. From (2.4) and (2.5) we have $\mathbf{E}[C_{40000}] = 400030000$, and $s_{40000} = 50\sqrt{2133413330/3}$. Hence

$$\begin{aligned}\mathbf{Prob}\{C_{40000} \le 400070003\} \\ &= \mathbf{Prob}\left\{\frac{C_{40000} - 400030000}{50\sqrt{2133413330/3}} \le \frac{400070003 - 400030000}{50\sqrt{2133413330/3}}\right\} \\ &= \mathbf{Prob}\{C^*_{40000} \le 0.0300016\ldots\}.\end{aligned}$$

Thus, from the Berry–Ésseen bound (2.7),

$$\begin{aligned}\Big|\big(1 - \mathbf{Prob}\big\{C_{40000} \le 400070003\big\}\big) - \Big(1 - \mathbf{Prob}\big\{\mathcal{N}(0, 1) \le 0.0300016\ldots\big\}\Big)\Big| \\ \le \frac{17.88 f(40000)}{s^3_{40000}}.\end{aligned}$$

From the bound (2.8) we can write

$$\begin{aligned}\Big|\mathbf{Prob}\{C_{40000} > 400070003\} - \mathbf{Prob}\big\{\mathcal{N}(0, 1) > 0.0300016\ldots\big\}\Big| \\ \le 0.1659527\ldots.\end{aligned}$$

Looking up the normal table we find

$$\mathbf{Prob}\{\mathcal{N}(0, 1) > 0.0300016\ldots\} = 0.488\ldots.$$

Thus

$$0.322 \ < \ \mathbf{Prob}\{C_{40000} > 400070003\} \ < \ 0.654.$$

Higher moments can be deduced from the well-known moments of the limiting normal distribution. Doberkat (1982a) finds the leading term in each moment directly by a combinatorial argument and Panny (1986) refines Doberkat's results and finds several lower order terms for the moments of the number of comparisons of LINEAR INSERTION SORT. Conversely, from their arguments, one can infer convergence in distribution of the normed number of comparisons to the random variable $\mathcal{N}(0, 1)$. The normal distribution is one of the distributions uniquely characterized by their moments, and it follows that $C_n^* \xrightarrow{\mathcal{D}} \mathcal{N}(0, 1)$. This route of proving asymptotic normality works here because moment calculations are particularly simple for LINEAR INSERTION SEARCH. For implementations of INSERTION SORT via some other search strategies, like the BINARY SEARCH strategy of the next section, the method of moments may prove to be a difficult route as the computation of the moments may be rather complex.

2.4 BINARY INSERTION SORT

This algorithm uses the binary search strategy, that is, it uses the BINARY SEARCH algorithm at each stage of the insertion. At the ith stage, BINARY SEARCH operates on a sorted array $A[1 \mathinner{..} i-1]$ to find a position for a new key K by probing the *middle* position $\lceil \frac{i}{2} \rceil$. At a later step in the algorithm, the search is narrowed down to $A[\ell \mathinner{..} u]$, a stretch of A between a lower index ℓ and an upper index u. The algorithm probes the middle position $m = \lceil (\ell + u)/2 \rceil$. If $K < A[m]$, the algorithm should only be concerned about finding an insertion position within $A[\ell \mathinner{..} m-1]$, because all the elements of $A[m \mathinner{..} u]$ are now known to be of values larger than K (as $A[1 \mathinner{..} i-1]$ is sorted). BINARY SEARCH then repeats its operation in $A[\ell \mathinner{..} m-1]$. Similarly, if $K > A[m]$, BINARY SEARCH repeats its operation in $A[m+1 \mathinner{..} u]$. The process continues to dispose of one of the two remaining halves until a gap between two elements of $A[1 \mathinner{..} i-1]$ is found. Data movements and insertion follow exactly as in LINEAR INSERTION SORT.

The ith BINARY SEARCH algorithm in this strategy has a complete insertion tree T_i of order $i-1$. The insertion tree of Figure 2.4 illustrates this insertion algorithm for $i = 6$; the root-to-leaf paths of the tree specify all possible probe sequences.

Let Z_i denote the number of leaves at level $\lceil \lg i \rceil = \lfloor \lg(i-1) \rfloor + 1$ in the tree T_i. Recall that an extended binary tree on $i-1$ internal nodes has i leaves and its height is $\lceil \lg i \rceil$. That is, unless the tree is perfect, the leaves appear on the two levels $\lfloor \lg(i-1) \rfloor + 1$ and $\lfloor \lg(i-1) \rfloor$. If the tree is perfect ($i = 2^k$, for some k) all its leaves lie on level $\lg i$. So, X_i (the number of comparisons taken to insert the ith key) is

$$X_i = \lfloor \lg(i-1) \rfloor + B_i, \tag{2.9}$$

where B_i is a BERNOULLI(Z_i/i) random variable that assumes the value 1 (with probability Z_i/i) or 0 (with probability $1 - Z_i/i$); in the case $i = 2^k$ this Bernoulli random variable is deterministically $B_i \equiv 1$. It is an easy combinatorial derivation (Exercise 1.6.2) to show that

$$Z_i = 2\big(i - 2^{\lfloor \lg(i-1) \rfloor}\big). \tag{2.10}$$

Hence

$$\mathbf{E}[B_i^2] = \mathbf{E}[B_i] = 0 \times \Big(1 - \frac{Z_i}{i}\Big) + 1 \times \frac{Z_i}{i} = \frac{Z_i}{i}.$$

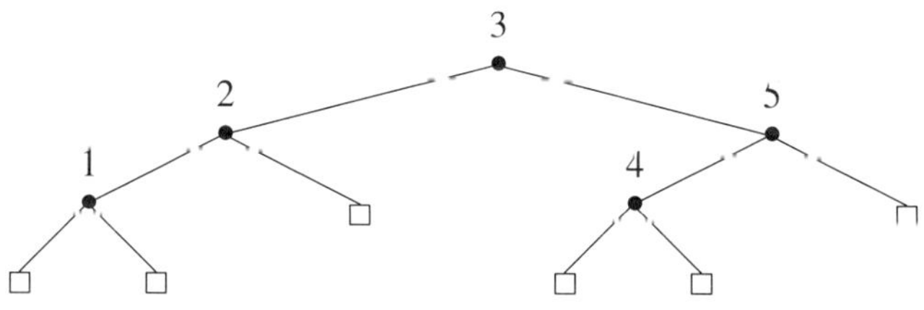

Figure 2.4. The insertion tree of BINARY INSERTION SORT for the insertion of the 6th key.

So, by (2.9) and (2.10)

$$\begin{aligned}\mathbf{E}[X_i] &= \lfloor \lg(i-1) \rfloor + \mathbf{E}[B_i]\\ &= \lfloor \lg(i-1) \rfloor + \frac{2\big(i - 2^{\lfloor \lg(i-1) \rfloor}\big)}{i}.\end{aligned}$$

Similarly we obtain for the variance

$$\begin{aligned}\mathbf{Var}[X_i] &= \mathbf{Var}\big[\lfloor \lg(i-1) \rfloor + B_i\big]\\ &= \mathbf{Var}[B_i]\\ &= \mathbf{E}[B_i^2] - \mathbf{E}^2[B_i]\\ &= \frac{Z_i}{i} - \frac{Z_i^2}{i^2}. \qquad (2.11)\end{aligned}$$

The mean of C_n, the total number of comparisons, is obtained by summing $\mathbf{E}[X_i]$ yielding

$$\begin{aligned}\mathbf{E}[C_n] &= \sum_{i=2}^{n} \lfloor \lg(i-1) \rfloor + \sum_{i=2}^{n} \frac{2\Big(i - 2^{\lfloor \lg(i-1) \rfloor}\Big)}{i}\\ &= \sum_{j=1}^{n-1} \big(\lg j + O(1)\big) + 2n - 2 - 2\sum_{j=1}^{n-1} \frac{2^{\lfloor \lg j \rfloor}}{j+1}.\end{aligned}$$

Note that the $O(1)$ term is uniform in i, because $x - \lfloor x \rfloor < 1$, for any real number x. The sum

$$\sum_{j=1}^{n-1} \frac{2^{\lfloor \lg j \rfloor}}{j+1} \le \sum_{j=1}^{n-1} \frac{2^{\lg j}}{j+1} = \sum_{j=1}^{n-1} \frac{j}{j+1} < n.$$

It follows that

$$\mathbf{E}[C_n] = O(n) + \sum_{j=1}^{n-1} \lg j.$$

By the Stirling approximation for factorials (which is developed in (1.6)) we have

$$\begin{aligned}\sum_{j=1}^{n-1} \lg j &= \lg n! - \lg n\\ &= n \lg n - \frac{n}{\ln 2} - \frac{1}{2} \lg n + \lg \sqrt{2\pi} + O\Big(\frac{1}{n}\Big).\end{aligned}$$

Hence,

$$\mathbf{E}[C_n] = n \lg n + O(n).$$

The calculation of the variance is more involved. It is easier to start with $n = 2^j$ so that the corresponding insertion tree T_{2^j}, of order $n - 1 = 2^j - 1$, is perfect; this perfect tree has height j. By the independence of the X_i's, the variance s_n^2 of the overall number of comparisons is the sum of the variances over all stages of insertion. The history of this insertion corresponds to a sequence of complete trees. The shape of the ith tree T_i can be obtained from the previous tree T_{i-1} by adjoining a new internal node to T_{i-1} at level $\lfloor \lg(i-1) \rfloor$ to replace a leaf, then adding two new leaves as children of this new node. (The labels, however, need to be recomputed.) This tree-growing property is discussed in Exercise 2.8. Another view of this is to consider the perfect tree as if it is obtained by a growth process that systematically fills out a binary tree by adding internal nodes at level k, until that level is saturated (for $k \in \{0, 1, \dots, j-1\}$). The process starts at $k = 0$ and gradually increases k to span the range $0, 1, \dots j-1$. The variance $s_n^2 = \mathbf{Var}[C_n]$ can thus be obtained by first considering the sum of the variances over all trees of height $k+1$, then taking the sum over $k = 0, \dots, j-1$. (The last tree T_{2^j} is perfect and consequently $\mathbf{Var}[X_{2^j}] = 0$.) From the variance computation (2.11) we have the representation:

$$s_n^2 = \sum_{k=0}^{j-1} \sum_{i=0}^{2^k-1} \mathbf{Var}[X_{2^k+i}] = \sum_{k=0}^{j-1} \sum_{i=0}^{2^k-1} \left\{ \frac{Z_{2^k+i}}{2^k+i} - \left(\frac{Z_{2^k+i}}{2^k+i} \right)^2 \right\}. \tag{2.12}$$

While the tree is filling out level k, the height of the insertion trees is held constant at $k+1$, and $\lfloor \lg(2^k + i - 1) \rfloor = k$, for $1 \le i \le 2^k$. Thus by (2.10),

$$Z_{2^k+i} = 2(2^k + i - 2^k) = 2i,$$

and the inner sum gives

$$\begin{aligned}\sum_{i=0}^{2^k-1} \left\{ \frac{Z_{2^k+i}}{2^k+i} - \left(\frac{Z_{2^k+i}}{2^k+i} \right)^2 \right\} &= \sum_{i=0}^{2^k-1} \left(\frac{2i}{2^k+i} \right) - \sum_{i=0}^{2^k-1} \left(\frac{2i}{2^k+i} \right)^2 \\ &= 6(2^k)\big(H_{2^{k+1}} - H_{2^k}\big) - 4^{k+1}\big(H_{2^{k+1}}^{(2)} - H_{2^k}^{(2)}\big) - 2^{k+1}.\end{aligned}$$

To simplify the latter expression, we use the asymptotic expansions

$$H_n = \ln n + \gamma + O\left(\frac{1}{n}\right);$$

$$H_n^{(2)} = \frac{\pi^2}{6} - \frac{1}{n} + O\left(\frac{1}{n^2}\right).$$

We obtain

$$\sum_{i=0}^{2^k-1} \left\{ \frac{Z_{2^k+i}}{2^k+i} - \left(\frac{Z_{2^k+i}}{2^k+i} \right)^2 \right\} = (6 \ln 2 - 4)2^k + O(1).$$

So, now

$$\begin{aligned}
s_n^2 &= (6\ln 2 - 4)\sum_{k=0}^{j-1} 2^k + O(\ln n) \\
&= (6\ln 2 - 4)2^j + O(\ln n) \\
&= (6\ln 2 - 4)n + O(\ln n).
\end{aligned}$$

Note that $h_n = \lg n$, whereas $s_n \sim \sqrt{(6\ln 2 - 4)n}$. So, for the perfect tree $T_n = T_{2^j}$,

$$h_n = o(s_n),$$

comporting to a requirement of Theorem 2.1.

For general n, the situation is not very much different from the special case $n = 2^j$. The insertion of the nth key for a general value $n \neq 2^j$, for any j, corresponds to a complete but not perfect tree of size $n - 1$ and height h_n. This value of n can be represented as $n = 2^{\lfloor \lg(n-1) \rfloor} + i$, for some i not exceeding $\frac{1}{2}(n + 1)$. The height $h_n = \lceil \lg n \rceil$. The growth process systematically saturates the levels $k = 0, 1, 2, \ldots, \lfloor \lg n \rfloor - 1$, in this order, with internal nodes, then inserts the remaining i internal nodes on level $\lfloor \lg n \rfloor$, thus placing $2i$ leaves on level $\lfloor \lg n \rfloor + 1 = \lceil \lg n \rceil$. We only need to add to our calculation of the variance of a perfect tree the contribution of the $2i$ leaves on level $\lceil \lg n \rceil$. That is,

$$s_n^2 = s^2_{2^{\lfloor \lg(n-1) \rfloor}} + \sum_{\ell=1}^{i} \mathbf{Var}\big[X_{2^{\lfloor \lg(n-1) \rfloor} + \ell}\big].$$

Each internal node added on level $\lfloor \lg n \rfloor$ increases the overall variance by a contribution of the two leaves introduced and reduces it by replacing an old leaf. The net gain can be easily computed (as in the inner sum of (2.12)). Adding up these gains we obtain

$$\begin{aligned}
\sum_{\ell=0}^{i} \mathbf{Var}[X_{2^{\lfloor \lg(n-1) \rfloor} + \ell}] = 4^{j+1}\Big(&\frac{1}{2^j + i + 1} - \frac{1}{2^j} \\
&+ 6(2^j)(H_{2^j + i + 1} - H_{2^j})\Big) - 2i + O(1).
\end{aligned}$$

Simplifying we obtain

$$\begin{aligned}
s_n^2 &= \Big[\frac{6(1 + f_n)\ln 2 - 6}{2^{f_n}} - 2 + \frac{4}{4^{f_n}}\Big] n + O(\ln n) \\
&\stackrel{\text{def}}{=} Q_n n + O(\ln n), \qquad (2.13)
\end{aligned}$$

where

$$f_n = \lg n - \lfloor \lg n \rfloor.$$

The sequence $f_1, f_2, \ldots$ is dense on the interval $[0, 1)$, that is, given any $\varepsilon > 0$, for any $x \in [0, 1)$ the interval $(x - \varepsilon, x + \varepsilon)$ contains points from the sequence, however small ε is. Thus, given any $x \in (0, 1]$, f_n is infinitely often in the interval $(x - \varepsilon, x + \varepsilon)$. That is, if we specify any $x \in [0, 1)$, the coefficient of n in the variance s_n^2 in (2.13) for infinitely many n comes arbitrarily close to the interpolating function

$$Q(x) \stackrel{\text{def}}{=} \frac{6(1 + x)\ln 2 - 6}{2^x} - 2 + \frac{4}{4^x}.$$

The function $Q(x)$ is the analytic continuation of Q_n (which is defined only at the positive integers), restricted to $[0, 1)$. When $n = 2^j$, the insertion tree is perfect and the variance $\mathbf{Var}[X_n]$ is 0. As n gradually increases $Q(n)$ goes through "cycles." First it decreases, then increases, then decreases again until $n = 2^{j+1}$, the next perfect tree, where the variance of the corresponding insertion goes back to 0 again. Figure 2.5 illustrates the cycle for $n = 4, 5, 6, 7$. At the beginning of this cycle, $Q_n = Q_4$ is identically $6 \ln 2 - 4$. As n increases, Q_n fluctuates till Q_8 "wraps around" to coincide again with $6 \ln 2 - 4$. The corresponding behavior in the total variance s_n^2 is that its leading term is asymptotic to $(6 \ln 2 - 4)n$ when n is a proper power of 2; it then increases at a small rate, because there is only a small proportion of nodes on the last level; the majority of the leaves are at level $\lfloor \lg n \rfloor$. Larger variability will appear when the proportion of nodes on that level are of an effective magnitude. When the majority of leaves are on level $\lfloor \lg n \rfloor + 1$, the total variance does not change much any more; each new insertion has a smaller variance than the preceding one; the increase

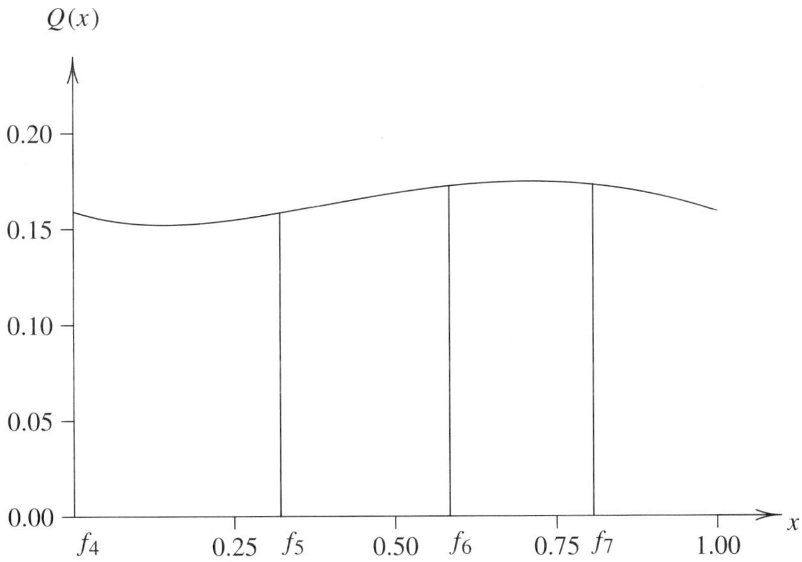

Figure 2.5. The cycle of BINARY INSERTION SORT for $n = 4, 5, 6, 7$.

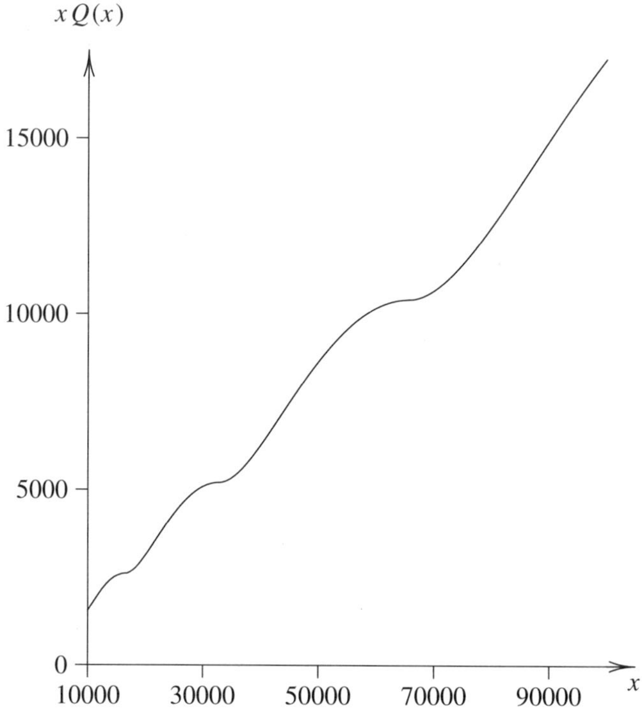

Figure 2.6. Oscillations in the leading term of the variance of the number of comparisons in BINARY INSERTION SORT.

in s_n^2 is small. The variance s_n^2 is an increasing function of the form

$$s_n^2 = Q_n n + O(\ln n).$$

Figure 2.6 illustrates the oscillating behavior of the leading term.

In particular $Q^* = \inf_n Q_n = 0.1519119\ldots$ and

$$s_n^2 \ge Q^* n + O(\ln n),$$

whereas $h_n \sim \lg n$. Hence $h_n = o(s_n)$ for general n, too. Theorem 2.1 asserts the Gaussian limit behavior

$$\frac{C_n - n \lg n}{\sqrt{n Q_n}} \xrightarrow{\mathcal{D}} \mathcal{N}(0, 1).$$

EXERCISES

2.1 How many comparisons does LINEAR INSERTION SORT (the sentinel version of Section 2.3) make to sort the list 10 9 8 7 6 5 4 3 2 1?

How many comparisons does BINARY INSERTION SORT make to sort this list? If the list is reversed, how many comparisons does each implementation make?

2.2 Suppose that the data file is available off-line, that is, the entire file is accessible prior to sorting. Assume the data are initially loaded in the (unsorted) array $A[1\,..\,n]$. Adapt LINEAR INSERTION SORT to sort A.

2.3 LINEAR INSERTION SORT is particularly suitable for linked lists. Write a LINEAR INSERTION SORT algorithm to progressively add data to a sorted linked data list resulting in a sorted list. Each insertion should be $O(1)$.

2.4 Consider the following possible speed-up method for FORWARD LINEAR SEARCH. Assume an array implementation with a sorted array $A[1\,..\,i-1]$ after $i-1$ insertions. At the ith stage, instead of going through the array $A[1\,..\,i-1]$ in increments of size 1 as in the usual FORWARD LINEAR SORT, a JUMP INSERTION SORT may compare the ith key by probing positions $c, 2c, 3c, \ldots$, until either a stopper (an array element larger than the new key) is encountered, or until the list is exhausted without finding such an element (that is, when the next candidate probe is at position $jc > n$). If a stopper is encountered at position jc, say, we start a search backward within the stretch $A[(j-1)c+1\,..\,jc-1]$. If no stopper is found, we make the insertion at position i. Find the mean number of comparisons made by JUMP INSERTION SORT. Show that $h_n = o(s_n)$, where h_n is the height of the insertion tree of this algorithm and s_n is the standard deviation of its number of comparisons. Deduce the Gaussian limit behavior of JUMP INSERTION SORT.

2.5 Refer to Exercise 2.4. Show that the choice $c = \lfloor\sqrt{n/2}\rfloor$ asymptotically optimizes the average behavior of JUMP INSERTION SEARCH.

2.6 Study Exercises 2.4 and 2.5. Derive a Gaussian law for JUMP INSERTION SORT at its asymptotically optimal choice of c.

2.7 Prove that the insertion tree for BINARY SEARCH is complete.

2.8 A search strategy is said to be tree-growing, when $T_1, T_2, \ldots$ is its sequence of insertion trees, and the shape of T_i is obtained from the shape of T_{i-1} by replacing a leaf in the latter tree to become an internal node and then adding two new leaves as children of this new internal node (considering only shapes and disregarding labels). Draw the sequence of the first eight trees in each of the following strategies:

(a) BACKWARD and FORWARD LINEAR INSERTION. Consider the sentinel version of both as well.

(b) BINARY SEARCH.

(c) 3-JUMP SEARCH (the strategy of JUMP INSERTION SEARCH of Exercise 2.4 with $c = 3$).

Prove that each of the above strategies is tree-growing.

2.9 Show that among all possible search strategies, BINARY SEARCH is the one that minimizes the average of the overall number of data comparisons.

2.10 Consider RANDOMIZED INSERTION SORT in which the probe for the ith datum is chosen at random from among the $i - 1$ sorted elements in the array (the ith probe position has the distribution of UNIFORM$[1 \mathinner{..} i-1]$). The algorithm is then applied recursively on the subfile chosen for the continuation of the search. Find the mean and variance of the overall number of comparisons to sort n elements. Prove that, with suitable norming, the overall number of comparisons of this randomized algorithm asymptotically has a normal distribution.

2.11 Assume an array storage for data. Let M_n be the number of data movements required while running any implementation of insertion sort. Argue that M_n does not depend on the search strategy. Prove that M_n has a Gaussian limit behavior.

3

Shell Sort

SHELL SORT is a generalization of the method of sorting by insertion. It is essentially several stages of INSERTION SORT designed with the purpose of speeding up INSERTION SORT itself. The algorithm was proposed by Shell in 1959. The algorithm is a rather practicable method of in-situ sorting and can be implemented with ease. From a theoretical standpoint, the interest is in that standard implementations of INSERTION SORT have an average of $\Theta(n^2)$ running time to sort n random keys, whereas the appropriate choice of the parameters of the stages of SHELL SORT can bring down the order of magnitude. For instance, by a certain choice of the structure of the stages, a 2-stage SHELL SORT can sort in $O(n^{5/3})$ average running time. Ultimately, an optimized choice of the parameters can come close to the theoretic lower bound of $\Theta(n \ln n)$ average and worst case; the best known sequence of parameters performs in $\Theta(n \ln^2 n)$ time in the average and worst case. In practice some SHELL SORT constructions perform competitively for the small and medium range of n. The analysis of the algorithm brings to bear some interesting stochastic processes like the Brownian bridge.

3.1 THE ALGORITHM

To sort by insertion, one progressively adds keys to an already sorted file. This is achieved by identifying the rank of the next key by searching the available sorted list. As discussed at length in Chapter 2, the search can be done in many different ways. We shall restrict our attention to LINEAR INSERTION SORT, since it has a search mechanism that integrates easily into SHELL SORT.

SHELL SORT performs several stages of LINEAR INSERTION SORT. It is well suited for arrays. We shall assume the data reside in a host linear array structure $A[1 \,..\, n]$ of size n. If the chosen SHELL SORT uses k stages, a k-long integer sequence decreasing down to 1 is chosen to achieve a faster sort than plain LINEAR INSERTION SORT as follows. Suppose the sequence is $t_k, t_{k-1}, \dots, t_1 = 1$. In sorting n keys, the first stage sorts keys that are t_k positions apart in the list. Thus t_k subarrays of length at most $\lceil n/t_k \rceil$ each are sorted by plain LINEAR INSERTION SORT. In the second stage, the algorithm uses the increment t_{k-1} to sort t_{k-1} subarrays of keys that are t_{k-1} positions apart (each subarray is of length at most $\lceil n/t_{k-1} \rceil$), and so on, down to the last stage, where an increment of 1 is used to sort the whole array by insertion. Thus, in the last stage the algorithm executes plain LINEAR INSERTION SORT. A SHELL SORT algorithm using the sequence $t_k, t_{k-1}, \ldots, 1$ will be referred to as $(t_k, t_{k-1}, \ldots, 1)$-SHELL SORT, and the stage that uses the increment t_j will be called the t_j-stage.

As an example, (2, 1)-SHELL SORT sorts the array

$$7 \quad 4 \quad 1 \quad 8 \quad 9 \quad 5 \quad 6 \quad 2 \quad 3$$

in two stages. In the first stage increments of 2 are used—the subarray of odd indexes is sorted by the regular LINEAR INSERTION SORT, and the subarray of even indexes is sorted by the regular LINEAR INSERTION SORT. The two interleaved arrangements

sorted odd positions:	1		3		6		7		9
sorted even positions:		2		4		5		8	

are then sorted by one run of the regular LINEAR INSERTION SORT on the whole array.

The code for the algorithm is a simple adaptation of LINEAR INSERTION SORT (see Figure 3.1). Assume the increment sequence comprises k increments stored in

```
for m ← k downto 1 do
    begin
        {insert sort with increment t[m]}
        h ← t[m];
        for s ← 0 to h − 1 do
            begin
                i ← h + 1;
                while i ≤ n do
                    begin
                        j ← i;
                        key ← A[j];
                        while (j − h > 0) and (key < A[j − h]) do
                            begin
                                A[j] ← A[j − h];
                                j ← j − h;
                            end;
                        A[j] ← key;
                        i ← i + h;
                    end;
            end;
    end;
```

Figure 3.1. The SHELL SORT algorithm.

an array $t[1\,..\,k]$. The algorithm executes its main loop k times, one iteration for each increment. An index, say m, records the stage number, which starts out at k. Toward a transparent code, the increment $t[m]$ of the m-stage is recorded in a variable, say h, to avoid double indexing (for instance use $A[h+1]$ instead of $A[t[k]+1]$). The algorithm then iterates LINEAR INSERTION SORT on h subarrays. The variable s is used to index these iterations. During the first iteration of the outer loop. The first iteration of the inner loop with $h = t[k]$ orders the subarray $A[1]$, $A[h+1]$, $A[2h+1], \dots, A[rh+1]$, where r is the largest integer so that $rh+1 \le n$, by LINEAR INSERTION SORT. In the second iteration of the inner loop (still with $h = t[k]$ as increment), the subarray $A[2]$, $A[h+2]$, $A[2h+2], \dots, A[rh+2]$, where r is reassigned the largest integer so that $rh+2 \le n$, is sorted by LINEAR INSERTION SORT, and so forth. During the sth iteration of the inner loop with $h = t[m]$, the array with beginning index $s+1$ and whose elements are h positions apart is the subject of LINEAR INSERTION SORT.

3.2 STREAMLINED STOCHASTIC ANALYSIS

There is a large variety of SHELL SORT algorithms according to the various sequences that can be chosen for the stages. For example, some flavors will choose a fixed number of stages with a fixed sequence of increments. Other flavors may choose a fixed number of stages, but the members of the sequence may grow with the number of keys, while some other flavors may choose an approach that allows the number of stages to increase with the number of keys.

We shall study the (2, 1)-SHELL SORT as a prototype for the analysis technique and the type of result one expects out of such analysis. Extensions of this prototype case may be possible. The analysis of $(h, 1)$-SHELL SORT may follow the same general lines and extensions to $(t_k, t_{k-1}, \dots, 1)$-SHELL SORT may be possible for some sequences of increments.

The probability model for the analysis is the random permutation model. The difficulty in the stochastic analysis of $(t_k, t_{k-1}, \dots, 1)$-SHELL SORT lies in that after the first stage the resulting data are no longer random. Instead, t_k *sorted* subarrays are interleaved. The second and subsequent stages may not then appeal to the results known for INSERTION SORT. For example, the 1-stage of the (2, 1)-SHELL SORT does not sort a random array of size n. The 2-stage somewhat orders the array as a whole, and many inversions are removed (some new ones may appear, though; see, for example, the positions of 5 and 6 before and after the 2-stage of the example of Section 3.1).

To illustrate the analysis in its simplest form, we shall consider in this section (2, 1)-SHELL SORT. We shall streamline the stochastic analysis of (2,1)-SHELL SORT by finding a probabilistic representation for its components. The forms arising in the analysis bear resemblance to the empirical distribution function, which has natural connections to the Brownian bridge. We shall say a quick word about each of these tools in individual units before we embark on a full stochastic analysis of (2,1)

SHELL SORT. A reader familiar with the notions of the Brownian bridge may skip directly to pick up the remainder of the discussion at Subsection 3.2.3.

3.2.1 The Empirical Distribution Function

Suppose we want to empirically determine a distribution function F. For that we take a sample $Z_1, \ldots, Z_n$ of points from that distribution. Let $F_n(t)$ be $\frac{1}{n}$ times the number of observations not exceeding t in our sample. Clearly $F_n(t)$ is a right-continuous nondecreasing staircase-like function that rises from 0 at $-\infty$ to 1 at $+\infty$. It therefore qualifies as a distribution function, which we shall call the *empirical distribution function*. Its rationale is to assign each point in the n observed data a probability of $1/n$, corresponding to a jump of this magnitude in the empirical distribution function at each sample point.

For example, suppose that it is not known to us that $X \overset{\mathcal{D}}{=} \text{BERNOULLI}(\frac{2}{3})$ and we want to assess its distribution function. We may decide to take a sample of size 100. Of these 100 points 35 may turn out 0, the remaining 65 turn out 1. For $0 \le t < 1$, the empirical distribution function is $F_{100}(t) = 0.35$. This is to be compared with the true distribution function $F(t) = \frac{1}{3}$.

The empirical distribution function is random; $F_n(t)$ is a random variable. For instance, repeating our assessment experiment, it may happen the second time around that only 27 points of our 100 sample points assume the value 0, the rest assume the value 1, in which case $F_{100}(t) = 0.27$, for $0 \le t < 1$. As we repeat the experiment a large number of times, we expect fluctuations in $F_n(t)$, but we also should believe its value is close to the true underlying distribution. Indeed, $F_n(t)$ has a representation as a sum of independent identically distributed random variables:

$$F_n(t) = \frac{1}{n} \sum_{j=1}^{n} \mathbf{1}_{\{Z_j \le t\}},$$

where $\mathbf{1}_{\mathcal{E}}$ is the indicator of the set $\mathcal{E}$. It follows from the strong law of large numbers that

$$F_n(t) \xrightarrow{\text{a.s.}} \mathbf{E}\big[\mathbf{1}_{\{Z_1 \le t\}}\big] = \mathbf{Prob}\{Z_1 \le t\} = F(t).$$

A strong law only states the behavior of the leading component of $F_n(t)$. More profoundly, the fluctuations $F_n(t) - F(t)$ normed by a suitable factor behave like a Brownian bridge, the second of our constructs, a connection established shortly after we introduce the Brownian bridge process.

3.2.2 The Brownian Bridge

The second of our constructs is the Brownian bridge, a Brownian motion over the time interval [0, 1] conditioned to return to 0 at time 1. We therefore define the more general Brownian motion first. The Brownian motion is a stochastic process proposed as an approximate model for a physical phenomenon first noted in 1827 by the botanist Robert Brown, who took interest in the then-unexplained random motion of particles submerged in a fluid. The observation did not draw much attention throughout the rest of the nineteenth century. With the advent of the twentieth century, molecular theory of the structure of matter settled in and Albert Einstein and independently Marian Smoluchowski around 1905 gave physical explanations based on the then-new theory of molecules. The spores immersed in a fluid experience rather frequent bombardment by the particles of the fluid. The result is to exhibit short intermittent displacements. Around 1918 Norbort Wiener wrote about a mathematical model that is a conceptual limit to this kind of random behavior.

It is sufficient for our purpose to understand the one-dimensional analog. The genesis is the usual *symmetric random walk*. Assume we have a particle starting out at 0 and at each time tick it makes a move either one step to the right with probability $1/2$ or one step to the left with probability $1/2$. Let X_i be the particle's displacement in the ith step. Then X_i is -1 or 1 with equal probability. After n discrete time ticks the particle has moved a distance of

$$S_n = X_1 + X_2 + \cdots + X_n.$$

Being a sum of independent identically distributed random variables, this sum follows standard probability theory and gives strong laws and central limit theorems in a well-developed standard theory for random walks. Figure 3.2 illustrates one possible realization of the standard symmetric random walk. The position of the particle at time n (integer) is S_n, the bullet corresponding to the nth time tick; however, the straight line segments connecting S_n and S_{n+1} are only inserted for visual convenience and may not have a physical meaning.

Now suppose that we speed up the process in a manner that emulates the rapid intermittent bombardment of spores in a fluid. We shall call the process so obtained the *incremental Brownian motion*, till we pass to a formal limit definition. Instead

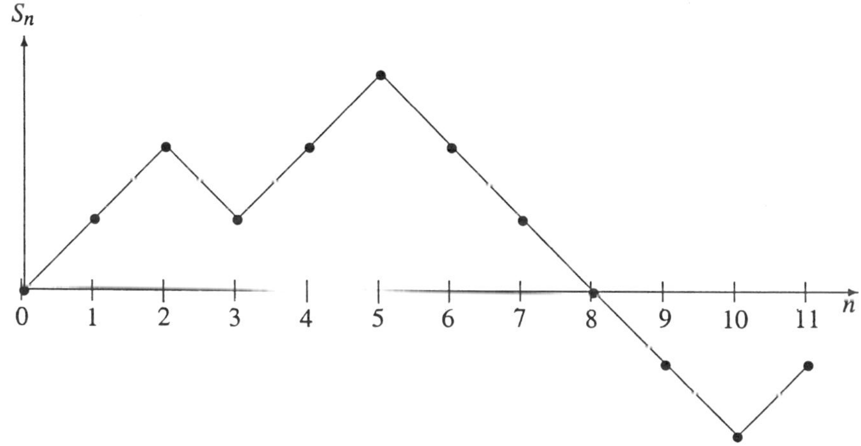

Figure 3.2. A realization of the standard symmetric random walk.

of waiting a whole time unit, the particle is forced to make its move every infinitesimal unit of time Δt. The size of the displacement is also made microscopic. Instead of jumps of magnitude 1, the particle is restricted to make displacements of the infinitesimal magnitude Δx. Where is the particle at time t and what kind of trajectory does it follow? If Δt and Δx are both very small, one does not expect the particle to change its position too much in a small time period approaching some notion of continuity.

Let the particle's displacement at time t be $D(t)$. The interval $[0, t]$ can be divided by time ticks $0, \Delta t, 2\Delta t, \dots, \lfloor t/\Delta t \rfloor \Delta t$. During the interval $(\lfloor t/\Delta t \rfloor \Delta t, t]$ the particle does not make any moves (but will move again at time $(\lfloor t/\Delta t \rfloor + 1)\Delta t$). The total displacement up to time t is

$$D(t) = Z_1 + Z_2 + \cdots + Z_{\lfloor t/\Delta t \rfloor},$$

with each Z_i being an independent random variable that takes the value $-\Delta x$ (with probability $1/2$) or $+\Delta x$ (with probability $1/2$).

The displacement $D(t)$ has mean 0 and, as $\Delta t \to 0$, has variance

$$\mathbf{Var}\big[D(t)\big] = \Big\lfloor \frac{t}{\Delta t} \Big\rfloor \mathbf{Var}[Z_1] = \Big(\frac{t}{\Delta t} + O(1)\Big)\,(\Delta x)^2.$$

For the incremental Brownian motion process to remain nontrivial at fixed t, as $\Delta t \to 0$, Δx must be of the order $\sigma\sqrt{\Delta t} + o(\sqrt{\Delta t}\,)$ for some positive constant σ (otherwise the process either dies out or blows up). Let us take $\Delta x = \sigma\sqrt{\Delta t}$, so that

$$\mathbf{Var}\big[D(t)\big] \to \sigma^2 t, \qquad \text{as } \Delta t \to 0.$$

If Δt is infinitesimally small, a finite interval $[0, t]$ will contain a very large number of time ticks at each of which a small displacement takes place, and $D(t)$ then becomes the sum of a large number of independent, identically distributed random variables. One can apply the standard central limit theorem to the statistical average of the incremental displacements Z_i, to obtain

$$\frac{\displaystyle\sum_{i=1}^{\lfloor t/\Delta t \rfloor} Z_i}{\sqrt{\mathbf{Var}\Big[\displaystyle\sum_{i=1}^{\lfloor t/\Delta t \rfloor} Z_i\Big]}} = \frac{D(t)}{\sqrt{\mathbf{Var}\big[D(t)\big]}} \xrightarrow{\mathcal{D}} \mathcal{N}(0, 1).$$

Under the chosen condition for speeding up the random walk, $\sqrt{\mathbf{Var}[D(t)]} \to \sigma\sqrt{t}$. The product of this deterministic convergence by the last relation yields

$$D(t) \xrightarrow{\mathcal{D}} \sigma\sqrt{t}\,\mathcal{N}(0, 1) \overset{\mathcal{D}}{=} \mathcal{N}(0, \sigma^2 t);$$

the displacement $D(t)$ is normally distributed with mean 0 and variance $\sigma^2 t$. Intuitively, the displacement averages to zero, as movement in both directions is equally likely, but as more time elapses, our uncertainty about the position of the particle increases (a larger variance).

The incremental Brownian motion process described above falls in a class of stochastic processes that are of independent increments. A process $Q(t)$ is said to have *independent increments* if for any time sequence $t_1 < t_2 < \ldots < t_n$, the increments $Q(t_1), Q(t_2) - Q(t_1), \ldots, Q(t_n) - Q(t_{n-1})$ are independent. The incremental Brownian motion process accrues an incremental amount of displacement that is independent of the particle's current position.

Moreover, a stochastic process $Q(t)$ is said to be a *stationary process* if the increment $Q(t+s) - Q(t)$ depends only on s (but not on t). If $s < t$, the displacement of the incremental symmetric random walk process at time t is $D(s) \oplus D(t-s)$, the convolution of $D(s)$ and a random variable distributed like $D(t-s)$. That is, the displacement over the interval $t-s$ is distributed like the process itself at time $t-s$ (as if the process were reset to start at time s).

The incremental Brownian motion is a stationary process that has independent increments. It may then be reasonable to require that the limiting process inherit these properties. We therefore use these properties and the normal limit as the basis for a formal definition of the Brownian motion process. The *Brownian motion*, $D(t)$, is a stochastic process that satisfies the following:

(i) $D(0) = 0$.

(ii) The process has independent and stationary increments.

(iii) The process at time t has a Gaussian distribution:

$$D(t) \stackrel{D}{=} \mathcal{N}(0, \sigma^2 t).$$

The *Brownian bridge* is a Brownian motion conditioned to return to the origin at time 1 (that is, it has the conditional distribution $D(t) \mid D(1) = 0$). We can standardize the bridge by considering the conditional Brownian motion with $\sigma = 1$ (if $\sigma \neq 1$, we can always study $D(t)/\sigma$.) We denote the *standard Brownian bridge* by $B(t)$. The distribution of the Brownian bridge process is obtained by conditioning the Brownian motion. Suppose $0 \le t_1 < t_2 < \cdots < t_n \le 1$. Let $f_{V_1,\ldots,V_n}(v_1, \ldots, v_n)$ denote the joint density of random variables $V_1, \ldots, V_n$. The joint distribution $f_{D(t_1),\ldots,D(t_n)}(d_1, \ldots, d_n)$ is more easily obtained from the increments $D(t_1), D(t_2) - D(t_1), \ldots, D(t_n) - D(t_{n-1})$. The increments amount to a linear transformation of random variables, introducing a new set:

$$\begin{aligned} I_1 &= D(t_1), \\ I_2 &= D(t_2) - D(t_1), \\ &\vdots \\ I_n &= D(t_n) - D(t_{n-1}). \end{aligned}$$

This linear transformation introduces a set of independent random variables because the Brownian motion is a stationary process of independent increments. From the standard theory of transformations, the set of increments and the set of actual displacements have joint distributions connected via:

$$f_{D(t_1),\ldots,D(t_n)}(d_1,\ldots,d_n) = \frac{1}{|J|} f_{I_1,\ldots,I_n}(d_1, d_2 - d_1, \ldots, d_n - d_{n-1}),$$

where J is the determinant of the Jacobian of transformation that is given by:

$$\begin{aligned} J &= \left|\frac{\partial D(t_i)}{\partial I_j}\right|, \qquad i, j = 1, \ldots, n \\ &= \begin{vmatrix} 1 & 0 & 0 & 0 & \ldots & 0 \\ 1 & 1 & 0 & 0 & \ldots & 0 \\ 1 & 1 & 1 & 0 & \ldots & 0 \\ \vdots & \vdots & \ddots & & & \vdots \\ 1 & 1 & \ldots & & 1 & 1 \end{vmatrix} \\ &= 1. \end{aligned}$$

By the independence of the increments, their joint density is the product of their individual normal densities:

$$\begin{aligned} f_{D(t_1),\ldots,D(t_n)}(d_1,\ldots,d_n) &= \frac{e^{-d_1^2/(2t_1)}}{\sqrt{2\pi t_1}} \times \frac{e^{-(d_2-d_1)^2/(2(t_2-t_1))}}{\sqrt{2\pi(t_2-t_1)}} \\ &\quad \times \cdots \times \frac{e^{-(d_n-d_{n-1})^2/(2(t_n-t_{n-1}))}}{\sqrt{2\pi(t_n-t_{n-1})}}. \end{aligned}$$

The Brownian bridge is a conditional Brownian motion:

$$B(t) = D(t) \mid D(1) = 0.$$

Its density $f_{B(t)}(x)$ is therefore obtained from the joint density of $D(t)$ and $D(1)$. For $0 \le s < t \le 1$, we are further interested in the covariance between $B(s)$ and $B(t)$. To save some effort, we develop the bivariate joint density $f_{B(s),B(t)}(x, y)$ first. The univariate density $f_{B(t)}$ is then one of the two marginal densities of the bivariate joint density. We have

$$\begin{aligned} f_{B(s),B(t)}(x, y) &= f_{D(s),D(t)}(x, y \mid D(1) = 0) \\ &= \frac{f_{D(s),D(t),D(1)}(x, y, 0)}{f_{D(1)}(0)} \\ &= \frac{\sqrt{2\pi}}{(2\pi)^{3/2}\sqrt{s(t-s)(1-t)}} \exp\Bigl(-\frac{x^2}{2s} - \frac{(y-x)^2}{2(t-s)} - \frac{y^2}{2(1-t)}\Bigr) \\ &= \frac{\exp\Bigl(-\frac{1}{2}\Bigl(\frac{t(1-t)x^2 - 2s(1-t)xy + s(1-s)y^2}{s(t-s)(1-t)}\Bigr)\Bigr)}{2\pi\sqrt{s(t-s)(1-t)}}. \end{aligned}$$

Put in matrix notation, this expression has a familiar compact form. Let $\mathbf{v} = \begin{pmatrix} x \\ y \end{pmatrix}$ and $\mathbf{v}'$ be its transpose. Then

$$f_{B(s),B(t)}(\mathbf{v}) = \frac{1}{2\pi\sqrt{|\boldsymbol{\Sigma}|}} e^{-\frac{1}{2}\mathbf{v}'\Sigma^{-1}\mathbf{v}},$$

where

$$\boldsymbol{\Sigma} = \begin{pmatrix} s(1-s) & s(1-t) \\ s(1-t) & t(1-t) \end{pmatrix};$$

the bivariate density of the Brownian bridge at times s and t is the same as that of a bivariate normal distribution with mean $\mathbf{0}$ and covariance matrix $\boldsymbol{\Sigma}$. The covariance of the Brownian bridge at times s and t is

$$\mathbf{Cov}\big[B(s), B(t)\big] = s(1-t).$$

The Brownian bridge $B(t)$ is the marginal distribution of the bivariate normal distribution discussed above. This marginal is a zero-mean normal distribution with variance $t(1-t)$, that is,

$$B(t) \overset{D}{=} \mathcal{N}\big(0, t(1-t)\big).$$

The Brownian bridge is closely tied to the empirical distribution function. To approach this intuitively, we consider first the connection between the Brownian bridge and the empirical distribution function $F(t) = t$, for $0 < t < 1$, of the UNIFORM(0, 1) random variable. This distribution function is assessed empirically by $F_n(t)$, that is, computed from a sample $U_1, \dots, U_n$ of n independent variates taken from that distribution:

$$F_n(t) = \frac{1}{n}\sum_{i=1}^{n} \mathbf{1}_{\{U_i \le t\}}.$$

For $0 \le s < t \le 1$, one finds

$$\begin{aligned} \mathbf{E}\big[F_n(t)\big] &= \mathbf{E}\Big[\frac{1}{n}\sum_{i=1}^{n} \mathbf{1}_{\{U_i \le t\}}\Big] \\ &= \frac{1}{n}\sum_{i=1}^{n} \mathbf{Prob}\{U_1 \le t\} \\ &= t. \end{aligned}$$

Interpreting the argument of the empirical distribution function as time, $F_n(t)$ is then a stochastic process with the covariance

$$\begin{aligned}\mathbf{Cov}\big[F_n(s), F_n(t)\big] &= \mathbf{E}\big[F_n(s)F_n(t)\big] - \mathbf{E}\big[F_n(s)\big]\mathbf{E}\big[F_n(t)\big] \\ &= \frac{1}{n^2}\mathbf{E}\Big[\Big(\sum_{i=1}^{n} \mathbf{1}_{\{U_i \le s\}}\Big)\Big(\sum_{j=1}^{n} \mathbf{1}_{\{U_j \le t\}}\Big)\Big] - st \\ &= \frac{1}{n^2}\mathbf{E}\Big[\sum_{i=1}^{n} \mathbf{1}_{\{U_i \le s\}} + 2\sum_{1\le i<j\le n} \mathbf{1}_{\{U_i \le s\}}\mathbf{1}_{\{U_j \le t\}}\Big] - st.\end{aligned}$$

The uniform variates are independent and identical in distribution, giving

$$\begin{aligned}\mathbf{Cov}\big[F_n(s), F_n(t)\big] &= \frac{1}{n^2} \times n\,\mathbf{Prob}\{U_1 \le s\} \\ &\qquad + \frac{1}{n^2} \times (n-1)n\,\mathbf{Prob}\{U_1 \le s\}\,\mathbf{Prob}\{U_2 \le t\} - st \\ &= \frac{s}{n} + \frac{(n-1)st}{n} - st \\ &= \frac{1}{n}s(1-t).\end{aligned}$$

The process $\sqrt{n}\,F_n(t)$ thus has the covariance structure

$$\mathbf{Cov}\big[\sqrt{n}\,F_n(s), \sqrt{n}\,F_n(t)\big] = s(1-t).$$

The centered process $\sqrt{n}\,\big(F_n(t) - t\big)$ is Markovian and has the same covariance structure of the standard Brownian bridge. We conclude that

$$\sqrt{n}\,\big(F_n(t) - t\big) \xrightarrow{\mathcal{D}} B(t).$$

Figure 3.3 shows the empirical distribution function F_{30} of 30 independent standard uniforms in comparison with the true distribution function common to these random variables, and Figures 3.4 illustrates $\sqrt{30}\,(F_{30}(t) - t)$, which has the general shape of a jagged Brownian bridge.

The discussion is not restricted to the empirical distribution function of the uniform random variable alone. The key point here is that the inversion process of a continuous distribution function (the so-called *probability integral transform*) always yields uniform random variables—if F is a continuous distribution function of X, we define $U = F(X)$ and then $U \overset{\mathcal{D}}{=} \text{UNIFORM}(0, 1)$; we leave it as an exercise to verify this. So,

$$\sqrt{n}\,\big(F_n(t) - F(t)\big) \xrightarrow{\mathcal{D}} B\big(F(t)\big),$$

where $F_n(t)$ is the empirical distribution function of $F(t)$.

The stochastic representations of (2, 1) SHELL SORT will soon be connected to the Brownian bridge. The area under the absolute Brownian bridge comes into this

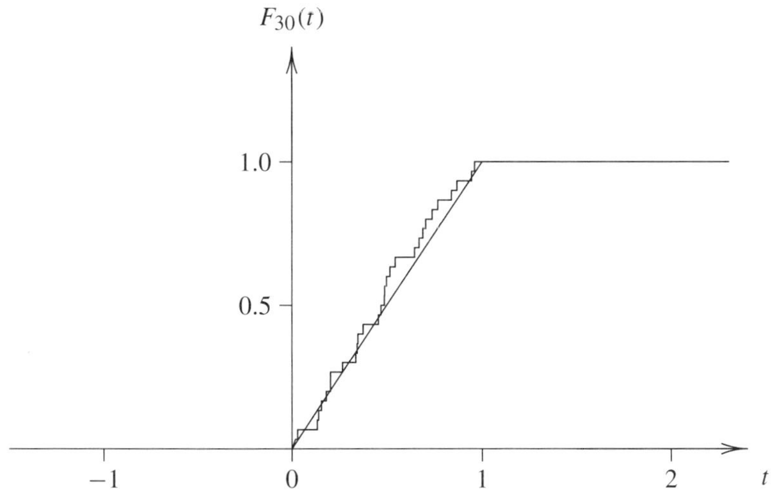

Figure 3.3. The empirical distribution function of the UNIFORM(0,1).

picture; the random variable

$$A = \int_0^1 |B(t)|\, dt$$

appears in the calculation. The distribution of A is complicated but known (Johnson and Killeen (1983)).

Many functionals of the Brownian bridge, including A, have been studied. All moments of A are known and the first few are tabulated in Shepp (1982). We shall only need up to the second moment. For the first moment of A, it is fairly simple to

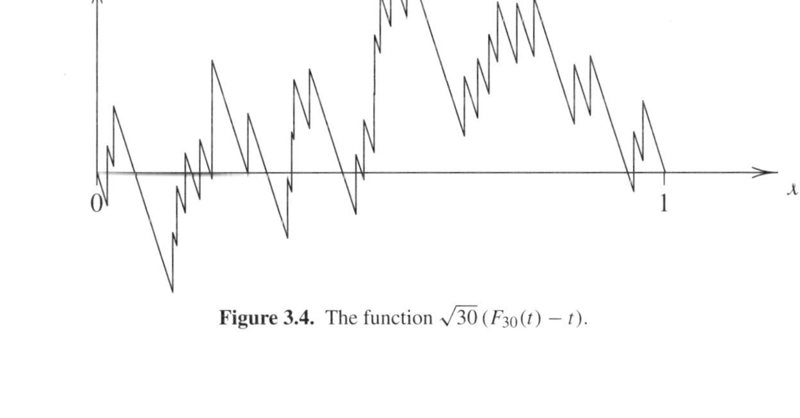

Figure 3.4. The function $\sqrt{30}\,(F_{30}(t) - t)$.

derive Shepp's first moment result from basic principles and standard distributions:

$$\begin{aligned}
\mathbf{E}\big|B(t)\big| &= \mathbf{E}\big|\mathcal{N}\big(0, t(1-t)\big)\big| \\
&= \frac{1}{\sqrt{2\pi t(1-t)}} \int_{-\infty}^{\infty} |x| \exp\Big(-\frac{x^2}{2t(1-t)}\Big)\, dx \\
&= 2\frac{t(1-t)}{\sqrt{2\pi t(1-t)}} \int_0^{\infty} v e^{-v^2/2}\, dv \\
&= \sqrt{\frac{2}{\pi} t(1-t)}\,.
\end{aligned}$$

Then

$$\begin{aligned}
\mathbf{E}[A] &= \mathbf{E}\left[\int_0^1 \big|B(t)\big|\, dt\right] \\
&= \int_0^1 \mathbf{E}\big|B(t)\big|\, dt \\
&= \sqrt{\frac{2}{\pi}} \int_0^1 \sqrt{t(1-t)}\, dt \\
&= \sqrt{\frac{2}{\pi}}\, \beta\left(\frac{3}{2}, \frac{3}{2}\right);
\end{aligned}$$

the Beta function $\beta(\frac{3}{2}, \frac{3}{2}) = \Gamma(\frac{3}{2})\,\Gamma(\frac{3}{2})/\,\Gamma(3) = \frac{\pi}{8}$, and

$$\mathbf{E}[A] = \sqrt{\frac{\pi}{32}}. \tag{3.1}$$

We can also do this for the second moment needed in the proof of the variance result below. One can compute Shepp's second moment result form basic principles involving the bivariate normal distribution. After somewhat heavy integral calculation one finds

$$\mathbf{E}[A^2] = \frac{7}{60}. \tag{3.2}$$

3.2.3 Using the Stochastic Tools

We have at our disposal enough tools to model (2, 1)-SHELL SORT. Suppose our data are n distinct real numbers (this is almost surely the case when the numbers come from a continuous probability distribution). Let us call the elements in odd positions X's and those in even positions Y's. Thus, if n is odd, initially our raw array prior to any sorting is

$$X_1, Y_1, X_2, Y_2, \ldots, Y_{\lfloor n/2 \rfloor}, X_{\lceil n/2 \rceil},$$

and if n is even the initial raw array is

$$X_1, Y_1, X_2, Y_2, \ldots, X_{n/2}, Y_{n/2}.$$

The 2-stage of the algorithm puts the X's in order among themselves and the Y's in order among themselves. The order statistic notation is handy here. Let $Z_{(j)}$ be the jth order statistic of $Z_1, \ldots, Z_m$. The 2-stage puts the array in the form

$$X_{(1)}, Y_{(1)}, X_{(2)}, Y_{(2)}, \ldots, Y_{(\lfloor n/2 \rfloor)}, X_{(\lceil n/2 \rceil)},$$

if n is odd, and

$$X_{(1)}, Y_{(1)}, X_{(2)}, Y_{(2)}, \ldots, X_{(n/2)}, Y_{(n/2)},$$

if n is even.

Let S_n be the number of comparisons that (2, 1)-SHELL SORT makes to sort n random keys, and let C_n be the number of comparisons that LINEAR INSERTION SORT makes to sort n random keys. The 2-stage of (2, 1)-SHELL SORT makes two runs of LINEAR INSERTION SORT on the subarrays $X_1, \ldots, X_{\lceil n/2 \rceil}$ and $Y_1, \ldots, Y_{\lfloor n/2 \rfloor}$, thus requiring

$$C_{\lceil n/2 \rceil} + \tilde{C}_{\lfloor n/2 \rfloor}$$

comparisons, where $\{C_j\}$ and $\{\tilde{C}_j\}$ are independent families.[1]

The 1-stage now comes in, requiring additional comparisons to remove the remaining inversions. We know that LINEAR INSERTION SORT makes

$$C(\Pi_n) = n + I(\Pi_n)$$

comparisons to sort a permutation Π_n with $I(\Pi_n)$ inversions. The overall number of comparisons, S_n, of (2, 1)-SHELL SORT is therefore given by the convolution

$$S_n = C_{\lceil n/2 \rceil} + \tilde{C}_{\lfloor n/2 \rfloor} + (n + I_n). \tag{3.3}$$

Gaussian laws have been developed in Chapter 2 for various forms of INSERTION SORT. In particular, C_n, the number of comparisons that LINEAR INSERTION SORT performs to sort n random keys, is asymptotically normally distributed:

$$\frac{C_n - \frac{1}{4}n^2}{n^{3/2}} \xrightarrow{\mathcal{D}} \mathcal{N}\Big(0, \frac{1}{36}\Big).$$

We only need the distribution of I_n to complete the analysis of (2, 1)-SHELL SORT.

[1]We shall use the term *independent families of random variables* to indicate that the members of one family are independent of each other as well as being independent of all the members of the other families.

Lemma 3.1 *(Smythe and Wellner, 2000). After the 2-stage of* (2, 1) *SHELL SORT, the remaining number of inversions, I_n, has the representation*

$$I_n \stackrel{\text{a.s.}}{=} \sum_{j=1}^{\lceil n/2 \rceil} \left| \sum_{i=1}^{\lfloor n/2 \rfloor} \mathbf{1}_{\{Y_i < X_j\}} - \sum_{i=1}^{\lceil n/2 \rceil} \mathbf{1}_{\{X_i < X_j\}} \right|,$$

where $\mathbf{1}_{\mathcal{E}}$ is the indicator of the set $\mathcal{E}$.

Proof. To avoid trivialities caused by data repetition, we shall discuss the case when all the data are distinct (which is the case with probability 1 under the random permutation model). For each of the X's we compute the number of inversions it will cause after the 2-stage. Suppose the rank of X_j in the entire data set is r. There are $r - 1$ data below X_j. If the rank of X_j among the X's is r_1, there are $r_1 - 1$ of the X's below X_j. To visualize data positions after the 2-stage, it is helpful to think of X_j as $X_{(r_1)}$ for the rest of the proof. The data points $X_{(1)}, \dots, X_{(r_1-1)}$ are below $X_{(r_1)}$. After the 2-stage, these items will be placed in this relative order to the left of $X_{(r_1)} = X_j$. Let $r_2 = (r - 1) - (r_1 - 1) = r - r_1$ be the number of Y's below X_j. For instance in the case n even and $r_1 \le r_2$ the arrangement after the 2-stage will look like:

$$\begin{array}{cccccccccccc} X_{(1)} & & X_{(2)} & & \dots & & X_{(r_1-1)} & & X_{(r_1)} & & \dots & & X_{(n/2)} \\ & Y_{(1)} & & Y_{(2)} & & \dots & & Y_{(r_1-1)} & \dots & Y_{(r_2)} & & \dots & & Y_{(n/2)}. \end{array}$$

After the 2-stage, the X's are sorted among themselves; $X_{(r_1)}$ causes no inversion with any other $X_{(r_1)}$. Among the Y's, $Y_{(1)}, \dots, Y_{(r_2)}$ are all less than $X_{(r_1)}$; all the other Y's are larger. Which of $Y_{(1)}, \dots, Y_{(r_2)}$ will be inverted with respect to $X_{(r_1)}$ will depend on the relation between r_1 and r_2. Two cases arise:

(a) The case $r_1 \le r_2$: In this case $Y_{(1)}, \dots, Y_{(r_1-1)}$ appear to the left of $X_{(r_1)}$; but $Y_{(r_1)}, \dots, Y_{(r_2)}$ appear to the right of $X_{(r_1)}$ and each of these latter Y's causes one inversion with $X_{(r_1)}$. Obviously the Y's above $X_{(r_1)}$ (that is, $Y_{(r_2+1)}, \dots, Y_{(\lfloor n/2 \rfloor)}$) all appear to the right of $X_{(r_1)}$, and cause no inversions with it. The number of inversions involving $X_{(r_1)}$ is therefore

$$r_2 - (r_1 - 1) = \sum_{i=1}^{\lfloor n/2 \rfloor} \mathbf{1}_{\{Y_i < X_j\}} - \sum_{i=1}^{\lceil n/2 \rceil} \mathbf{1}_{\{X_i < X_j\}}.$$

(b) The case $r_1 > r_2$: In this case $Y_{(1)}, \dots, Y_{(r_2)}$ appear to the left of $X_{(r_1)}$ and $Y_{(r_2+1)}, \dots, Y_{(r_1-1)}$ are the only Y's that are out of position with respect to $X_{(r_1)}$; the number of inversions involving $X_{(r_1)}$ is therefore

$$(r_1 - 1) - r_2 = \sum_{i=1}^{\lceil n/2 \rceil} \mathbf{1}_{\{X_i < X_j\}} - \sum_{i=1}^{\lfloor n/2 \rfloor} \mathbf{1}_{\{Y_i < X_j\}}.$$

In either case, to I_n the key X_j contributes

$$\left| \sum_{i=1}^{\lfloor n/2 \rfloor} \mathbf{1}_{\{Y_i < X_j\}} - \sum_{i=1}^{\lceil n/2 \rceil} \mathbf{1}_{\{X_i < X_j\}} \right|$$

inversions. The total number of inversions after the 2-stage is then obtained by summing over j. ■

Let $F(t)$ be the distribution function common to the data items, and let $\hat{F}_n(t)$ be the corresponding empirical distribution function computed from a sample of size n. Let U_i and V_i be the "projections" of the X_i's and Y_i's via the probability integral transform, that is,

$$U_i = F(X_i), \qquad i = 1, \dots, \lceil n/2 \rceil,$$
$$V_i = F(Y_i), \qquad i = 1, \dots, \lfloor n/2 \rfloor.$$

The random variables U_i and V_i are independent and each is distributed like a UNIFORM(0, 1) random variable. The representation of the number of comparisons in the 1-stage of (2, 1) SHELL SORT as a sum of indicators (Lemma 3.1) admits expressions via the empirical distribution function—we have the terms

$$\begin{aligned}
\sum_{i=1}^{\lfloor n/2 \rfloor} \mathbf{1}_{\{Y_i < X_j\}} &\overset{\text{a.s.}}{=} \sum_{i=1}^{\lfloor n/2 \rfloor} \mathbf{1}_{\{Y_i \le X_j\}} \\
&= \sum_{i=1}^{\lfloor n/2 \rfloor} \mathbf{1}_{\{F(Y_i) \le F(X_j)\}} \\
&= \sum_{i=1}^{\lfloor n/2 \rfloor} \mathbf{1}_{\{V_i \le U_j\}} \\
&= \lfloor n/2 \rfloor \hat{F}_{\lfloor n/2 \rfloor}(U_j),
\end{aligned}$$

and

$$\begin{aligned}
\sum_{i=1}^{\lceil n/2 \rceil} \mathbf{1}_{\{X_i < X_j\}} &\overset{\text{a.s.}}{=} \sum_{\substack{i=1 \\ i \ne j}}^{\lceil n/2 \rceil} \mathbf{1}_{\{X_i \le X_j\}} + \mathbf{1}_{\{X_j < X_j\}} \\
&= \sum_{i=1}^{\lceil n/2 \rceil} \mathbf{1}_{\{X_i \le X_j\}} - \mathbf{1}_{\{X_j \le X_j\}} + 0 \\
&= \lceil n/2 \rceil \tilde{F}_{\lceil n/2 \rceil}(U_j) - 1;
\end{aligned}$$

where $\tilde{F}_j(t) \overset{D}{=} \hat{F}_j(t)$, and $\hat{F}_j(t)$ and $\tilde{F}_j(t)$ are independent (the Y's are independent of the X's).

For compactness, let $T_n^{(j)}$ denote the inner summation of the expression of Lemma 3.1, namely

$$T_n^{(j)} \stackrel{\text{def}}{=} \left| \sum_{i=1}^{\lfloor n/2 \rfloor} \mathbf{1}_{\{Y_i < X_j\}} - \sum_{i=1}^{\lceil n/2 \rceil} \mathbf{1}_{\{X_i < X_j\}} \right|.$$

For uniform data, we have discussed the convergence

$$\sqrt{n}\left(F_n(t) - t\right) \xrightarrow{\mathcal{D}} B(t),$$

where $B(t)$ is the (standard) Brownian bridge. Thus

$$\begin{aligned} \frac{T_n^{(j)}}{\sqrt{n}} &= \left| \sqrt{\frac{1}{n}\left\lfloor \frac{n}{2} \right\rfloor} \sqrt{\left\lfloor \frac{n}{2} \right\rfloor} \hat{F}_{\lfloor n/2 \rfloor}(U_j) - \sqrt{\frac{1}{n}\left\lceil \frac{n}{2} \right\rceil} \sqrt{\left\lceil \frac{n}{2} \right\rceil} \tilde{F}_{\lceil n/2 \rceil}(U_j) + O\left(\frac{1}{\sqrt{n}}\right) \right| \\ &\xrightarrow{\mathcal{D}} \frac{1}{\sqrt{2}} \left| B(U_j) - \tilde{B}(U_j) \right|, \end{aligned}$$

where $B(t)$ and $\tilde{B}(t)$ are two independent standard Brownian bridges. The difference $B(t) - \tilde{B}(t)$ is clearly a Markovian stationary process of independent increments and has a simple normal distribution because

$$B(t) - \tilde{B}(t) \stackrel{D}{=} \mathcal{N}\big(0, t(1-t)\big) - \tilde{\mathcal{N}}\big(0, t(1-t)\big) \stackrel{D}{=} \mathcal{N}\big(0, 2t(1-t)\big),$$

($\tilde{\mathcal{N}}(0, t(1-t)) \stackrel{D}{=} \mathcal{N}(0, t(1-t))$, and is independent of $\mathcal{N}(0, t(1-t))$). Therefore $(B(t) - \tilde{B}(t))/\sqrt{2}$ is distributed like the standard Brownian bridge. We have

$$\frac{T_n^{(j)}}{\sqrt{n}} \xrightarrow{\mathcal{D}} |B(U_j)|. \tag{3.4}$$

Theorem 3.1 *(Knuth, 1973). The average number of comparisons made by the 2-stage of* (2, 1)*-SHELL SORT to sort n random keys is asymptotically equivalent to* $\frac{1}{8}n^2$*; the 1-stage then performs a number of comparisons asymptotic to* $\sqrt{\frac{\pi}{128}}\, n^{3/2}$*. The overall number of comparisons is therefore asymptotically equivalent to* $\frac{1}{8}n^2$*.*

Proof. Knuth (1973) proves this result by a combinatorial argument that counts certain kinds of paths in a two-dimensional lattice. We present a proof due to Louchard (1986) based on the stochastic elements introduced in this chapter. Let S_n be the number of comparisons that (2, 1)-SHELL SORT performs to sort n random keys. Recall (3.3):

$$S_n = C_{\lceil n/2 \rceil} + \tilde{C}_{\lfloor n/2 \rfloor} + (n + I_n).$$

Take the expectation

$$\mathbf{E}[S_n] = \mathbf{E}\big[C_{\lceil n/2\rceil}\big] + \mathbf{E}\big[\tilde{C}_{\lfloor n/2\rfloor}\big] + \mathbf{E}[I_n] + n.$$

Two of the three components in the average of the convolution of S_n are the average number of comparisons of LINEAR INSERTION SORT on $\lfloor n/2\rfloor$ and $\lceil n/2\rceil$ keys (each is asymptotically equivalent to $\frac{1}{16}n^2$; this was developed in Chapter 2; see (2.4)). The 2-stage asymptotically makes $n^2/16 + n^2/16 = n^2/8$ comparisons.

As in the development of Lemma 3.1, the third component (comparisons of the 1-stage) is $n + I_n$, where the random variable I_n takes the form

$$I_n \stackrel{\text{a.s.}}{=} \sum_{j=1}^{\lceil n/2\rceil} T_n^{(j)},$$

and $T_n^{(1)}, \dots, T_n^{(\lceil n/2\rceil)}$ are identically distributed. So, $\mathbf{E}[I_n] = \lceil n/2\rceil\, \mathbf{E}[T_n^{(1)}]$. By the connection (3.4) to Brownian bridge:

$$\mathbf{E}\big[T_n^{(1)}\big] \sim \sqrt{n}\, \mathbf{E}\big|B(U_1)\big|.$$

Conditioning on the uniform variate U_1, together with Shepp's first moment result (3.1), yield the expectation

$$\begin{aligned}\mathbf{E}\big[T_n^{(1)}\big] &\sim \sqrt{n} \int_0^1 \mathbf{E}\big|B(t)\big|\, dt \\ &\sim \sqrt{\frac{\pi}{32}n}. \qquad (3.5)\end{aligned}$$

The 1-stage makes $\lceil n/2\rceil\, \mathbf{E}\big[T_n^{(1)}\big] \sim \sqrt{\frac{\pi}{128}}\, n^{3/2}$ comparisons. ∎

The stochastic representation of (2, 1)–SHELL SORT allows us to go forward with the variance calculation and ultimately with the limiting distribution.

Theorem 3.2 *(Knuth, 1973). The variance of the number of comparisons of* (2, 1)-*SHELL SORT for sorting n random keys is asymptotically equivalent to* $(\frac{13}{360} - \frac{\pi}{128})n^3$.

Proof. Recall again the convolution (3.3):

$$S_n = C_{\lceil n/2\rceil} + \tilde{C}_{\lfloor n/2\rfloor} + (n + I_n);$$

and take its variance to get from the independence of the parts

$$\mathbf{Var}[S_n] = \mathbf{Var}\big[C_{\lceil n/2\rceil}\big] + \mathbf{Var}\big[C_{\lfloor n/2\rfloor}\big] + \mathbf{Var}[I_n].$$

The random variables $C_{\lceil n/2 \rceil}$ and $\tilde{C}_{\lfloor n/2 \rfloor}$ are the number of comparisons of LINEAR INSERTION SORT on $\lceil n/2 \rceil$ and $\lfloor n/2 \rfloor$ keys (each has variance that is asymptotically equivalent to $\frac{1}{288}n^3$, which was developed in Chapter 2; see (2.5)).

For I_n, we have

$$\mathbf{Var}[I_n] = \sum_{i=1}^{\lceil n/2 \rceil} \mathbf{Var}\big[T_n^{(i)}\big] + 2 \sum_{1 \le i < j \le \lceil n/2 \rceil} \mathbf{Cov}\big[T_n^{(i)}, T_n^{(j)}\big].$$

The variables $T_n^{(1)}, \dots, T_n^{(\lceil n/2 \rceil)}$ are identically distributed, which simplifies the calculation to

$$\mathbf{Var}[I_n] = \lceil n/2 \rceil \mathbf{Var}\big[T_n^{(1)}\big] + \lceil n/2 \rceil \Big(\lceil n/2 \rceil - 1\Big) \mathbf{Cov}\big[T_n^{(1)}, T_n^{(2)}\big].$$

Recall the statement

$$\frac{T_n^{(j)}}{\sqrt{n}} \xrightarrow{\mathcal{D}} \big|B(U_j)\big|.$$

The variance of $T_n^{(j)}$ is thus $O(n)$. The covariance can be obtained from

$$\mathbf{Cov}\big[T_n^{(1)}, T_n^{(2)}\big] = \mathbf{E}\big[T_n^{(1)} T_n^{(2)}\big] - \mathbf{E}\big[T_n^{(1)}\big]\mathbf{E}\big[T_n^{(2)}\big].$$

The latter component in this covariance was worked out in (3.5); it is

$$\mathbf{E}^2\big[T_n^{(1)}\big] \sim \frac{\pi}{32} n.$$

The former part is obtained by conditioning on the uniform random variates U_1 and U_2:

$$\begin{aligned}
\mathbf{E}\Big[T_n^{(1)} T_n^{(2)}\Big] &\sim n\mathbf{E}\left[\big|B(U_1)\big|\,\big|B(U_2)\big|\right] \\
&= n \int_0^1 \int_0^1 \mathbf{E}\left[\big|B(s)\big|\,\big|B(t)\big|\right] ds\, dt \\
&= n\mathbf{E}\left[\int_0^1 \int_0^1 \big|B(s)\big|\,\big|B(t)\big|\, ds\, dt\right] \\
&= n\mathbf{E}\left[\left(\int_0^1 \big|B(t)\big|\, dt\right)^2\right] \\
&= \frac{7}{60} n,
\end{aligned}$$

by Shepp's (1982) calculation (cf. (3.2)). So,

$$\mathbf{Cov}\big[T_n^{(1)} T_n^{(2)}\big] \sim \frac{7}{60} n - \frac{\pi}{32} n.$$

We can combine the variances of the various stages to get an overall result:

$$\begin{aligned}\mathbf{Var}[S_n] &= \mathbf{Var}\big[C_{\lceil n/2\rceil}\big] + \big[\tilde{C}_{\lfloor n/2\rfloor}\big] + \mathbf{Var}[I_n] \\ &\sim \frac{1}{288}n^3 + \frac{1}{288}n^3 + \Big[O(n^2) + \frac{n^2}{4}\Big(\frac{7}{60}n - \frac{\pi}{32}n\Big)\Big] \\ &\sim \Big(\frac{1}{144} + \frac{7}{240} - \frac{\pi}{128}\Big)n^3. \qquad \blacksquare\end{aligned}$$

Theorem 3.3 *(Louchard, 1986). Let S_n be the number of comparisons of (2,1)-SHELL SORT to sort n random keys. Let $A \equiv \int_0^1 |B(t)|\,dt$. Then*

$$\frac{S_n - \frac{1}{8}n^2}{n^{3/2}} \xrightarrow{\mathcal{D}} \mathcal{N}\Big(0, \frac{1}{144}\Big) + A.$$

Proof. Recall again the convolution (3.3):

$$S_n = C_{\lceil n/2\rceil} + \tilde{C}_{\lfloor n/2\rfloor} + (n + I_n);$$

whose average is

$$\mathbf{E}[S_n] = \mathbf{E}\big[C_{\lceil n/2\rceil}\big] + \mathbf{E}\big[C_{\lfloor n/2\rfloor}\big] + \mathbf{E}[I_n] + n.$$

Center S_n by subtracting off the mean and scale it by the order of its standard deviation; the standard deviation is found in Theorem 3.2:

$$\frac{S_n - \mathbf{E}[S_n]}{n^{3/2}} = \frac{C_{\lceil n/2\rceil} - \mathbf{E}\big[C_{\lceil n/2\rceil}\big]}{n^{3/2}} + \frac{\tilde{C}_{\lfloor n/2\rfloor} - \mathbf{E}\big[\tilde{C}_{\lfloor n/2\rfloor}\big]}{n^{3/2}} + \frac{I_n - \mathbf{E}[I_n]}{n^{3/2}}.$$

The three components on the right-hand side are independent. Hence, if we identify their three corresponding limits, the convolution of the three limits will be the limiting distribution.

For the third component $(I_n - \mathbf{E}[I_n])/n^{3/2}$, Theorem 3.1 gives us the limit

$$\frac{\mathbf{E}[I_n]}{n^{3/2}} \to \sqrt{\frac{\pi}{128}}.$$

The stochastic part, $I_n/n^{3/2}$, converges weakly—this can be seen by going back to the form

$$I_n \text{a.s.} \sqrt{n} \sum_{j=1}^{\lceil n/2\rceil} \left| \sqrt{\frac{1}{n}\Big\lfloor\frac{n}{2}\Big\rfloor}\sqrt{\Big\lfloor\frac{n}{2}\Big\rfloor}\,\hat{F}_{\lfloor n/2\rfloor}(U_j) - \sqrt{\frac{1}{n}\Big\lceil\frac{n}{2}\Big\rceil}\sqrt{\Big\lceil\frac{n}{2}\Big\rceil}\,\tilde{F}_{\lceil n/2\rceil}(U_j) + O\Big(\frac{1}{\sqrt{n}}\Big)\right|$$

$$= n^{3/2} \sum_{j=1}^{\lceil n/2\rceil} \frac{1}{\sqrt{2}}\left|\sqrt{\Big\lfloor\frac{n}{2}\Big\rfloor}\,\hat{F}_{\lfloor n/2\rfloor}(U_j) - \sqrt{\Big\lceil\frac{n}{2}\Big\rceil}\,\tilde{F}_{\lceil n/2\rceil}(U_j)\right| \times \frac{2}{\lceil n/2\rceil} + O\Big(\frac{1}{\sqrt{n}}\Big).$$

Let $U_{(1)}, \ldots, U_{(\lceil n/2 \rceil)}$ be the order statistics of the uniform random variates $U_1, \ldots, U_{\lceil n/2 \rceil}$. This enables us to view the sum (in the limit) as a Stieltjes integral:

$$\frac{I_n}{n^{3/2}} \xrightarrow{\mathcal{D}} \lim_{n\to\infty} \int_{j=1}^{\lceil n/2 \rceil} \sqrt{2}\left|\sqrt{\left\lfloor \frac{n}{2} \right\rfloor}\, \hat{F}_{\lfloor n/2 \rfloor}(U_{(j)}) \right.$$
$$\left. -\sqrt{\left\lceil \frac{n}{2} \right\rceil}\, \tilde{F}_{\lceil n/2 \rceil}(U_{(j)})\right| \, d\hat{F}_{\lceil n/2 \rceil}(U_{(j)}).$$

The empirical measure $\hat{F}_{\lceil n/2 \rceil}(U_{(j)})$ converges to the uniform measure on (0, 1) on the time scale t. Thus the incremental rise $dF_{\lceil n/2 \rceil}(U_{(j)})$ approaches the differential element dt. Furthermore, $\sqrt{2}\left|\sqrt{\lfloor \frac{n}{2} \rfloor}\, \hat{F}_{\lfloor n/2 \rfloor}(U_{(j)}) - \sqrt{\lceil \frac{n}{2} \rceil}\, \tilde{F}_{\lceil n/2 \rceil}(U_{(j)})\right|$ converges to the absolute standard Brownian bridge $B(t)$ as already discussed. So,

$$\frac{I_n - \mathbf{E}[I_n]}{n^{3/2}} \xrightarrow{\mathcal{D}} \int_0^1 |B(t)|\, dt - \sqrt{\frac{\pi}{128}}.$$

As for the first two components coming from the contribution of the 2-stage, the limit laws we obtained in Chapter 2 for INSERTION SORT provide us with the two (independent) normal limits $\mathcal{N}(0, \frac{1}{288})$ and $\tilde{\mathcal{N}}(0, \frac{1}{288})$. Hence

$$\frac{S_n - \mathbf{E}[S_n]}{n^{3/2}} \xrightarrow{\mathcal{D}} \mathcal{N}\Big(0, \frac{1}{288}\Big) + \tilde{\mathcal{N}}\Big(0, \frac{1}{288}\Big) + A - \sqrt{\frac{\pi}{128}}. \tag{3.6}$$

The leading terms of $\mathbf{E}[S_n]$ are determined in Theorem 3.1:

$$\mathbf{E}[S_n] = \frac{1}{8}n^2 + \sqrt{\frac{\pi}{128}}\, n^{3/2} + o(n^{3/2}).$$

Now apply Slutsky's additive form—add the deterministic convergence relation

$$\frac{\mathbf{E}[S_n] - \frac{1}{8}n^2}{n^{3/2}} \to \sqrt{\frac{\pi}{128}}$$

to the limit law (3.6) to obtain the required result. ■

3.3 OTHER INCREMENT SEQUENCES

We analyzed at length the (2,1)-SHELL SORT in the previous sections. The methods may extend to $(h, 1)$-Shell SORT and even to some longer sequences of fixed length, as is done thoroughly in Smythe and Wellner (2000).

Several natural questions pose themselves here. What is an optimal choice (in the average sense) of h in $(h, 1)$-SHELL SORT? Would the methods that lead to optimizing $(h, 1)$-SHELL SORT extend naturally to optimizing the choices in longer sequences, like $(k, h, 1)$-SHELL SORT, for example? Do sequences of length increasing with n, the number of items sorted, perform better than fixed-length sequences? What is an optimal sequence? Given an optimal sequence, does SHELL SORT achieve the theoretic $\Theta(n \ln n)$ bound?

At present, a satisfactory answer to all these questions is not yet known. However, we know some partial answers that may shed some light on the big picture. The lack of a universal answer and methods for analyzing SHELL SORT has stirred quite a bit of intellectual curiosity. Many sequences have been proposed and many conjectures have been generated. We shall not try to portray the whole picture of every development that took place in the investigation of SHELL SORT. We shall try only to bring about some significant results in the area.

Most sorting algorithms in this book have an average behavior of either $\Theta(n^2)$ or $\Theta(n \ln n)$. Curiously, SHELL SORT with various choices of the increment sequences fills that spectrum with interesting orders of magnitude in between these two typical extremes. It is instructive to look at a special construction that gives one of these atypical orders of magnitude. Let us try to optimize the two-stage $(h, 1)$-SHELL SORT.

Theorem 3.4 *(Knuth, 1973). The choice* $h = \lfloor \sqrt[3]{16n/\pi} \rfloor$ *optimizes* $(h, 1)$*-SHELL SORT. The* $(\lfloor \sqrt[3]{16n/\pi} \rfloor, 1)$*-SHELL SORT sorts n keys in* $\Theta(n^{5/3})$ *average time.*

Proof. Knuth (1973) presents an argument based on an exact count of Hunt (1967). For asymptotics, it is sufficient to consider a simpler asymptotic estimate. The h-stage of $(h, 1)$-SHELL SORT creates h files of size $n/h + O(1)$ each. On average, the creation of these blocks requires h times the average amount of time LINEAR INSERTION SORT spends on sorting $n/h + O(1)$ random keys, an asymptotic total of $h \times \frac{(n/h)^2}{4} = \frac{n^2}{4h}$.

There are $\binom{h}{2}$ pairs of these subfiles; any two of them are interleaved like a permutation of size $2n/h + O(1)$ after it has been subjected to the 2-stage of $(2, 1)$-SHELL SORT. Let I_n be the number of inversions that $(2, 1)$-SHELL SORT leaves after the 2-stage. The 1-stage needs $n + I_n$ comparisons. We studied the average of I_n in Theorem 3.1. Accordingly, on average the 1-stage of $(h, 1)$-SHELL SORT asymptotically requires

$$n + \mathbf{E}[I_n] \sim \sqrt{\frac{\pi}{128}} \left(\frac{2n}{h}\right)^{3/2} \binom{h}{2}$$

comparisons.

If h grows with n, over the two stages the algorithm requires an asymptotic average of

$$\frac{n^2}{4h} + \frac{1}{8}\sqrt{\pi n^3 h}.$$

Differentiate this quantity with respect to h and set the derivative to 0 to obtain $h = \sqrt[3]{16n/\pi}$ as the value of h that asymptotically minimizes the number of comparisons. The $(\lfloor \sqrt[3]{16n/\pi} \rfloor, 1)$-SHELL SORT runs asymptotically in $\Theta(n^{5/3})$ average time. ∎

We see in the proof of Theorem 3.4 that on average the h-stage makes $\frac{n^2}{4h}$ comparisons. If h is fixed with respect to n, the average number of comparisons of the h-stage alone will grow quadratically. This is a feature observed in fixed sequences. To do better, one needs to let the number of the increments depend on n.

It is not known definitively whether SHELL SORT is optimal; no one knows whether there exists some sequence of increments that achieves $\Theta(n \ln n)$ time on average to sort n keys. Several reasonable sequences give $\Theta(n^a)$, with $a > 1$, average running time; $(\lfloor \sqrt[3]{16n/\pi} \rfloor, 1)$-SHELL SORT instantiates this (Theorem 3.4). Obviously, sequences with $\Theta(n^a)$ average running time, with $a > 1$, cannot achieve the theoretic lower bound. Several sequences are known to give reasonably fast performance. The general spirit in these sequences is that they do not have fixed length. Such sequences are usually derived from an increasing numeric sequence $1 = a_1, a_2, \ldots$, to be called the *generator sequence* in what follows. For any n, we determine the largest k for which $a_k \le n$; we then apply $(a_k, a_{k-1}, \ldots, a_1)$-SHELL SORT. Thus the length of the sequence grows with n, and a large number of increments will be used for large n.

Sparse breakthroughs have taken the form of discovery of clever generator sequences that improve the average order of magnitude. When Shell introduced his algorithm in 1959, he suggested the sequence $\lfloor n/2 \rfloor$, $\lfloor n/4 \rfloor$, $\lfloor n/8 \rfloor, \ldots, 1$, which becomes a diminishing sequence of proper powers of 2 in the case when n itself is a power of 2. However, this sequence has a bad worst-case behavior. To remedy the difficulty several authors suggested perturbations that take the form of an offset by ± 1 from Shell's essentially powers-of-2 sequence. Hibbard (1963) suggests the generator sequence $2^k - 1$, with $k = 1, 2, \ldots$. Lazarus and Frank (1960) suggest using Shell's sequence but adding 1 whenever an increment from Shell's sequence is even. Papernov and Stasevich (1965) suggest the generator sequence $2^k + 1$, with $k = 1, 2, \ldots$ (Lazarus and Frank's sequence when n is a proper power of 2), giving $O(n^{3/2})$ running time in the worst case; a proof of this and several other results related to the choice of this sequence are discussed in a chain of exercises (Exercises 3.2–3.4).

Sedgewick (1986) suggests deriving the increments from the generator sequence $1, 8, 23, 77, \ldots$, that is, $a_1 = 1$, and $a_k = 4^{k-1} + 3 \times 2^{k-2} + 1$, for $k \ge 2$. Sedgewick (1986) proves that a SHELL SORT construction based on this generator sequence requires $O(n^{4/3})$ running time for all inputs of size n. Weiss and Sedgewick (1990a) proved that $\Theta(n^{4/3})$ is the running time for all inputs of size n, still an order of magnitude higher than the $\Theta(n \ln n)$. Incerpi and Sedgewick (1985) argue that for any fixed k there are SHELL SORT constructions with $O(\ln n)$ increments for which the running time is $O(n^{1+1/k})$; Selmer (1989) gives a proof based on the a number-theoretic problem of Frobenius (introduced shortly). On the other hand, Poonen (1993) shows that a worst-case input for k increments requires at least $n^{1+c/\sqrt{k}}$, $c > 0$, comparisons (Plaxton and Suel (1997) provide a simple proof). Thus SHELL SORT cannot be worst-case optimal with a fixed number of increments.

Theoretically, one of the best known numeric sequences for generating increments is $1, 2, 3, 4, 6, 8, 9, 12, 16, \ldots$; numbers of the form $2^a \times 3^b$, for $a, b \ge 0$. In practice, a SHELL SORT construction based on this sequence is slow as it must involve $\Theta(\ln^2 n)$ increments, too large a number of increments giving an impractical amount of overhead. This sequence was suggested by Pratt in 1971. From an asymptotic viewpoint, this sequence produces the best known to-date increment sequence. To intuitively understand why Pratt's $2^a \times 3^b$ sequence performs well, whereas a SHELL SORT construction based on Shell's essentially 2^k generator sequence performs poorly, it is worth looking into what happens to the input after it has passed a stage. We shall call a file *k-ordered* if all k subarrays of k keys apart are sorted. For example, consider again the input instance of Section 3.1:

$$7 \quad 4 \quad 1 \quad 8 \quad 9 \quad 5 \quad 6 \quad 2 \quad 3.$$

The 2-stage of (2, 1)-SHELL SORT produces the 2-ordered permutation:

$$1 \quad \mathbf{2} \quad 3 \quad \mathbf{4} \quad 6 \quad \mathbf{5} \quad 7 \quad \mathbf{8} \quad 9.$$

It is curious to observe the following result

Lemma 3.2 *(Knuth, 1973). Suppose the decreasing sequence of increments $t_k, t_{k-1}, \ldots, 1$ is used in the SHELL SORT algorithm. At the end of the t_j-stage, the input is t_k-ordered, t_{k-1}-ordered, ..., and t_j-ordered.*

Proof. Let $X_1, \ldots, X_m$ and $Y_1, \ldots, Y_m$ be sequences of numbers such that

$$X_1 \le Y_1, \qquad X_2 \le Y_2, \qquad \ldots \qquad , \qquad X_m \le Y_m. \tag{3.7}$$

We show that their order statistics $X_{(1)}, \ldots, X_{(m)}$ and $Y_{(1)}, \ldots, Y_{(m)}$ satisfy

$$X_{(1)} \le Y_{(1)}, \qquad X_{(2)} \le Y_{(2)}, \qquad \ldots \qquad , \qquad X_{(m)} \le Y_{(m)}. \tag{3.8}$$

This is true because each of the Y's is at least as large as some X. But $Y_{(j)}$ is as large as j of the Y's, which are known to be at least as large as j of the X's. That is, $Y_{(j)} \ge X_{(j)}$, for $j = 1, \ldots, m$.

Let $h < k$. Consider an array $A[1\,..\,n]$ of numbers to be the subject of SHELL SORT algorithm that uses k and h as successive increments. After the k-stage the array is k-ordered. Now the h-stage is about to take over. We want to show that after the h-stage the file remains k-ordered, that is, $A[i] \le A[i+k]$, for all $1 \le i \le n-k$.

For any position $1 \le i \le n-k$, we can find the starting position μ of the subarray of h keys apart to which $A[i]$ belongs by going as far back in the array in steps of h. We can then shift the position up by k to find the starting position of a second subarray whose elements are k away from their counterpart in the first subarray. The h stage will sort each of these subarrays. Formally, let $1 \le \mu \le h$ be such that $i \equiv \mu \pmod h$. Appeal to the relations (3.7) and (3.8) with $X_j = A[\mu + (j-1)h]$, $Y_j = A[\mu + k + (j-1)h]$, with j running from 1 to $(i - \mu + h)/h$. Right before the h stage, we have $X_j \le Y_j$, for $j = 1, \ldots, (i - \mu + h)/h$, by the sorting effect of the k stage which puts in order $A[s]$ and $A[s+k]$ for any $s \le n-k$. The relations (3.7) and (3.8) guarantee that the largest among the X's will be at most as large as the largest among the Y's. The h ordering will now bring the largest of the X's into $A[\mu + h[(i - \mu + h)/h - 1]] = A[i]$ and will bring the largest of the Y's into $A[\mu + k + h[(i - \mu + h)/h - 1]] = A[i + k]$. The resulting array (which is now h-ordered) will remain k-ordered. ∎

Lemma 3.2 shows that, for example, after the 2-stage of (4, 2, 1)-SHELL SORT performs on some input, the input will be 4-ordered and 2-ordered. This is not particularly useful—interaction between odd and even positions takes place only at the 1-stage. An adversary can construct an input which leaves in $\Theta(n^2)$ inversions after the 2-stage of (4, 2, 1)-SHELL SORT. One will have to wait until the 1-stage to remove these inversions in $\Theta(n^2)$ time. By contrast, (3, 2, 1)-Shell SORT mixes in the 2-stage subarrays handled in the 3-stage.

A number-theoretic property is needed to complete the analysis of the construction known to be theoretically best for SHELL SORT. The aspect we need is the definition of a special number.

A country has issued only k different stamps of distinct values $s_1, \ldots, s_k$. The *Frobenius number* $R(s_1, \ldots, s_k)$ is the largest value that cannot be exactly composed of these stamps, that is the largest integer that cannot be expressed as a sum $\alpha_1 s_1 + \alpha_2 s_2 + \cdots + \alpha_k s_k$, with nonnegative coefficients $\alpha_1, \ldots, \alpha_k$. For instance, if $k = 2$ and the two stamps issued are of values $s_1 = 3$ and $s_2 = 5$, then

$$
\begin{aligned}
1 &= ?\\
2 &= ?\\
3 &= 1 \times 3\\
4 &= ?\\
5 &= 1 \times 5\\
6 &= 2 \times 3\\
7 &= ?\\
8 &= 1 \times 3 + 1 \times 5\\
9 &= 3 \times 3\\
10 &= 2 \times 5\\
11 &= 2 \times 3 + 1 \times 5\\
&\ \vdots
\end{aligned}
$$

and one can show in this case that $R(3, 5) = 7$ from the more general result of Exercise 3.2 for the case of two stamps. Even though it is an old problem, determining $R(s_1, \ldots, s_k)$ by an explicit formula of $s_1, \ldots, s_k$ is not solved except in some special cases. For the cases where $s_1, \ldots, s_k$ are pairwise relatively prime, $R(s_1, \ldots, s_k)$ is known to exist finitely.

Lemma 3.3 *(Poonen, 1993). Let $s_1, \ldots, s_k$ be positive integers for which the Frobenius number $R(s_1, \ldots, s_k)$ is well-defined. The number of steps required to h-order an array $A[1 \mathinner{..} n]$ that is already $s_1 h$-ordered, $\ldots$, $s_k h$-ordered by an h-stage of SHELL SORT is at most $[1 + R(s_1, \ldots, s_k)](n + h)$.*

Proof. Let $\alpha_1, \ldots, \alpha_k$ be any nonnegative integers. Since the array is $s_1 h$-ordered, we have $A[j] \le A[j + \alpha_1 s_1 h]$; since the array is also $s_2 h$-ordered, we have $A[j + \alpha_1 s_1 h] \le A[j + \alpha_1 s_1 h + \alpha_2 s_2 h], \ldots$; and since the array is $s_k h$-ordered, we have $A[j + \alpha_1 s_1 h + \cdots + \alpha_{k-1} s_{k-1} h] \le A[j + \alpha_1 s_1 h + \cdots + \alpha_k s_k h]$. By transitivity,

$$A[j] \le A[j + \alpha_1 s_1 h + \cdots + \alpha_k s_k h], \tag{3.9}$$

whenever the indexing is feasible.

The definition of the Frobenius number $R = R(s_1, \ldots, s_k)$ implies that $R + 1$ can be expressed as $\alpha_1 s_1 + \cdots + \alpha_k s_k$ for nonnegative coefficients $\alpha_1, \ldots, \alpha_k$. Hence. The partial order (3.9) guarantees

$$A[j] \le A[j + (R + 1)h].$$

Running the h-stage on an input with this partial order drags any key at most $R + 1$ steps back before finding a "stopper." During the h-stage, the algorithm sorts h subarrays of size at most $\lceil n/h \rceil$ each. Inserting an element at its correct position in its respective subarray involves at most $R + 1$ comparisons. It is, of course, assumed that $h \le n$; at most $h \times \lceil \frac{n}{h} \rceil (R + 1) < (n + h)(R + 1) = O(n)$ comparisons are needed at the h-stage. ■

Theorem 3.5 *(Pratt, 1971). Generate the numbers of the sequence $2^a \times 3^b$, $a, b \ge 0$ in natural increasing order. Let t_j be the jth member of this sequence. Let k_n be the largest index for which $t_{k_n} \le n$. The $(t_{k_n}, t_{k_n - 1}, \ldots, 1)$-SHELL SORT requires $\Theta(n \ln^2 n)$ average and worst-case running time.*

Proof. This proof is due to Poonen (1993). Consider an increment h. It is, of course, assumed that $h \le n$. Take up the large increments first. For any $h \ge \frac{1}{6} n$ from Pratt's generator sequence, the worst case is that of sorting h files each having at most $\binom{\lceil n/h \rceil}{2}$ inversions. This requires at most

$$h \times \left(\lceil n/h \rceil + \binom{\lceil n/h \rceil}{2} \right) = O(n)$$

comparisons in the worst case. The argument for the average case is the same; the average number of inversions has the same order of magnitude as the worst-case number of inversions.

For $h < \frac{1}{6}n$, both $2h$ and $3h$ are less than n. By the h-stage, the input is $3h$-ordered and $2h$-ordered. By Lemma 3.3 the algorithm requires at most $(n+h)(R(2, 3)+1) = 2(n + h)$ comparisons for this stage.

Any increment (large or small) requires $O(n)$ comparisons. There are $\lceil \log_2 n \rceil$ powers of 2 that are less than n and $\lceil \log_3 n \rceil$ powers of 3 that are less than n—there are at most $\lceil \log_2 n \rceil \lceil \log_3 n \rceil = \Theta(\ln^2 n)$ increments.[2] The number of comparisons is $O(n \ln^2 n)$.

We now turn to the lower bound. The number of comparison that LINEAR INSERTION SORT makes on an input of size n is $C_n \geq n$. Any increment h then requires at least $h \times \lfloor \frac{n}{h} \rfloor = \Omega(n)$ comparisons for any input. There are $\Theta(\ln^2 n)$ increments. The total number of comparisons is also $\Omega(n \ln^2 n)$ in the worst and average case. ∎

Theorem 3.5 tells us that asymptotically SHELL SORT can come close to the theoretic $\Theta(n \ln n)$ bound on average and in the worst case, but is not quite there. Yet, various flavors of SHELL SORT with several (but a fixed number of) increments are relatively fast in the small and mid-range number of input keys. These flavors of SHELL SORT remain popular sorting methods in practice.

EXERCISES

3.1 Suppose X is a continuous random variable whose distribution function $F(x)$ is strictly increasing. Show that $F(X) \overset{D}{=}$ UNIFORM(0, 1).

3.2 Let $s_1, \dots, s_k$ be $k \geq 1$ given distinct numbers and let $R(s_1, \dots, s_k)$ be their corresponding Frobenius number.

(a) What is the Frobenius number $R(1, s_2, \dots, s_k)$?

(b) Argue that if each of s_i successive integers can be expressed as a linear combination $\alpha_1 s_1 + \cdots + \alpha_k s_k$ (with nonnegative coefficients $\alpha_1, \dots, \alpha_k$), then any higher integer can be expressed as a linear combination of $s_1, \dots, s_k$ with nonnegative coefficients.

3.3 Let $k > h > g$ be three successive increments of SHELL SORT. Prove that if k and h are relatively prime, the number of comparisons that SHELL SORT uses at the g-stage is $O(khn/g)$. (Hint: It is known (Sharp and Curran (1884)) that the Frobenius number $R(s_1, s_2)$ is $s_1 s_2 - s_1 - s_2$, when s_1 and s_2 are relatively prime.)

3.4 (Papernov and Stasevich, 1965) Let $k_n = \lfloor \lg n \rfloor$ and $h_s = 2^s - 1$. Prove that $(h_{k_n}, h_{k_n - 1}, \dots, 1)$-SHELL SORT sorts any file of size n in $O(n^{3/2})$ time.

[2]Elsewhere in the book we use the notation $\lg n$ for $\log_2 n$; in this proof both $\log_2 n$ and $\log_3 n$ appear. We emphasize the base of each logarithm for clarity.

4

Bubble Sort

A natural way to sort is to make local exchanges among neighbors that are out of order. Because the exchanges are done locally between adjacent neighbors, an exchange can move a key at most one position closer to its final destination. A key distant from its final position may therefore require many exchanges before it gets to its correct position. This makes the algorithm slow. Indeed, as the analysis will reveal, BUBBLE SORT is a naive algorithm, requiring $\Theta(n^2)$ running time on average to sort n random keys. Albeit being too natural an algorithm, unfortunately it has a slow performance.

4.1 THE ALGORITHM

The BUBBLE SORT algorithm goes down a data array $A[1 \mathinner{..} n]$ and exchanges two adjacent keys $A[i]$ and $A[i+1]$ if they are not in their natural order, that is, if $A[i] > A[i+1]$. BUBBLE SORT thinks of such a pair as a "bad pair." A "good pair" is a data pair $A[i]$ and $A[i+1]$ in natural order, that is, $A[i] \le A[i+1]$. Starting at the top of the array ($i = 1$), BUBBLE SORT goes down considering the pairs $A[i]$ and $A[i+1]$, for $i = 1, \dots, n-1$, fixing bad pairs and leaving good pairs unchanged. The small keys "bubble up" to the surface and large keys "sink down" to the bottom. Performing these local exchanges for $i = 1, \dots, n-1$, completes the first *pass* over the input. One pass orders the array somewhat, but may not be sufficient to sort. More passes are performed, as many as needed to completely sort the array. During the first pass, the largest element, wherever it is, will sink all the way down to the last position. If it is at position $j < n$, when $i = j$ the largest key, $A[i]$, is compared to $A[i+1]$. It will prove to be at least as large as $A[i+1]$ and if it is strictly larger, it will sink down one position to be at $A[j+1]$ by the swap, and if it is not larger, then it is equal to $A[j+1]$; in either case $A[j+1]$ now holds the largest data value. But then, when i is advanced to position $j+1$, the pair $A[j+1]$ and $A[j+2]$ is considered, and again the largest key will sink further down to position $j+2$, and so on, till it ends up at position n. In the process, several other large elements may be at their natural position by the end of the first pass. The second time around, BUBBLE SORT need not go all the way down. During the first pass it detects the position of the last swap, possibly several positions above n, and that defines a reduced range for the second pass. The algorithm proceeds like this in passes, fixing some bad pairs in each pass and handing over to the next pass a reduced range of indexes, till at some pass no exchanges are made at all. That is when BUBBLE SORT gets to know that sorting has been achieved. At least one key is moved to its correct position in each pass and it is evident that at most n passes are needed to sort any arrangement of n keys.

To illustrate the action of BUBBLE SORT, consider the data set

$$32 \quad 54 \quad 27 \quad 13 \quad 85 \quad 44 \quad 63 \quad 68.$$

Figure 4.1 depicts the step-by-step exchanges of the first pass, with the pair being considered at each step for an exchange indicated. Bad pairs are flagged with an asterisk.

Figure 4.2 shows the status of the array after each pass. In Figure 4.2 *last* is the last index of a key involved in an exchange during the pass. Thus $A[1 \,..\, last]$ is the object of sorting in the next pass. Note that *last* may move in increments larger than one. For instance, relative to its position at the end of the first pass, *last* moved three steps up at the end of the second pass. Note also that during the fourth pass the algorithm did not detect any bad pairs, and it halted at the end of that pass.

The above ideas may be implemented by two nested loops, one to initiate passes, and an inner one that does the bookkeeping of a pass. To initiate passes, BUBBLE SORT keeps track of a logical variable, *more*, which tells the next iteration whether more exchanges are needed. At the very beginning of execution, *more* is of course assigned the value **true**, for the algorithm does not have any prior information on the input. Even if the data are sorted, BUBBLE SORT will have to go through them at least once to realize that. The number of passes is not known in advance; a **while** loop is appropriate for the situation. This loop runs till no more passes are required. When the algorithm enters a new pass, it wishfully hopes that the pass is the last; it

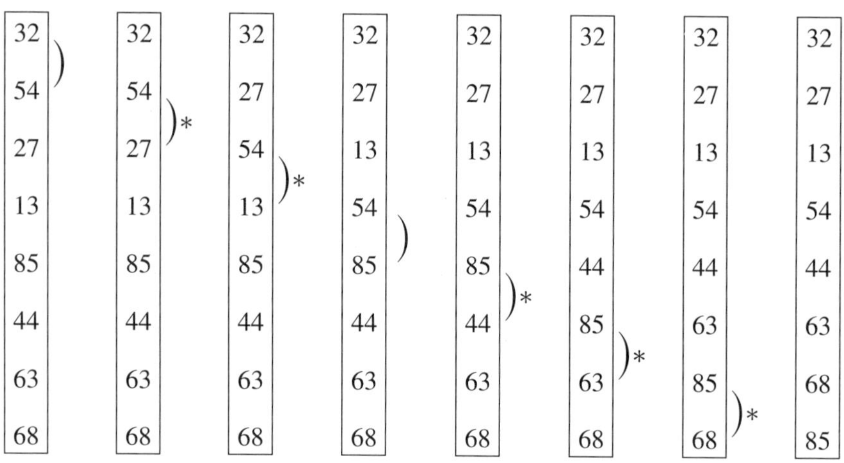

Figure 4.1. Pair comparisons in the first pass of BUBBLE SORT.

The array after

pass 0 (input)	pass 1	pass 2	pass 3	pass 4 (sorted)
32	32	27	13 ← *last*	13
54	27	13	27	27
27	13	32	32	32
13	54	44 ← *last*	44	44
85	44	54	54	54
44	63	63	63	63
63	68 ← *last*	68	68	68
68 ← *last*	85	85	85	85

Figure 4.2. The state of an input array to BUBBLE SORT after each pass.

assigns *more* the negated value **false**. The algorithm then considers the pairs $A[i]$ and $A[i+1]$, fixing bad pairs, starting at $i = 1$ and working through the range of indexes of the pass. If the algorithm detects a bad pair, it realizes that its assumption that this was going to be the last pass is untrue; BUBBLE SORT takes back its assumption and reverts *more* to the **true** status. BUBBLE SORT also records i, the lower index of the two keys involved in the exchange in a variable, say *last*. By the end of the inner loop the lower of the two indexes of the last two keys involved in an exchange sticks in *last*; $A[last]$ and $A[last+1]$ were exchanged, after which point the algorithm did not detect any other keys below *last* that are out of position. In between passes, a pass tells the next that its range may be reduced; the next pass need only concern itself with $A[1 .. last]$, where *last* picks up the position of the last key involved in an exchange (the last value assigned to *last* in the inner loop). The variable *last* controls the iterations of the next pass, which needs to consider pairs $A[i]$ and $A[i+1]$, only for $i = 1, \ldots, last - 1$. Because a definite number of steps is known in advance to a pass, the working of the pass can be done by a **for** loop. The very first pass must consider all pairs $A[i]$ and $A[i+1]$, for $i = 1, \ldots, n-1$; at the very beginning *last* must be initialized to n.

4.2 A LIMIT LAW FOR PASSES

We can analyze the number of comparisons that BUBBLE SORT makes via the number of passes the algorithm executes. So, we shall take up this latter parameter first.

```
last ← n;
more ← true;
while more do
    begin
        more ← false;
        for i ← 1 to last − 1 do
            if A[i] > A[i + 1] then
                begin
                    swap(A[i], A[i + 1]);
                    last ← i;
                    more ← true;
                end;
    end;
```

Figure 4.3. The BUBBLE SORT algorithm.

Let P_n be the number of passes that BUBBLE SORT makes on a random permutation of size n. Suppose that initially $K = A[i]$ is a key preceded by a certain number V_i of larger members of the permutation. In its various passes, the algorithm moves exactly one of the keys that are larger than K past K. Thus V_i passes will be sufficient to remove all the inversions caused by the initial entry at position i. To remove all the inversions, $\max\{V_1, \dots, V_n\}$ passes are needed. One last pass will then be needed to scan the array and recognize that the array has been completely sorted. Hence,

$$P_n = 1 + \max\{V_1, \dots, V_n\}.$$

We studied the distribution of the number of inversions caused by an element of a permutation. Recalling a discussion from Subsection 1.10.2,

$$V_i \overset{D}{=} \text{UNIFORM}[0 \mathinner{..} i - 1],$$

and $V_1, \dots, V_n$ are independent. The simple distributions and independence of V_i, $i = 1, \dots, n$, admit an easy exact distribution for P_n. Of course, P_n is a discrete random variable and it is sufficient to study its staircase distribution function at the integer points $k = 1, \dots, n$. We have

$$\begin{aligned}
\mathbf{Prob}\{P_n \le k\} &= \mathbf{Prob}\big\{1 + \max\{V_1, \dots, V_n\} \le k\big\} \\
&= \mathbf{Prob}\{V_1 \le k - 1, \dots, V_n \le k - 1\} \\
&= \mathbf{Prob}\{V_1 \le k - 1\} \dots \mathbf{Prob}\{V_n \le k - 1\},
\end{aligned}$$

a decomposition by independence. Because $V_i = \text{UNIFORM}[0 \mathinner{..} i - 1] \le k - 1$, for $i = 1, \dots, k$, the probabilities $\mathbf{Prob}\{V_i \le k - 1\}$, for $i = 1, \dots, k$, are all equal to 1. That is,

$$\mathbf{Prob}\{P_n \le k\} = \prod_{j=k+1}^{n} \mathbf{Prob}\{V_j \le k-1\} = \prod_{j=k+1}^{n} \frac{k}{j}.$$

We have thus arrived at the exact distribution stated next.

Theorem 4.1 *(Knuth, 1973). Let P_n be the number of passes that BUBBLE SORT makes on a random permutation of size n. The exact distribution function of this random variable is*

$$\mathbf{Prob}\{P_n \le k\} = \frac{k!\,k^{n-k}}{n!},$$

for $k = 1, \ldots, n$.

The exact distribution is simple enough to allow a direct passage to the limit distribution. As a byproduct, we shall get the asymptotic mean and variance of the number of passes.

Theorem 4.2 *The number of passes, P_n, that BUBBLE SORT makes while sorting a random permutation of size n, satisfies*

$$\frac{P_n - n}{\sqrt{n}} \xrightarrow{\mathcal{D}} -R,$$

where R is a Rayleigh random variable (a nonnegative continuous random variable with density function $xe^{-x^2/2}$, for $x > 0$).

Proof. We shall seek center and scale factors, μ_n and σ_n, so that $P_n^* = (P_n - \mu_n)/\sigma_n$ converges in distribution. Consider the probability

$$\mathbf{Prob}\{P_n^* \le x\} = \mathbf{Prob}\{P_n \le \mu_n + x\sigma_n\} = \mathbf{Prob}\{P_n \le \lfloor \mu_n + x\sigma_n \rfloor\}.$$

For an appropriate range of x, Theorem 4.1 applies giving

$$\mathbf{Prob}\{P_n^* \le x\} = \lfloor \mu_n + x\sigma_n \rfloor^{n - \lfloor \mu_n + x\sigma_n \rfloor} \times \frac{\lfloor \mu_n + x\sigma_n \rfloor!}{n!}.$$

We have $\lfloor \mu_n + x\sigma_n \rfloor = \mu_n + x\sigma_n - \beta_n$, for some $0 \le \beta_n < 1$. Then

$$\mathbf{Prob}\{P_n^* \le x\} = \big(\mu_n + x\sigma_n - \beta_n\big)^{n - (\mu_n + x\sigma_n - \beta_n)} \times \frac{(\mu_n + x\sigma_n - \beta_n)!}{n!}.$$

The factorials can be asymptotically handled by Stirling's approximation, yielding

$$\begin{aligned}\mathbf{Prob}\{P_n^* \le x\} &\sim \big(\mu_n + x\sigma_n - \beta_n\big)^{n-\mu_n - x\sigma_n+\beta_n} \\ &\quad\times \frac{\left(\dfrac{\mu_n + x\sigma_n - \beta_n}{e}\right)^{\mu_n+x\sigma_n-\beta_n} \times \sqrt{2\pi(\mu_n + x\sigma_n - \beta_n)}}{\left(\dfrac{n}{e}\right)^n \times \sqrt{2\pi n}} \\ &= \frac{\big(\mu_n + x\sigma_n - \beta_n\big)^n}{n^n} e^{n-\mu_n-x\sigma_n+\beta_n} \sqrt{\frac{\mu_n + x\sigma_n - \beta_n}{n}} \\ &= \frac{\mu_n^n}{n^n}\left(1 + \frac{x\sigma_n - \beta_n}{\mu_n}\right)^n e^{n-\mu_n-x\sigma_n+\beta_n} \sqrt{\frac{\mu_n + x\sigma_n - \beta_n}{n}}.\end{aligned}$$

The form suggests use of the standard calculus relation

$$\left(1 + \frac{f_n}{n}\right)^n \sim e^{f_n},$$

when $f_n = O(n^\varepsilon)$, with $\varepsilon < 1$; we take $\mu_n = n$. Having decided to try centering by n, we can also determine the appropriate range of x for that centering. As $P_n \le n$, the probability $\mathbf{Prob}\{P_n^* \le x\} = 1$, for any nonnegative x. We need only consider negative x. Any fixed negative x is in the support, because $P_n \ge 1$, that is, $P_n^* \ge -(n-1)/\sigma_n$, and x enters the support as $-(n-1)/\sigma_n \le x < 0$, if $\sigma_n = o(n)$, and n is large enough. We can write

$$\mathbf{Prob}\{P_n^* \le x\} \sim \left(1 + \frac{x\sigma_n - \beta_n}{n}\right)^n e^{-x\sigma_n+\beta_n} \sqrt{\frac{n + x\sigma_n}{n}}.$$

Now,

$$\left(1 + \frac{x\sigma_n - \beta_n}{n}\right)^n = \exp\left\{n \ln\left(1 + \frac{x\sigma_n - \beta_n}{n}\right)\right\}.$$

Expanding the logarithm as $(x\sigma_n/n - \beta_n/n) - x^2\sigma_n^2/(2n^2) + O(\sigma_n/n^2)$, we arrive at

$$\mathbf{Prob}\{P_n^* \le x\} \sim e^{-x^2\sigma_n^2/(2n)+O(\sigma_n/n)} \sqrt{\frac{n + x\sigma_n}{n}}.$$

The right-hand side of the last relation converges if $\sigma_n = c\sqrt{n}$, for any constant c. Let us take $c = 1$ (and subsume it in the limit R), to finally obtain

$$\mathbf{Prob}\left\{\frac{P_n - n}{\sqrt{n}} \le x\right\} \to e^{-x^2/2}, \qquad \text{for } x < 0.$$

The latter distribution function is that of a negative random variable, $-R$, where R has the distribution function

$$F_R(x) = \begin{cases} 1 - e^{-x^2/2}, & \text{for } x > 0; \\ 0, & \text{otherwise.} \end{cases}$$

The density of R follows by differentiation. ■

Knuth (1973) computes the average and variance from the exact distribution by considering asymptotic equivalents of sums involving the exact distribution function. The asymptotic route considered in Knuth (1973) alludes to an argument by De Bruijn that connects the exact distribution function to one of Ramanujan's functions. We shall follow an alternate route to derive the asymptotic mean and variance from the limit distribution.

Only mild conditions are required to show that moment calculations follow from convergence in distribution. In pathological cases one may not be able to claim, for example, that if $X_n \xrightarrow{\mathcal{D}} X$, then $\mathbf{E}[X_n] \to \mathbf{E}[X]$. However, random variables arising in algorithmics are usually well behaved. Often all their moments exist. Our case is no exception; it can be checked that it satisfies a uniform integrability condition sufficient for convergence of the first two moments to the first two moments of the limit random variable.

Theorem 4.3 *(Knuth, 1973). The average number of passes that BUBBLE SORT makes on a random permutation of size n is*

$$n - \sqrt{\frac{\pi n}{2}} + o(\sqrt{n}).$$

The variance is asymptotically equivalent to

$$\Bigl(2 - \frac{\pi}{2}\Bigr)n.$$

Proof. Owing to the presence of suitable conditions (uniform integrability), we have two straightforward asymptotic calculations. For the first moment, we have

$$\mathbf{E}\Bigl[\frac{P_n - n}{\sqrt{n}}\Bigr] \to \mathbf{E}[-R] = -\int_0^\infty x^2 e^{-x^2/2}\, dx = -\sqrt{\frac{\pi}{2}},$$

implying

$$\mathbf{E}[P_n] = n - \sqrt{\frac{\pi n}{2}} + o(\sqrt{n}).$$

For the second moment, we have

$$\mathbf{Var}\Bigl[\frac{P_n - n}{\sqrt{n}}\Bigr] \to \mathbf{Var}[-R],$$

or

$$\begin{aligned}\frac{1}{n}\mathbf{Var}[P_n] &\to \mathbf{Var}[R]\\ &= \mathbf{E}[R^2] - \mathbf{E}^2[R]\\ &= \int_0^\infty x^3 e^{-x^2/2}\,dx - \Big(-\sqrt{\frac{\pi}{2}}\Big)^2\\ &= 2 - \frac{\pi}{2}.\end{aligned}$$ ■

The orders of the mean and variance of the number of passes indicate a narrow concentration around the mean value.

Corollary 4.1 *Let P_n be the number of passes that BUBBLE SORT makes in sorting a random permutation of $\{1, \ldots, n\}$. Then*

$$\frac{P_n}{n} \xrightarrow{P} 1.$$

Proof. Fix $\varepsilon > 0$. By Chebyshev's inequality

$$\begin{aligned}\mathbf{Prob}\Big\{\Big|\frac{P_n}{n} - 1\Big| \ge \varepsilon\Big\} &\le \frac{\mathbf{Var}[P_n/n - 1]}{\varepsilon^2}\\ &= \frac{\mathbf{Var}[P_n]}{\varepsilon^2 n^2}\\ &= \frac{(2 - \pi/2)n + o(n)}{\varepsilon^2 n^2}\\ &\to 0, \qquad \text{as } n \to \infty.\end{aligned}$$ ■

4.3 A LIMIT LAW FOR COMPARISONS

Let C_n be the number of comparisons that BUBBLE SORT makes to sort a random permutation of size n. If the algorithm makes n passes, necessarily after each the range of sorting diminishes by only 1, forcing the range delimiter *last* to decrement by 1 (refer to the algorithm of Figure 4.3). The ranges of the sort are then successively $n, n - 1, \ldots, 0$; in this case the algorithm exerts $(n - 1) + (n - 2) + \cdots + 1 \sim \frac{1}{2}n^2$ comparisons. But indeed, the number of passes is narrowly concentrated around n—in view of Corollary 4.1, the number of passes behaves almost deterministically like n. Therefore we expect C_n to be concentrated around $\frac{1}{2}n^2$.

The number of comparisons can be sandwiched by bounds involving the square of the number of passes. This will be sufficient to derive a law of large numbers for the comparisons of BUBBLE SORT.

The variable *last* in the algorithm of Figure 4.3 diminishes from n to 0 along a sequence $n = L_1, L_2, \ldots, L_{P_n}, 0$. Within the jth pass the range of BUBBLE SORT is $A[1 \mathinner{..} L_j]$. Given the number of passes P_n, a perturbation (preserving the number of passes) of only one of the values L_j by Δ will change the overall number of comparisons by Δ, as can be seen from the following straightforward argument. Suppose P_n is given and the sequence $L_1, L_2, \ldots, L_{P_n}$ is specified. The number of comparisons is

$$C_n = \sum_{j=1}^{P_n} (L_j - 1).$$

A sequence $(L'_1, \ldots, L'_{P_n}) = (L_1, L_2, \ldots, L_{j-1}, L_j + \Delta, L_{j+1}, \ldots, L_{P_n})$ corresponds to a permutation for which BUBBLE SORT makes the same number of passes, but requires

$$C'_n = \sum_{k=1}^{P_n} (L'_k - 1) = \Delta + \sum_{k=1}^{P_n} (L_k - 1) = C_n + \Delta$$

comparisons. When P_n and $L_1, L_2, \ldots, L_{P_n}$ are given, the sequence $n, n-1, n-2, n - P_n + 1$ will correspond to at least the same number of comparisons; all the bounding values $L_1, L_2, \ldots, L_{P_n}$ are perturbed as far as possible in the positive direction. We have the upper bound

$$\begin{aligned} C_n &\le (n-1) + (n-2) + \cdots + (n - P_n) \\ &= nP_n - (1 + 2 + \cdots + P_n) \\ &= nP_n - \frac{1}{2} P_n (P_n + 1). \end{aligned} \tag{4.1}$$

On the other hand, if all the numbers in the given sequence are perturbed in the negative direction as far as possible, the result will be the sequence $n, P_n - 1, \ldots, 1$, which corresponds to a lower bound on the number of comparisons

$$\begin{aligned} C_n &> (n-1) + (P_n - 2) + (P_n - 3) + \cdots + 1 \\ &= n - 1 + \frac{1}{2}(P_n - 2)(P_n - 1). \end{aligned} \tag{4.2}$$

Combining the two bounds (4.1) and (4.2),

$$\frac{1}{2n^2}(2n - 3P_n) + \frac{P_n^2}{2n^2} \le \frac{C_n}{n^2} \le \frac{nP_n - \frac{1}{2}P_n^2 - \frac{1}{2}P_n}{n^2} = \frac{P_n}{n} \times \left(1 - \frac{P_n}{2n}\right) - \frac{P_n}{2n^2}.$$

Corollary 4.1 asserts that $P_n/n \xrightarrow{P} 1$. Thus, $P_n/n^2 \xrightarrow{P} 0$. In probability, C_n is sandwiched from above and below by 1/2.

Theorem 4.4 *The number of passes, C_n, that BUBBLE SORT makes in sorting a random permutation of size n, satisfies*

$$\frac{C_n}{n^2} \xrightarrow{P} \frac{1}{2}.$$

As discussed in the paragraph preceding Theorem 4.3, random variables arising is algorithmics are usually well-behaved and we have sufficient conditions to warrant that the moments of C_n/n^2 behave like the degenerate random variable $1/2$. In particular we have the first moment convergence

$$\frac{\mathbf{E}[C_n]}{n^2} \rightarrow \frac{1}{2}.$$

BUBBLE SORT's average number of comparisons $\mathbf{E}[C_n]$ is asymptotic to $\frac{1}{2}n^2$, which is close to the algorithm's asymptotic worst-case performance. Even though it is a very intuitive sorting algorithm, its behavior is narrowly concentrated around its worst case, and will generally be rather slow.

EXERCISES

4.1 Does BUBBLE SORT (as implemented in Figure 4.3) handle repeated data correctly or does it require some modification to deal with repeats?

4.2 Is BUBBLE SORT a stable sorting algorithm?

4.3 Refer to the algorithm of Figure 4.3. Suppose you have a permutation of $\{1, \ldots, n\}$ that assigns the numbers $n = L_1, L_2, \ldots, L_{P_n}$ to the variable *last* in P_n passes. Describe a permutation of $\{1, \ldots, n\}$ that also requires P_n passes, but corresponds to the sequence of assignments $n, n-1, n-2, \ldots, n - P_n + 1$. Describe a permutation of $\{1, \ldots, n\}$ that requires P_n passes, and corresponds to the sequence of assignments n, $P_n - 1$, $P_n - 2, \ldots, 2, 1$.

4.4 (Dobosiewicz, 1980) As presented in the chapter, BUBBLE SORT makes local exchanges among adjacent keys. It may be possible to speed up BUBBLE SORT if we let a key move a long distance. In the spirit of $(t_k, t_{k-1}, \ldots, 1)$-SHELL SORT one can design an algorithm to go through a "bubbling pass" on subarrays whose keys are t_k positions apart, then go through another bubbling pass on a subarray whose keys are t_{k-1} positions apart, and so forth until the usual BUBBLE SORT (the algorithm of Figure 4.3) is applied at the last stage. Implement such a $(t_k, t_{k-1}, \ldots, 1)$-BUBBLE SORT.

5

Selection Sort

SELECTION SORT is an intuitive but slow sorting method. The method is in the $O(n^2)$ naive class of sorting algorithms. It is based on successive selection of minima (or maxima). To sort n items from an ordered set, the algorithm proceeds in stages: It first isolates the smallest element by scanning the list of n elements using $n - 1$ data comparisons. This minimal element is then moved to the top of the list (in practical implementation this minimum is swapped with the element at the top of the array containing the list). Now that the smallest element is in its correct place, in the second stage the algorithm is concerned with the bottom $n - 1$ elements. The algorithm finds the second smallest in the array (the smallest among the remaining $n - 1$ elements) by making $n - 2$ comparisons, moves it to the second position from the top, then finds the third smallest (using $n - 3$ comparisons), moves it to the third position from the top, and so forth. Obviously the algorithm is bound to be slow because at every new stage the algorithm is entirely oblivious to any partial information obtained while scanning during the successive selection of minima.

5.1 THE ALGORITHM

The algorithm is presented in Figure 5.1. Right before the selection of the ith minimum, $A[1 \mathinner{..} i - 1]$ contains the smallest $i - 1$ data elements in sorted order. The ith order statistic will then be found in $A[i \mathinner{..} n]$ and stored in the variable min. The selection of the ith minimum is done by presuming $A[i]$ to be the minimum, then checking this hypothesis against the elements of $A[i + 1 \mathinner{..} n]$. Every time an element smaller than min is encountered, the algorithm picks it as a better candidate for min and records its position in pos. Right before the end of the ith stage pos will hold the position of the ith order statistic. The content of positions i and pos are swapped (possibly swapping $A[i]$ with itself).

Figure 5.2 shows the various stages of SELECT SORT on the input data set 22, 8, 16, 5, 14, 17. The decreasing-size boxes show the diminishing ranges of SELECTION SORT. The minimum at a stage (a stage is a column in the figure) is indicated by a pointing arrow. At the end of a stage, the minimum of the stage at the position pointed to is swapped with the element at the beginning of the sorting range. In the next column over, that minimum appears at the top of the previous sorting range, and the range of sorting is shrunk by one position. On one occasion, the

```
for i ← 1 to n − 1do
    begin
        pos ← i;
        min ← A[i];
        for j ← i + 1 to n do
            if A[j] < min then
                begin
                    min ← A[j];
                    pos ← j;
                end;
        call swap(A[i], A[pos]);
    end;
```

Figure 5.1. The SELECTION SORT algorithm.

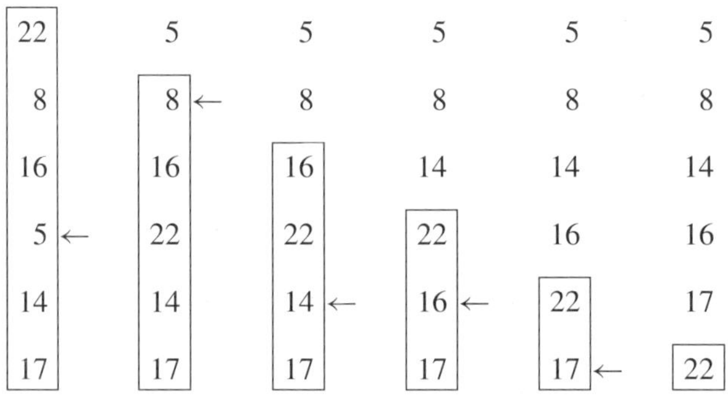

Figure 5.2. Successive selection of minima by SELECTION SORT.

minimum of a range was the first element in the range (the second column from the left) and it is swapped with itself.

5.2 ANALYSIS

The number of comparisons made by SELECTION SORT is deterministic. The first stage always takes $n - 1$ comparisons, the second $n - 2$, and so on. Thus the number of comparisons is exactly

$$(n - 1) + (n - 2) + \cdots + 3 + 2 + 1;$$

that is, $\frac{1}{2}n(n-1)$. The algorithm makes $n-1$ swaps (sometimes swapping an element with itself, when the element nearest to the top in the remaining list happens to be the smallest among the remaining elements).

The only random aspect of the algorithm is the number of times the variables *min* and *pos* are updated to reflect the discovery of a better choice. This quantity is of secondary importance relative to comparisons, as generally an assignment is done much faster than a comparison on a modern computer. There is one update for each of these variables right before the inner loop of the algorithm, then the body of that inner loop executes a random number of updates on them. Let U_n be the number of updates performed on *min* and *pos* while sorting a random list of size n. During the first stage, each time an array element smaller than *min* is encountered, *min* and *pos* are updated. Together with the initializing update outside the inner loop, during the first stage the number of updates is the number of left-to-right minima or the number of records.[1] Denote the number of records in a random list of size n by R_n. Supposing the minimum appears at (random) position K_n, the relative ranks of the list $A[1 \mathinner{..} K_n - 1]$ form a random permutation of $\{1, \dots, K_n - 1\}$. The number of updates during the first stage is one plus the number of records in $A[1 \mathinner{..} K_n - 1]$, that is $1 + R_{K_n - 1}$, where the 1 accounts for the condition that $A[K_n]$ is the minimal element in the array, and thus will always force an update. At the end of the scan of the first stage, $A[K_n]$ is swapped with $A[1]$, leaving a random list in $A[2 \mathinner{..} n]$, with U_{n-1} updates to be performed. We have a recurrence for the conditional expectation:

$$\mathbf{E}[U_n \mid K_n] = U_{n-1} + R_{K_n - 1} + 1.$$

Unfortunately U_{n-1} and R_{K_n} are dependent random variables, which complicates the analysis of U_n considerably. No complete analysis beyond variance is known to this day. Even the variance is found by a rather lengthy and sophisticated calculation that embeds the algorithm in a two-dimensional stochastic process (Yao, 1988). Nonetheless the last recurrence is good enough for the computation of the average. Taking expectations of the last recurrence, we obtain:

$$\mathbf{E}[U_n] = \mathbf{E}[U_{n-1}] + \mathbf{E}[R_{K_n - 1}] + 1.$$

To find $\mathbf{E}[R_{K_n - 1}]$, we condition on K_n. The minimum during the first stage is equally likely to appear at any position, that is, K_n has the UNIFORM$[1 \mathinner{..} n]$ distribution with $\mathbf{Prob}\{K_n = k\} = 1/n$. We find

$$\begin{aligned}\mathbf{E}[R_{K_n - 1}] &= \sum_{k=1}^{n} \mathbf{E}[R_{k-1}]\, \mathbf{Prob}\{K_n = k\} \\ &= \frac{1}{n} \sum_{k=1}^{n} H_{k-1} \\ &= \frac{1}{n}\big(nH_{n-1} - (n-1)\big);\end{aligned}$$

[1]Records were defined in Section 1.10.1 in the sense of "record large." In the present context we are working with the symmetric notion of "record small," which has the same distributional characteristics as record large.

in the last display we have used the property that $\mathbf{E}[R_j] = H_j$ as we found in our discussion on records (see (1.10)), and the identity

$$\sum_{j=1}^{n} H_j = (n+1)H_n - n. \tag{5.1}$$

We get an unconditional recurrence for the desired average:

$$\begin{aligned}\mathbf{E}[U_n] &= \mathbf{E}[U_{n-1}] + \frac{1}{n}\big(nH_{n-1} - (n-1)\big) + 1\\ &= \mathbf{E}[U_{n-1}] + H_n.\end{aligned}$$

Unwinding the recurrence with the initial condition $\mathbf{E}[U_1] = 0$, we obtain

$$\begin{aligned}\mathbf{E}[U_n] &= \sum_{j=2}^{n} H_j\\ &= (n+1)(H_n - 1)\\ &\sim n \ln n,\end{aligned}$$

where again we used (5.1) to reduce the combinatorial sum.

EXERCISES

5.1 If SELECTION SORT is stopped after the selection of k minima, the algorithm finds only the first k order statistics. Adapt SELECTION SORT to find only the first $k \le n$ order statistics. How many comparisons does the adapted algorithm make to find the first k order statistics?

5.2 Adapt SELECTION SORT to be an algorithm for order statistics. Given a set of ranks, $i_1, \ldots, i_k$, the adapted algorithm should find the keys with these ranks. How many comparisons does your adapted algorithm make?

5.3 Let a set of k ranks be chosen at random from $\{1, \ldots, n\}$ so that all $\binom{n}{k}$ sets of k ranks are equally likely, and then given to your multiple order statistic finding algorithm of the previous exercise.

(a) What is the average the number of comparisons?

(b) What is the variance of the number of comparisons?

(c) As an indication of the class of distributions involved determine the limiting distribution of a suitably normed version of the number of comparisons for the case $k = 1$.

5.4 Illustrate by an instance that the variables $R_{K_n - 1}$ and U_{n-1} (counting the number of records and the number of updates of pos and min in the algorithm of Figure 5.1) are not independent.

5.5 We have a list of n numeric keys and we are going to perform m search operations on them. Each key is equally likely to be the subject of any search. When is it cheaper on average to perform these m search operations by linear search than to sort first by SELECTION SORT? That is, what is the largest value for m for which the average number of comparisons in m such searches is less than the number of comparisons in the SELECTION SORT of n keys?

6

Sorting by Counting

Simple counting can be used effectively to sort data by comparison or by non-comparison-based algorithms. Suppose we have a list $A[1 \mathinner{..} n]$ of data elements. The total order of this data set can be discovered by determining for each element the number of list elements not exceeding it. More precisely, assuming our list is made up of *distinct* data, if we determine that there are j elements less than or equal to $A[i]$, then $A[i]$ must have rank j in the set. Only minor adjustments are needed if our data are not distinct, a case that arises with 0 probability in the random permutation model.

6.1 COUNT SORT

We shall dub the sorting algorithm based on straightforward counting COUNT SORT. As outlined in the introductory paragraph of this chapter, sorting a data set $A[1 \mathinner{..} n]$ by counting is basically a determination of the count of the number elements not exceeding $A[i]$ for every i. We shall assume that these counts are kept in an array of counters, to be declared *count*$[1 \mathinner{..} n]$. After running the algorithm, *count*$[i]$ will contain the number of elements that do not exceed $A[i]$; the rank of $A[i]$ will be this counter's value. Starting *count*$[i]$ at 0, every element below $A[i]$ will contribute 1 to the count, as well as $A[i]$ itself because, of course, any element is less than or equal to itself. Subsequently, we can cut the work by about half. We can generate only the indexes $1 \le i < j \le n$, and if $A[j] \le A[i]$, then right there we can also account for the outcome of the comparison $A[i] \le A[j]$, which we would have made, had we generated all possible pairs of indexes. We can also ignore all the comparisons $A[i]$ with $A[i]$, for $i = 1, \dots, n$, by initializing all counters to 1.

At the end of executing this algorithm, the array *count*$[1 \mathinner{..} n]$ will hold a permutation of $\{1, \dots, n\}$ that carries the required complete sorting information. Figure 6.1 illustrates an input data set and the result in *count* after running the algorithm. For example, $A[3] = 12$ is the smallest input key; after executing COUNT SORT, the

A:	40	16	12	86	60	40	19	50	40	26
count:	5	2	1	10	9	6	3	8	7	4

Figure 6.1. The counts after running COUNT SORT on some data.

```
for i ← 1 to n do
    count[i] ← 1;
for i ← 1 to n do
for j ← i + 1 to n do
    if A[j] < A[i] then
        count[i] ← count[i] + 1
    else count[j] ← count[j] + 1;
```

Figure 6.2. The COUNT SORT algorithm.

corresponding counter, *count*[3], contains 1, indicating that this key has rank 1 or is the smallest among the data presented to the algorithm. Three keys are equal to 40. There are four keys of lower value. The algorithm chooses to rank these three repeat keys 5, 6, and 7, with the least indexed repeat key receiving the least rank. Of course, other interpretations of the ranks of equal keys are also possible and will amount to only minor modifications in the basic algorithm. This version of the algorithm is presented in Figure 6.2.

The analysis of COUNT SORT is very straightforward. The first loop makes n initializations. The algorithm then executes two nested loops. Embedded in the two loops is the comparison statement. One comparison is executed for every pair of distinct indexes $1 \le i < j \le n$. Deterministically $\binom{n}{2}$ comparisons are made. Thus the dominant component of the algorithm is deterministic and is of quadratic order, putting the algorithm in the class of naive sorting algorithms. Nothing is really stochastic until one looks into some of the very fine details of the algorithm, such as machine-level implementation, which will then introduce a little variability around the chief component. For example, at the level of assembly language the **if**-clause is implemented by a set of instructions followed by a JUMP instruction to bypass the **else**-clause. This instruction has no counterpart after the assembly translation of the **else**-clause as there is nothing to bypass. The number of times this JUMP instruction is executed is the number of times the **if** path is chosen, that is, the number of times the algorithm encounters two indexes i and j such that $A[j] < A[i]$. This number is, of course, random and introduces the only stochastic element of the algorithm. Corresponding to every inversion of the permutation underlying an input of n distinct keys, an extra JUMP instruction in the **if**-clause is executed. Therefore J_n, the total number of jumps until a file of n random data is sorted, is the total number of inversions of that random permutation. A result for inversions in random permutations is given in Section 1.10.2. We repeat that result here, cast in the context of COUNT SORT. The limit result is still valid if our data are not distinct, a case that arises with probability zero in the random permutation model anyway.

Theorem 6.1 *Let J_n be the total number of jumps in the* **if**-*clause of COUNT SORT to sort n random keys following the random permutation probability model. In the limit, as $n \to \infty$,*

$$\frac{J_n - n^2/4}{n^{3/2}} \xrightarrow{\mathcal{D}} \mathcal{N}\Big(0, \frac{1}{36}\Big).$$

6.2 SORTING BY COUNTING FREQUENCIES

The general idea of counting can be used very selectively to design efficient sorting algorithms for discrete data from finite ranges. The algorithm so derived counts frequencies of occurrence of values and henceforth will be called FREQUENCY COUNTING SORT. Supposing our data are drawn (possibly with replacement) from a finite set $S = \{s_1, s_2, \ldots, s_k\}$, whose listing is according to a strict total order, that is, $s_1 < s_2 < \ldots < s_k$. Our n keys (some possibly are repeats) come from S. To sort these n keys we can set up a simple frequency counting scheme. We can count how many occurrences there are of s_1, how many of s_2, and so on. We can keep these frequencies in an array indexed by $s_1, \ldots, s_k$. At the end of the tally, we can figure out exactly how many times an element below s_j appeared in the set and sort accordingly.

For example, the parent data set may be the range of integers $b \mathinner{..} c$. Adapted to this particular data type, the algorithm is very straightforward. The data are kept in $A[1 \mathinner{..} n]$; the algorithm initializes all counters to 0, then every time it finds an element $A[j]$ equal to i, it updates $count[i]$ by 1 (that is, it updates $count[A[j]]$ by 1). At the end, it is known that the integer i appears in the data set $count[i]$ times; thus we can print (or add to a new data structure like an array or a linked list) $count[i]$ copies of i after the integers that precede i and before the integers that are larger than i. For example, suppose our range of integers is $6 \mathinner{..} 13$. We set up $count[6 \mathinner{..} 13]$ to receive the final count. If our raw data set is the one given in Figure 6.3; after we tally the frequencies, $count[6] = count[7] = count[8] = 1$, indicating that each of the integers 6, 7 and 8 appears only once, whereas $count[9] = 3$, indicating that 9 appears three times. We can then print the keys 6, 7, 8, 9, 9, 9, ... or adjoin them (in this order) to a secondary data structure.

Figure 6.4 gives the formal algorithm for sorting a set of integers from the range $a \mathinner{..} b$ (with the $count[a \mathinner{..} b]$ serving as the counting data structure). The version cho-

A:	9	12	6	8	9	13	9	7	12	13

indexes:	6	7	8	9	10	11	12	13
count:	1	1	1	3	0	0	2	2

Figure 6.3. The counts after running FREQUENCY COUNT SORT on a set of integer data.

```
for i ← b to c do
        count[i] ← 0;
for i ← 1 to n do
        count[A[i]] ← count[A[i]] + 1;
for i ← b to c do
for j ← 1 to count[i] do
        print(i);
```

Figure 6.4. The FREQUENCY COUNT SORT algorithm for integers in the range $a \mathinner{..} b$.

sen for presentation is the one that prints, but equally simple is a version that adjoins the data sequentially in the same order either to a secondary array of size n or to a linked list that starts out empty then grows to size n linking integers in sorted order.

The algorithm just discussed can be applied to any discrete data type from a set of reasonable size like, for example, a small finite set of integers or other objects. Technically speaking, real data are "discretized" in internal computer operation. Owing to finite precision of all modern digital computers, real numbers are not genuinely real inside a computer but rather approximations of these numbers are used. The finite precision of data in a computer forces us to work with equally spaced close "ticks" on the real line. These ticks are close enough for almost all practical applications, such as financial records, to consider these approximations quite satisfactory. The ticks are typically of order 10^{-38} on personal computers and 10^{-75} or less on mainframes. Therefore, if need be, we can even consider a version of FREQUENCY COUNT SORT to sort computer "real" data. Even for a small real interval, the number of ticks contained in the interval will be enormous. Therefore the algorithm cannot be practical unless the data come from a very small range, such as physics data under very accurate measurements in a laboratory experiment.

There is not a single conditional statement in FREQUENCY COUNT SORT. The algorithm is therefore purely deterministic and takes the exact same amount of computing time for all subsets of the same size coming from the same parent set. The first loop runs in $O(1)$ time. The second loop performs n counting steps. After the count, the sum of all counts must obviously be n. Collectively the two nested loops perform n **print** operations (or **adjoin** operations of whatever kind in the version that builds a secondary data structure). This algorithm, whenever the nature of the data permits, runs in linear time.

EXERCISES

6.1 Is the COUNT SORT algorithm (Figure 6.2) stable?

6.2 Argue that the COUNT SORT algorithm (Figure 6.2) handles repeated data correctly.

6.3 What is the rate of convergence to the normal distribution of Theorem 6.1?

7

Quick Sort

QUICK SORT is one of the fastest in-situ sorting algorithms. The algorithm was invented by Hoare in 1961. Since then the method has enjoyed high popularity and several implementations are in operation as the sorting method of choice in computer systems such as the UNIX operating system.

QUICK SORT is a parsimonious divide-and-conquer algorithm. Its basic idea is as follows. A list of n distinct keys is given in the unsorted array $A[1\,..\,n]$. We select an element, called the *pivot*, and locate its position in the final sorted list by comparing it to all the other elements in the list. In the process, the remaining $n - 1$ elements are classified into two groups: Those that are less than the pivot are moved to the left of the pivot's final position, and those that are greater than the pivot are moved to the right of the pivot's final position. The pivot itself is then moved between the two groups to its correct and final position. This stage of the algorithm is called the *partitioning stage*. QUICK SORT is then applied recursively to the left and right sublists until small lists of size 1 or less are reached; these are left intact as they are already sorted.

Simple modifications can be introduced to handle lists with key repetitions, a case that occurs with probability zero in a sample from a continuous distribution (i.e., when the random permutation model applies). We shall therefore assume in the following text that all n keys in the list are distinct and their actual values are assimilated by a random permutation of $\{1, \ldots, n\}$ and will leave other variations to the exercises.

7.1 THE PARTITIONING STAGE

It is obvious from the above layout that QUICK SORT only does the bookkeeping necessary for monitoring which array segments to be sorted next. The actual data movement to accomplish sorting is done within the partitioning stage (to be set up as a procedure called PARTITION).

During the partitioning stage the pivot is compared to the remaining $n - 1$ keys; this takes at least $n-1$ comparisons. We shall present an implementation that actually takes $n - 1$ comparisons. A popular implementation of the partitioning algorithm due to Sedgewick makes $n + 1$ comparisons. Sedgewick's partitioning algorithm is discussed in Exercise 7.2.

QUICK SORT recursively invokes itself to sort the two sublists. At a general stage QUICK SORT may need to sort the particular segment $A[\ell\,..\,u]$, with ℓ and u being the lower and upper limits of the sublist to be sorted. Therefore the partitioning procedure must be able to handle the partitioning of any general segment $A[\ell\,..\,u]$. We shall consider an implementation of PARTITION that takes in ℓ and u and returns p, the final position of the pivot; the array A is accessed globally from within PARTITION. A side effect of the call to PARTITION is to rearrange A so that $A[\ell\,..\,p-1] < pivot$ and $A[p+1\,..\,u] > pivot$; the pivot itself is moved to its correct and final position. A typical call from within QUICK SORT is

$$\textbf{call } Partition(\ell, u, p).$$

The procedure PARTITION considered here is based on a simple idea. In partitioning the segment $A[\ell\,..\,u]$, the element $A[u]$ is used as a pivot. PARTITION scans the remaining segment $A[\ell\,..\,u-1]$ sequentially, from left to right, keeping an invariant property. At the ith stage of the scan, within the segment scanned so far we keep track of a position p to the right of which we place all the elements encountered so far that are larger than the pivot, but up to p all the elements of $A[\ell\,..\,p]$ are smaller; see Figure 7.1.

Assume this has been observed up to position $i-1$, $\ell \le i \le u-1$. We now consider the next element $A[i]$. If $A[i] > pivot$, then $A[i]$ is on the correct side of p, we *extend* the range for which the invariant property holds on the right of p by simply advancing i, and now all the keys from position ℓ up to p are smaller than the pivot, and all the elements to the right of position p up to i are greater. If $A[i] < pivot$, we increment p by 1, then swap $A[i]$ with $A[p]$; we thus bring an element smaller than the pivot to the position indexed by the new incremented value of p to replace an element greater than *pivot*. This greater element is bumped to the end of the segment scanned so far; the invariant property is maintained. At the end of the scan, $A[\ell\,..\,p]$ contains keys smaller than *pivot*, and $A[p+1\,..\,u-1]$ contains elements greater than *pivot*. We can bring *pivot* to its correct position $p+1$ by swapping the pivot $A[u]$ with $A[p+1]$. The algorithm is presented in procedure form in Figure 7.2.

The diagram of Figure 7.3 illustrates the operation of PARTITION on a stretch of six keys. In the figure the pivot is circled and the top line depicts the selection of the pivot. Each subsequent line represents some ith stage and depicts the addition of $A[i]$ to one of the two growing parts of the partition. The two growing parts of the partition are boxed; only the second line has one box with an element larger than the pivot; the other part (with keys less than the pivot) is still empty at this stage.

We shall discuss next the effect of the above PARTITION procedure on the randomness of the two subarrays produced. The hypothesis is that at the end of PAR-

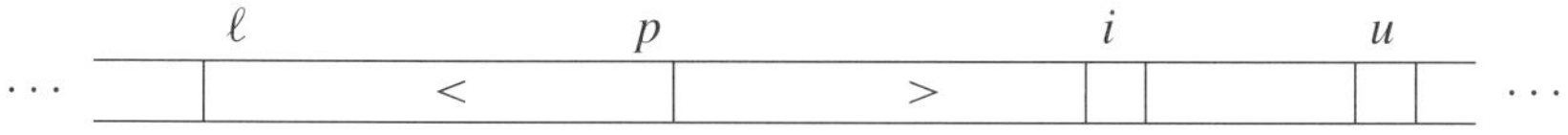

Figure 7.1. An invariant property of PARTITION.

```
procedure Partition(ℓ, u: integer; var p: integer);
    local i, pivot: integer;
    begin
        {maintain the following invariant:
            A[ℓ .. p] < pivot and A[p + 1 .. i] ≥ pivot}
        pivot ← A[u];
        p ← ℓ − 1;
        for i ← ℓ to u − 1 do
            if A[i] < pivot then
                begin
                    p ← p + 1;
                    call swap(A[p], A[i]);
                end;
        p ← p + 1;
        call swap(A[p], A[u]);
    end;
```

Figure 7.2. A partitioning algorithm that takes $n - 1$ comparisons.

TITION the relative ranks of the keys in the subarray $A[1 \mathinner{..} p-1]$ form a *random permutation* of the integers $\{1, \dots, p-1\}$. Likewise the relative ranks of the keys in $A[p+1 \mathinner{..} n]$ form a random permutation of the integers $\{1, \dots, n-p\}$. We prove this hypothesis by induction on i (the stage number). We start with keys in $A[1 \mathinner{..} n]$ whose (absolute) ranks form a random permutation of $\{1, \dots, n\}$. At the ith stage of PARTITION, $A[i]$ is considered. The sequential rank of $A[i]$, $Seqrank(A[i])$, is distributed as discrete UNIFORM$[1 \mathinner{..} i]$, according to the random permutation model (see Proposition 1.6).

Assume that at the ith stage of the scan, the relative ranks of $A[1 \mathinner{..} p]$ form a random permutation of $\{1, \dots, p\}$ and those in $A[p+1 \mathinner{..} i-1]$ form a random per-

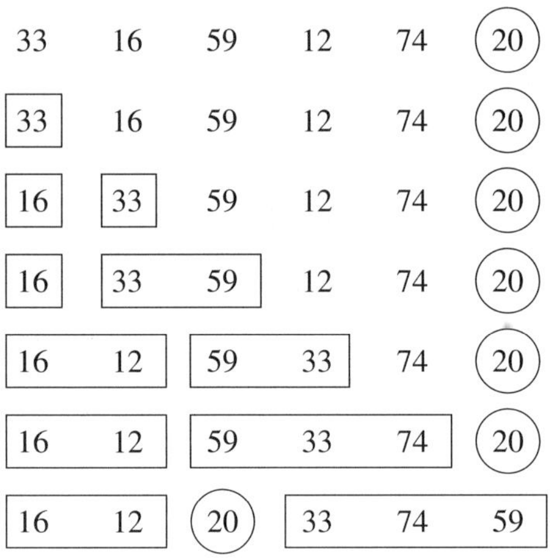

Figure 7.3. The action of PARTITION.

mutation of $\{1, \ldots, i-p-1\}$. Thus, the conditional distribution of $Absrank(A[i])$, given $A[i] < pivot$, is discrete UNIFORM$[1 \mathinner{..} p+1]$. In this case, swapping $A[i]$ and $A[p+1]$ is the same as juxtaposing an element at the end of $A[1 \mathinner{..} p]$ at position $A[p+1]$ and the sequential rank of the new element is equally likely to be any of the numbers $1, 2, \ldots, p+1$. Hence after this swap the relative ranks of $A[1 \mathinner{..} p+1]$ will be a random permutation of $\{1, \ldots, p+1\}$; see Proposition 1.6. As for the other segment, the subarray $A[p+1 \mathinner{..} i]$, the effect of the swap between $A[i]$ and $A[p+1]$ is like a circular shuffle of a permutation on $\{1, \ldots, i-p-1\}$ moving its first element to the end; the result is a random permutation on $\{1, \ldots, i-p-1\}$; see Exercise 1.10.1.

The argument for the case $A[i] > pivot$ mirrors the one we used for $A[i] < pivot$. In either case, the induction hypothesis implies that when we are done with $A[i]$ the two subarrays $A[1 \mathinner{..} p]$ and $A[p+1 \mathinner{..} i]$ contain elements whose relative ranks are random permutations of their respective sizes, completing the induction.

The last swap, moving the pivot to its correct position, preserves the randomness, too. The element $A[p+1]$, whose sequential rank a priori has a UNIFORM$[1 \mathinner{..} n]$ distribution, is now known to have rank exceeding p, so relative to the elements now contained in $A[p+1 \mathinner{..} u-1]$ it has relative rank distributed like UNIFORM$[1 \mathinner{..} n-p]$. According to Proposition 1.6, the relative ranks of the entire segment $A[p+1 \mathinner{..} u]$ form a random permutation. Swapping the pivot with $A[p+1]$ is a circular shuffle that preserves this randomness (Exercise 1.10.1).

In summary, at the end of PARTITION, the relative ranks of the elements in $A[1 \mathinner{..} p-1]$ are a random permutation of $\{1, \ldots, p-1\}$ and the relative ranks of the elements in $A[p+1 \mathinner{..} n]$ are a random permutation of $\{1, \ldots, n-p\}$. This is instrumental in forming recurrence relations that assume that the sublists are probabilistic copies (on their respective sizes) of the original problem.

7.2 BOOKKEEPING

As discussed above, QUICK SORT itself acts as a manager directing PARTITION to rearrange the elements of the list. The recursive algorithm is specified in Figure 7.4.

```
procedure QuickSort(ℓ,u: integer);
    local p: integer;
    begin
        if ℓ < u then
            begin
                call Partition(ℓ, u, p);
                call QuickSort(ℓ, p − 1);
                call QuickSort(p + 1, u);
            end;
    end;
```

Figure 7.4. QUICK SORT algorithm.

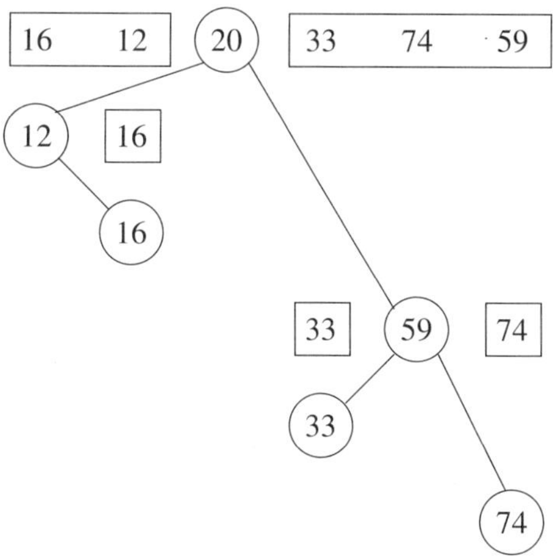

Figure 7.5. Operation of QUICK SORT by repeated calls to PARTITION.

In the algorithm shown in the figure, the array $A[1\,..\,n]$ is accessed globally; the algorithm may be initiated by an external call like **call** *QuickSort*$(1, n)$.

Figure 7.5 depicts the action of QUICK SORT on the same data set used for illustrating PARTITION (Figure 7.3). Each line of Figure 7.5 shows a stretch of the array at the end of a call to QUICK SORT to handle a segment; the pivot chosen for that step is circled.

7.3 QUICK SORT TREE

It is useful to make a connection between QUICK SORT and random binary search trees whose properties are well understood. Let us construct the *Quick Sort tree* as follows. For every pivot encountered create a node in the tree labeled with that pivot. The first pivot labels the root of the tree. As the first pivot splits the list into two parts (of which one or both may be empty), a pivot will be found in each nonempty list and will father a subtree. If the list of smaller elements is nonempty, the pivot found in it is attached to the root as a left child. Likewise, if there are larger elements, the pivot found there will be attached to the root as a right child. As QUICK SORT continues recursively, nodes are added in this fashion to the tree until files of size 0 are the subject of sorting; those are represented by leaves. As a visual aid for this construction consider connecting the pivots (circled numbers) in Figure 7.5. Redrawing this tree with nodes at the same distance from the root appearing at the same level we get the tree of Figure 7.6.

Our discussion of the randomness of the subfiles created by PARTITION indicates that the QUICK SORT tree is a random binary search tree. Indeed, as the ranks of the n keys of the input form a random permutation, the pivot is equally likely to be any of the input keys. It is then an easy induction argument—given the absolute rank

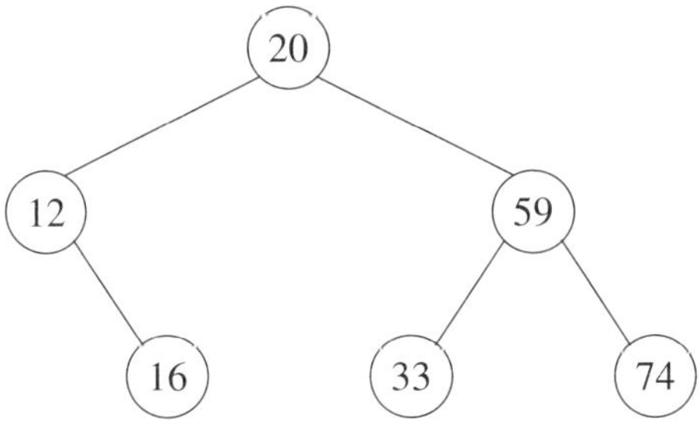

Figure 7.6. The Quick Sort tree of the running example.

of the first pivot is $r \in \{1, \dots, n\}$, the two remaining lists, of sizes $r - 1$ and $n - r$, are random as in our discussion above. If we assume the assertion true for any list of size less than n, then the two sublists give rise to two random binary search trees of sizes $r - 1$ and $n - r$, which are attached as left and right subtree. This construction coincides with the inductive definition of the random binary search tree (see Section 1.6); the resulting tree is a random binary search tree; the induction is complete. A node at level k in the QUICK SORT tree is involved in k comparisons with its k predecessors, one comparison for each predecessor when this predecessor is chosen as the pivot. The total number of comparisons is therefore the sum of the depths of all the nodes of the tree, that is, the internal path length of the Quick Sort tree (recall that the root is at level 0); whence C_n, the number of comparisons made by QUICK SORT to sort a list of n elements, is the internal path length of the Quick Sort tree and is related to X_n, the external path length of the Quick Sort tree, by the relation

$$X_n = C_n + 2n$$

(review Proposition 1.3).

7.4 PROBABILISTIC ANALYSIS OF QUICK SORT

We demonstrate in this section that QUICK SORT has good probabilistic behavior that makes it one of the best known sorting algorithms. We shall see that QUICK SORT makes $2n \ln n$ comparisons, on average, to sort a list of n elements. This only means that the majority of sample space points have good behavior. Nonetheless, QUICK SORT can occasionally exhibit slow performance on certain (rare) inputs. For example, QUICK SORT performs badly on sorted arrays! If the given list is already sorted (say the input is the permutation $\Pi^{(n)}_{ascend} = (1, 2, \dots, n)$), QUICK SORT will just waste time at each partitioning stage trying to find a good location for the pivot, only to discover at the end that the pivot is already in its correct place. This takes $n - 1$ comparisons at the first level of recursion, then recursively the subarrays considered will each be sorted in increasing order. At the second recursive step $n - 2$ comparisons will be made, and so on, and in this case

$$C_n\big(\Pi^{(n)}_{ascend}\big) = (n-1) + (n-2) + \cdots + 1$$
$$= \frac{n(n-1)}{2}.$$

This $\Theta(n^2)$ behavior is the worst-case number of data comparisons (and on the other hand is the best case for the number of swaps). The point here is that, although QUICK SORT has an $O(n \log n)$ average number of comparisons, there are cases in which QUICK SORT performs like naive $\Theta(n^2)$ algorithms. One can construct other deterministic permutations on which QUICK SORT has other orders of magnitude between the best $O(n \log n)$ and the worst $O(n^2)$; see Exercise 7.3.

In the previous section we discussed PARTITION, an algorithm for the partitioning stage with $n - 1$ data comparisons when performing on a list of size n. The bookkeeping algorithm recursively applies PARTITION; the number of data comparisons will build up. Let C_n be the total number of comparisons used by QUICK SORT to sort the given data array of size n. PARTITION moves the pivot to a random position $P_n \overset{D}{=}$ UNIFORM$[1\,..\,n]$. The recursive structure of QUICK SORT is then reflected by the recurrence

$$C_n \overset{D}{=} C_{P_n-1} + \tilde{C}_{n-P_n} + n - 1, \tag{7.1}$$

where $\tilde{C}_k \overset{D}{=} C_k$, and the families $\{\tilde{C}_j\}_{j=1}^{\infty}$ and $\{C_j\}_{j=1}^{\infty}$ are independent; the boundary conditions are $C_0 = C_1 = 0$. It should be noted that even though C_j and $\tilde{C}_k$ are independent for every j, k, the pair C_{P_n-1} and $\tilde{C}_{n-P_n}$ are dependent (through their connection via P_n). It should also be noted that the random variables $C_1, \ldots, C_n$ are all well defined on the sample space of permutations of $\{1, \ldots, n\}$. The average behavior of QUICK SORT can be directly obtained by taking expectation of (7.1), yielding the basic recurrence for averages:

$$\mathbf{E}[C_n] = \mathbf{E}[C_{P_n-1}] + \mathbf{E}[\tilde{C}_{n-P_n}] + n - 1.$$

The uniform distribution of P_n induces symmetry in the algorithm and $C_{P_n-1} \overset{D}{=} \tilde{C}_{n-P_n}$. That is, C_{P_n-1} and $\tilde{C}_{n-P_n}$ have the same probabilistic behavior (hence the same average). We can write the average number of comparison in the form

$$\mathbf{E}[C_n] = n - 1 + 2\mathbf{E}[C_{P_n-1}].$$

By conditioning on P_n, the random landing position of the pivot, we can compute the average of C_{P_n-1} as follows:

$$\mathbf{E}[C_{P_n-1}] = \sum_{p=1}^{n} \mathbf{E}[C_{P_n-1} \mid P_n = p]\, \mathbf{Prob}\{P_n = p\}$$
$$= \frac{1}{n} \sum_{p=1}^{n} \mathbf{E}[C_{p-1}].$$

We thus get the recurrence

$$nE[C_n] = (n-1)n + 2\sum_{p=1}^{n} E[C_{p-1}]. \tag{7.2}$$

We can solve this equation by differencing as follows. A version of (7.2) for $n-1$ is

$$(n-1)E[C_{n-1}] = (n-2)(n-1) + 2\sum_{p=1}^{n-1} E[C_{p-1}]. \tag{7.3}$$

Subtracting (7.3) from (7.2) the resulting equation can be reorganized in a form suitable for direct iteration:

$$\begin{aligned}
E[C_n] &= \frac{2(n-1)}{n} + \frac{n+1}{n}E[C_{n-1}] \\
&= \frac{2(n-1)(n+1)}{n(n+1)} + \frac{2(n-2)(n+1)}{(n-1)n} + \frac{n+1}{n-1}E[C_{n-2}] \\
&\ \ \vdots \\
&= 2(n+1)\sum_{j=2}^{n} \frac{j-1}{j(j+1)} + \frac{1}{2}(n+1)E[C_1] \\
&= 2(n+1)\sum_{j=1}^{n} \frac{j-1}{j(j+1)}.
\end{aligned}$$

(In the last line, we used the boundary condition $E[C_1] = 0$.) Writing $(j-1)/(j(j+1))$ as $2/(j+1) - 1/j$ simplifies the average to

$$\begin{aligned}
E[C_n] &= 2(n+1)\sum_{j=1}^{n}\left(\frac{2}{j+1} - \frac{1}{j}\right) \\
&= 2(n+1)\left[2\left(H_n + \frac{1}{n+1} - 1\right) - H_n\right].
\end{aligned}$$

Theorem 7.1 *(Hoare, 1962). The average number of comparisons that QUICK SORT makes to sort a random list of n keys is*

$$E[C_n] = 2(n+1)H_n - 4n.$$

As $n \to \infty$,

$$E[C_n] \sim 2n \ln n.$$

The variance of C_n can be found by a similar technique. The principle of the calculation is the same, though the algebraic manipulation is quite a bit more involved;

it may be best to reduce the combinatorial sums therein by a symbolic manipulation computer system such as Maple or Mathematica. (Exercise 7.5 outlines a fast way of evaluating the leading term in this variance.)

We first compute the second moment by conditioning on P_n, the pivot's final position to get

$$\mathbf{E}[C_n^2] = \sum_{p=1}^{n} \mathbf{E}[C_n^2 \mid P_n = p] \, \mathbf{Prob}\{P_n = p\}.$$

The probabilistic recurrence (7.1) gives

$$\begin{aligned}\mathbf{E}[C_n^2 \mid P_n = p] &= \mathbf{E}\big[(C_{p-1} + \tilde{C}_{n-p} + n - 1)^2\big] \\ &= \mathbf{E}[C_{p-1}^2] + \mathbf{E}[\tilde{C}_{n-p}^2] + (n-1)^2 + 2(n-1)\mathbf{E}[C_{p-1}] \\ &\qquad + 2(n-1)\mathbf{E}[\tilde{C}_{n-p}] + 2\mathbf{E}[C_{p-1}]\,\mathbf{E}[\tilde{C}_{n-p}].\end{aligned}$$

We can replace every $\mathbf{E}[\tilde{C}_{n-p}]$ by $\mathbf{E}[C_{n-p}]$, because for every k, C_k and $\tilde{C}_k$ are identically distributed. Unconditioning, we obtain a recurrence for the second moment:

$$\begin{aligned}\mathbf{E}[C_n^2] = (n-1)^2 + \frac{1}{n}\sum_{p=1}^{n}\Big\{&\mathbf{E}[C_{p-1}^2] + \mathbf{E}[C_{n-p}^2] + 2(n-1)\mathbf{E}[C_{p-1}] \\ &+ 2(n-1)\mathbf{E}[C_{n-p}] + 2\mathbf{E}[C_{p-1}]\,\mathbf{E}[C_{n-p}]\Big\}.\end{aligned}$$

The term $\mathbf{E}[C_{n-p}^2]$ can be replaced by $\mathbf{E}[C_{p-1}^2]$ and the term $2(n-1)\mathbf{E}[C_{n-p}]$ can be replaced by $2(n-1)\mathbf{E}[C_{p-1}]$ because they occur within sums over p (so the former gives sums like the latter but written backward). Letting

$$f(n) = n(n-1)^2 + 4(n-1)\sum_{p=1}^{n} \mathbf{E}[C_{p-1}] + 2\sum_{p=1}^{n} \mathbf{E}[C_{p-1}]\,\mathbf{E}[C_{n-p}],$$

we can combine the latter notation with the unconditional second moment to get a recurrence for the second moment:

$$\mathbf{E}[C_n^2] = \frac{2}{n}\sum_{p=1}^{n} \mathbf{E}[C_{p-1}^2] + \frac{f(n)}{n},$$

where the averages involved are given by Theorem 7.1. To solve this recurrence, we can employ the differencing technique we used for the average. Using the backward difference operator notation $\nabla f(n) = f(n) - f(n-1)$, we subtract from the latter equation a version of itself with n replaced by $n-1$ and unwind the recurrence:

$$\begin{aligned}
\mathbf{E}[C_n^2] &= \frac{\nabla f(n)}{n} + \frac{n+1}{n}\mathbf{E}[C_{n-1}^2] \\
&= \frac{(n+1)\nabla f(n)}{(n+1)n} + \frac{(n+1)\nabla f(n-1)}{n(n-1)} + \frac{(n+1)n}{n(n-1)}\mathbf{E}[C_{n-2}^2] \\
&\ \ \vdots \\
&= \sum_{j=1}^{n} \frac{(n+1)\nabla f(j)}{j(j+1)};
\end{aligned}$$

we used the boundary condition $C_1 \equiv 0$. As mentioned above, a symbolic manipulation system can be programmed to reduce the last sum.

Theorem 7.2 *(Knuth, 1973).*

$$\begin{aligned}
\mathbf{Var}[C_n] &= 7n^2 - 4(n+1)^2 H_n^{(2)} - 2(n+1)H_n + 13n \\
&\sim \Big(7 - \frac{2\pi^2}{3}\Big)n^2, \qquad \textit{as } n \to \infty.
\end{aligned}$$

We have found the mean and the variance of the number of comparisons that QUICK SORT makes to sort a random list of n elements. If a limit distribution exists, the proper centering factor should be $\mathbf{E}[C_n] \sim 2n \ln n$, the asymptotic mean, and the norming factor should be of order n, the same order of the square root of the variance of C_n. This suggests studying the normalized random variable

$$C_n^* \stackrel{\text{def}}{=} \frac{C_n - 2n \ln n}{n}.$$

We next discuss the existence of a limit law for C_n^*. In fact we prove a stronger almost-sure result in the next theorem.

Theorem 7.3 *(Régnier, 1989). There exists a limiting square-integrable random variable C such that*

$$C_n^* \xrightarrow{\text{a.s.}} C.$$

Proof. Let us grow a binary search tree T_n from a random permutation of $\{1, \ldots, n\}$. Let $\mathcal{F}_n$ be the sigma field generated by the trees $T_1, T_2, \ldots, T_{n-1}$. Suppose the n leaves of T_{n-1} are indexed $1, \ldots, n$, from left to right, say. Let D_i be the depth of the ith leaf of T_{n-1}. When we insert the last entry into T_{n-1} to obtain T_n, it replaces some leaf of T_{n-1}, say leaf i, at depth D_i. Two new nodes are created at depths $D' = D'' = D_i + 1$. Suppose the level of insertion of the nth key is L_n. Hence, given that $L_n = D_i$, the new external path length is

$$\begin{aligned}
X_n &= D_1 + \cdots + D_{i-1} + D' + D'' + D_{i+1} + \cdots + D_n \\
&= D_1 + \cdots + D_{i-1} + 2(D_i + 1) + D_{i+1} + \cdots + D_n \\
&= D_1 + \cdots + D_n + D_i + 2 \\
&= X_{n-1} + D_i + 2.
\end{aligned}$$

Then

$$\begin{aligned}
\mathbf{E}[X_n \mid \mathcal{F}_{n-1}] &= \sum_{i=1}^{n} \mathbf{E}\big[X_n \mid \mathcal{F}_{n-1} \text{ and } L_n = D_i\big] \times \mathbf{Prob}\{L_n = D_i | \mathcal{F}_{n-1}\} \\
&= \frac{1}{n} \sum_{i=1}^{n} (X_{n-1} + 2 + D_i) \\
&= X_{n-1} + 2 + \frac{1}{n} \sum_{i=1}^{n} D_i \\
&= X_{n-1} + 2 + \frac{1}{n} X_{n-1} \\
&= \frac{n+1}{n} X_{n-1} + 2. \qquad (7.4)
\end{aligned}$$

Taking expectations gives us

$$\mathbf{E}[X_n] = \frac{n+1}{n} \mathbf{E}[X_{n-1}] + 2. \qquad (7.5)$$

We recall Proposition 1.3, which gives the relationship between the external path length, and the internal path length, which is also C_n:

$$X_n = C_n + 2n.$$

Equations (7.4) and (7.5) can be expressed in terms of C_n and one gets

$$\mathbf{E}\Big[\frac{C_n - \mathbf{E}[C_n]}{n+1} \,\Big|\, \mathcal{F}_{n-1}\Big] = \frac{C_{n-1} - \mathbf{E}[C_{n-1}]}{n}.$$

That is, $(C_n - \mathbf{E}[C_n])/(n+1)$ is a martingale. This centered martingale has zero mean. According to Theorem 7.2,

$$\begin{aligned}
\mathbf{Var}\Big[\frac{C_n - \mathbf{E}[C_n]}{n+1}\Big] &= \frac{1}{(n+1)^2} \mathbf{Var}[C_n] \\
&\le 7 + \frac{13}{n} \\
&\le 20, \qquad \text{for all } n \ge 1.
\end{aligned}$$

Therefore this zero-mean martingale has uniformly bounded variance . It follows from the martingale convergence theorem that almost surely this martingale converges to a limit. That is, there is a square-integrable random variable C', such that

$$\frac{C_n - \mathbf{E}[C_n]}{n+1} \xrightarrow{\text{a.s.}} C'.$$

In Theorem 7.1 the asymptotic approximation of the harmonic number gives

$$\frac{\mathbf{E}[C_n] - 2n\ln n}{n+1} \xrightarrow{\text{a.s.}} 2\gamma - 4.$$

From the rules of almost-sure convergence, we can add the last two relations to obtain

$$\frac{C_n - 2n\ln n}{n+1} \xrightarrow{\text{a.s.}} C' + 2\gamma - 4 \stackrel{\text{def}}{=} C,$$

and C is square-integrable, too. Multiplying the last relation by the obvious deterministic convergence relation $(n+1)/n \to 1$, we obtain the statement of the theorem. ■

The distribution of C is not known explicitly. However, it can be characterized implicitly as the solution of a distributional equation as we discuss in the next paragraphs. Several attempts have been made toward an explicit characterization of the limiting distribution. Hennequin (1991) found all its cumulants, Rösler (1991) proved that the limit law of C is the fixed point of a contraction mapping, McDiarmid and Hayward (1992) empirically obtained its skewed shape, and Tan and Hadjicostas (1995) proved that it possesses (an unknown) density with respect to Lebesgue measure.

An implicit characterization of the distribution of C is obtained by first expressing all the random variables in the basic probabilistic recurrence (7.1) in the normalized form—we subtract $2n\ln n$ from both sides and divide by n to obtain

$$\begin{aligned}
C_n^* &= \frac{C_n - 2n\ln n}{n} \\
&\stackrel{D}{=} \frac{C_{P_n-1}}{n} + \frac{\tilde{C}_{n-P_n}}{n} + \frac{(n-1) - 2n\ln n}{n} \\
&= \frac{P_n - 1}{n} \times \frac{C_{P_n-1} - 2(P_n-1)\,\ln(P_n-1)}{P_n-1} \\
&\quad + \frac{n-P_n}{n} \times \frac{\tilde{C}_{n-P_n} - 2(n-P_n)\,\ln(n-P_n)}{n-P_n} \\
&\quad + \frac{2(P_n-1)\,\ln(P_n-1) + 2(n-P_n)\,\ln(n-P_n) - 2n\ln n}{n} \\
&\quad + \frac{n-1}{n},
\end{aligned}$$

which can be written as

$$C_n^* \stackrel{D}{=} \frac{P_n - 1}{n} C^*_{P_n - 1} + \frac{n - P_n}{n} \tilde{C}^*_{n - P_n} + G_n(P_n), \tag{7.6}$$

where

$$G_n(x) = \frac{1}{n}\big[2(x-1)\,\ln(x-1) + 2(n-x)\,\ln(n-x) - 2n\ln n\big] + \frac{n-1}{n},$$

and for every k, $\tilde{C}_k^* \stackrel{D}{=} C_k^*$, and the families $\{C_k^*\}_{k=1}^\infty$ and $\{\tilde{C}_k^*\}_{k=1}^\infty$ are independent.

As ascertained by Theorem 7.3, a limit distribution exists for C_n^*; we can characterize it by passing the last equation to the limit. However, this is a delicate operation and must be handled with care. We have three additive terms in the distributional functional equation (7.6), and all three converge to limiting random variables. Convergence in distribution is a weak mode of convergence and we cannot immediately infer from this that the sum of the limiting random variables is the limit of the right-hand side of (7.6), even though it will turn out to be, but this needs some technical work. Recall that if $X_n \xrightarrow{\mathcal{D}} X$ and $Y_n \xrightarrow{\mathcal{D}} Y$, it is not necessarily the case that $X_n + Y_n \xrightarrow{\mathcal{D}} X + Y$. This is true when X_n and Y_n satisfy some additional conditions like independence or when at least one of the two limits is a constant.

The ingredients on the right-hand side of the functional equation (7.6) are not independent. For instance, the random variables $C^*_{P_n-1}$ and $\tilde{C}^*_{n-P_n}$ are dependent through their mutual dependence on P_n, even though C_{j-1} and $\tilde{C}_{n-j}$ are independent. Curiously, the dependence is asymptotically weak enough, as $n \to \infty$. That is, for large n the three additive factors of (7.6) are "almost independent"; and eventually when n becomes infinite they become so.

The terms on the right-hand side converge as follows. Let U be a UNIFORM(0, 1) random variable and C be the limit of C_n^*, as in Theorem 7.3. It is easy to see that

$$\frac{P_n - 1}{n} \xrightarrow{\mathcal{D}} U,$$

$$P_n \xrightarrow{\text{a.s.}} \infty.$$

From the latter, we have

$$C^*_{P_n - 1} \xrightarrow{\text{a.s.}} C.$$

Therefore,

$$\frac{P_n - 1}{n} C^*_{P_n - 1} \xrightarrow{\mathcal{D}} UC.$$

Similarly

$$\frac{n - P_n}{n} \tilde{C}^*_{n - P_n} \xrightarrow{\mathcal{D}} (1 - U)\tilde{C},$$

where $\tilde{C} \overset{D}{=} C$, and C and C are independent. Moreover, in $G_n(P_n)$ we write $2n \ln n = 2(P_n - 1 + n - P_n + 1) \ln n = 2(P_n - 1) \ln n + 2(n - P_n) \ln n + 2 \ln n$. In the regrouped $G_n(P_n)$ the term

$$\frac{1}{n}\big[2(P_n - 1)\,\ln(P_n - 1) - 2(P_n - 1)\ln n\big] = 2\frac{P_n - 1}{n}\,\ln\Big(\frac{P_n - 1}{n}\Big)$$
$$\xrightarrow{\mathcal{D}} 2U \ln U,$$

and

$$\frac{1}{n}\big[2(n - P_n)\,\ln(n - P_n) - 2(n - P_n)\ln n\big] = 2\frac{n - P_n}{n}\,\ln\Big(\frac{n - P_n}{n}\Big)$$
$$\xrightarrow{\mathcal{D}} 2(1 - U)\ln(1 - U),$$

and, of course,

$$\frac{n-1}{n} - \frac{2\ln n}{n} \to 1.$$

These convergence relations suggest a form of a distributional functional equation for the limit:

$$C \overset{D}{=} UC + (1 - U)\tilde{C} + G(U), \tag{7.7}$$

where U is a UNIFORM(0, 1) random variable that is independent of C, and

$$G(u) \overset{\text{def}}{=} 2u \ln u + 2(1 - u)\ln(1 - u) + 1.$$

However, as discussed, we cannot immediately claim the right-hand side of (7.7) as the limit of the right-hand side of (7.6).

The dependence among the three additive terms in (7.6) is weak enough to permit the right-hand side of (7.7) to be the limit and, after all, the conjectured functional equation (7.7) is the valid form for the distribution, a statement that is proved in what follows. The proof is based on convergence in the Wasserstein metric space.

The *Wasserstein distance of order k* between two distribution functions F and G is defined by

$$d_k(F, G) = \inf ||W - Z||_k,$$

where the infimum is taken over all random variables W and Z having the respective distributions F and G (with $||\,.\,||_k$ being the usual L^k norm). If F_n is a sequence of distribution functions of the random variables W_n, it is known (Barbour, Holst, and Janson (1992)) that convergence in the second-order Wasserstein distance implies weak convergence, as well as convergence of the first two moments. A few exercises are given at the end of this chapter to familiarize the reader with this metric space.

Because the input is a random permutation, we have the representations

$$P_n = \lceil nU \rceil \stackrel{D}{=} \text{UNIFORM}[1..n],$$

for some UNIFORM(0, 1) random variable U.

Let F_n and F be the distribution functions of C_n^* and the limit C, respectively. Further, let

$$a_n = \mathbf{E}\big[(C_n^* - C)^2\big].$$

Since the second-order Wasserstein distance $d_2(F_n, F)$ is given by an infimum taken over all random variables with distributions F_n and F, we must have

$$d_2^2(F_n, F) \le a_n.$$

If we manage to show that $a_n \to 0$, as $n \to \infty$, we shall have proved the required convergence in distribution. We have

$$\begin{aligned} a_n = \mathbf{E}\Big[\Big\{\Big(&\frac{\lceil nU \rceil - 1}{n} C^*_{\lceil nU \rceil - 1} - UC\Big) \\ &+\Big(\frac{n - \lceil nU \rceil}{n} \tilde{C}^*_{n-\lceil nU \rceil} - (1-U)\tilde{C}\Big) \\ &+\big(G_n(\lceil nU \rceil) - G(U)\big)\Big\}^2\Big]. \end{aligned}$$

As we square the argument of the expectation operator, only the squared terms stay because all cross-product terms are 0.

We have the equality

$$\begin{aligned} a_n = \mathbf{E}\Big[\Big(&\frac{\lceil nU \rceil - 1}{n} C^*_{\lceil nU \rceil - 1} - UC\Big)^2 \\ &+\Big(\frac{n - \lceil nU \rceil}{n} \tilde{C}^*_{n-\lceil nU \rceil} - (1-U)\tilde{C}\Big)^2 \\ &+\Big(G_n\big(\lceil nU \rceil\big) - G(U)\Big)^2\Big]. \end{aligned}$$

If $\lceil nU \rceil = k$, then $k = nU + \alpha_{kn}$, where α_{kn} is a nonnegative function of k and n, and is of a uniformly bounded magnitude; for all $n \ge 1$, $0 \le \alpha_{kn} < 1$. Condition on $\lceil nU \rceil = k$ and write

$$a_n = \Bigg(\frac{1}{n}\sum_{k=1}^{n} \mathbf{E}\Big[\Big(\frac{k-1}{n} C^*_{k-1} - \frac{k - \alpha_{kn}}{n} C\Big)^2\Big]$$

$$
\begin{aligned}
&+\mathbf{E}\Big[\Big(\frac{n-k}{n}C^*_{n-k}-\frac{n-k+\alpha_{kn}}{n}C\Big)^2\Big]\Big)\\
&+\mathbf{E}\Big[\Big(G_n(\lceil nU\rceil)-G(U)\Big)^2\Big]\\
=&\Big\{R_n+\frac{1}{n}\sum_{k=1}^{n}\mathbf{E}[(C^*_{k-1}-C)^2]\Big(\frac{k-1}{n}\Big)^2\Big\}\\
&+\Big\{\tilde{R}_n+\frac{1}{n}\sum_{k=1}^{n}\mathbf{E}[(C^*_{n-k}-C)^2]\Big(\frac{n-k}{n}\Big)^2\Big\}\\
&+\mathbf{E}\Big[\Big(G_n(\lceil nU\rceil)-G(U)\Big)^2\Big];
\end{aligned} \tag{7.8}
$$

the remainder terms R_n, $\tilde{R}_n$, and $\mathbf{E}\big[\big(G_n(\lceil nU\rceil)-G(U)\big)^2\big]$ are uniformly bounded in n—to see this, we use the uniform bound $0 \le \alpha_{kn} < 1$ and the variance calculation of C (Exercise 7.5) to develop the following:

$$
\begin{aligned}
R_n &= \frac{2}{n}\sum_{k=1}^{n}\frac{(k-1)(1-\alpha_{kn})}{n^2}\mathbf{E}[C^2]+\frac{1}{n}\sum_{k=1}^{n}\frac{(1-\alpha_{kn})^2}{n^2}\mathbf{E}[C^2]\\
&< \frac{2}{n}\sum_{k=1}^{n}\frac{k-1}{n^2}\mathbf{E}[C^2]+\frac{1}{n}\sum_{k=1}^{n}\frac{\mathbf{E}[C^2]}{n^2}\\
&< \Big(7-\frac{2\pi^2}{3}\Big)\Big[\frac{1}{n}+\frac{1}{n^2}\Big]\\
&\le 2\Big(7-\frac{2\pi^2}{3}\Big)\frac{1}{n}.
\end{aligned}
$$

Similarly, the remainder $\tilde{R}_n$ is also bounded by $\frac{2}{n}(7-\frac{2}{3}\pi^2)$.

Again, manipulating $2n\ln n$ as $2(P_n-1+n-P_n+1)\ln n = 2(P_n-1)\ln n + 2(n-P_n)\ln n + 2\ln n$ in $G_n(P_n)$, we bound the difference between $G_n\big(\lceil nU\rceil\big)$ and $G(U)$ as follows:

$$
\begin{aligned}
G_n\big(\lceil nU\rceil\big)-G(U) = &\frac{2\big(\lceil nU\rceil-1\big)}{n}\ln\Big(\frac{\lceil nU\rceil-1}{n}\Big)\\
&+\frac{2\big(n-\lceil nU\rceil\big)}{n}\ln\Big(\frac{n-\lceil nU\rceil}{n}\Big)\\
&-\frac{2\ln n}{n}+\frac{n-1}{n}\\
&-\big(2U\ln U+2(1-U)\ln(1-U)+1\big)
\end{aligned}
$$

$$\leq 2U \ln U + 2(1-U)\ln(1-U) - \frac{2\ln n}{n} + \frac{n-1}{n}$$
$$-\big(2U \ln U + 2(1-U)\ln(1-U) + 1\big).$$

It follows that

$$\mathbf{E}\big[\{G_n(\lceil nU \rceil) - G(U)\}^2\big] \leq \Big(\frac{2\ln n + 1}{n}\Big)^2 \leq \frac{4}{n},$$

for all $n \geq 1$. Together, all remainder terms in (7.8) are less than $\frac{4}{n}(7 - \frac{2}{3}\pi^2) + \frac{4}{n} < \frac{6}{n}$, for all $n \geq 1$. By a change of the second summation variable in (7.8), that summation is made identical to the first and we can represent (7.8) as an inequality:

$$\begin{aligned} a_n &\leq \frac{2}{n^3}\sum_{k=1}^{n}(k-1)^2 \mathbf{E}\big[(C_{k-1}^* - C)^2\big] + \frac{6}{n} \\ &\leq \frac{2}{n^3}\sum_{k=1}^{n}(k-1)^2 a_{k-1} + \frac{6}{n}. \end{aligned}$$

Theorem 7.4 *(Rösler, 1991). Let C_n be the number of comparisons that QUICK SORT makes to sort an input of size n following the random permutation model. The normed number of comparisons $C_n^* = (C_n - 2n\ln n)/n$ converges almost surely to a random variable C satisfying the distributional functional equation*

$$C \stackrel{D}{=} UC + (1-U)\tilde{C} + G(U),$$

where U is UNIFORM(0, 1), *C, $\tilde{C}$, and U are independent, and $\tilde{C} \stackrel{D}{=} C$ and $G(u) = 2u\ln u + 2(1-u)\ln(1-u) + 1$.*

Proof. Let C^* be the random variable on the right-hand side in the functional equation in the theorem. Let $F_n(x)$ and $F(x)$ be respectively the distribution functions of C_n^* and C^*. Let $d_2^2(F_n, F)$ be the second-order Wasserstein distance between these two distributions.

In the preparatory computation of the Wasserstein distance preceding the theorem, we have shown that

$$d_2^2(F_n, F) \leq a_n = \mathbf{E}\big[(C_n^* - C)^2\big] \leq \frac{2}{n^3}\sum_{j=0}^{n-1} j^2 a_j + \frac{6}{n}, \tag{7.9}$$

for all $n \geq 1$. We next show by induction on n that $a_n \to 0$, as $n \to \infty$. The argument works by first treating (7.9) as an equality to guess the general form of a solution to a recurrence of this type; we then take a simple bound on the nth term and use it as an upper bound at the nth step of the induction.

A recurrence of the form

$$b_n = \frac{2}{n^3} \sum_{j=0}^{n-1} j^2 b_j + \frac{6}{n}$$

can be treated by differencing the nth and the $(n+1)$st terms to obtain

$$b_{n+1} = \frac{n^2(n+2)}{(n+1)^3} b_n + \frac{6(2n+1)}{(n+1)^3}.$$

This recurrence is easily linearized by setting $x_n = n^2 b_n/(n+1)$. This transforms the equation into

$$x_{n+1} = x_n + \frac{6(2n+1)}{(n+2)(n+1)}.$$

Iterating this formula (with initial condition $b_1 = 6$) we find

$$b_n = \frac{6(n+1)}{n^2}\left(2H_n + \frac{3}{n+1} - 3\right) < \frac{24\ln(n+1)}{n}.$$

We proceed with our induction to show that $a_n \le \frac{24}{n}\ln(n+1)$. Using this induction hypothesis in inequality (7.9) we have

$$\begin{aligned}
a_{n+1} &\le \frac{2}{(n+1)^3} \sum_{j=1}^{n} 24j\ln(j+1) + \frac{6}{n+1} \\
&\le \frac{48}{(n+1)^3} \int_1^{n+1} x\ln(x+1)\,dx + \frac{6}{n+1} \\
&= \frac{48}{(n+1)^3}\left[\frac{1}{2}n^2\ln(n+2) + n\ln(n+2) - \frac{1}{4}n^2\right] + \frac{6}{n+1} \\
&= \left\{\frac{24\ln(n+2)}{n+1} - \frac{24\ln(n+2)}{n+1}\right\} + \frac{24n^2\ln(n+2)}{(n+1)^3} \\
&\qquad + \frac{48n\ln(n+2)}{(n+1)^3} - \frac{12n^2}{(n+1)^3} + \frac{6}{n+1} \\
&= \frac{24\ln(n+2)}{n+1} - \frac{24\ln(n+2)}{(n+1)^3} - \frac{12n^2}{(n+1)^3} + \frac{6}{n+1}.
\end{aligned}$$

The quantity

$$-\frac{24\ln(n+2)}{(n+1)^3} - \frac{12n^2}{(n+1)^3} + \frac{6}{n+1}$$

is negative for all $n \geq 1$, and we have

$$a_{n+1} \leq \frac{24 \ln(n+2)}{n+1};$$

the induction is complete.

It follows that, as $n \to \infty$,

$$d_2^2(F_n, F) \leq a_n \to 0,$$

or

$$C_n \xrightarrow{\mathcal{D}} C^*;$$

in view of Theorem 7.3, C^* must be equal to C almost surely. ■

7.5 QUICK SELECTION

QUICK SORT can easily be modified to find a collection of order statistics in a given list of numbers without completely sorting the list. We shall call this variant MULTIPLE QUICK SELECT (MQS). Obviously, if we want to use the idea of partitioning to find p order statistics in a list, we need not continue partitioning the two sublists as in QUICK SORT; we need only identify the relevant sublists containing the desired order statistics and proceed recursively with those sublists while truncating the other irrelevant sublist. Occasionally the algorithm will be one sided and its speed must be somewhat faster than plain QUICK SORT. The version of MQS for only a single order statistic ($p = 1$) was introduced in Hoare (1961) and was called FIND. Chambers (1971) extends the algorithm for general $p \leq n$.

When MQS operates to find p order statistics, we can think of the algorithm as a hierarchy of levels (the *level* being the number of order statistics sought). The algorithm begins at level p, and recursively moves down to level 0. At level 0, the algorithm is asked to search for no order statistics. No work is really required for this and the algorithm terminates its recursion. At every stage only segments containing desired order statistics are searched and the rest are ignored.

For each *fixed* set of p order statistics one should anticipate a limit distribution that depends on the particular set chosen, for a suitably normed version of the random number of comparisons (possibly the norming factors are different for different sets of p order statistics). These limit distributions are rather difficult to obtain.

The analysis is far more tractable under an averaging technique introduced in Mahmoud, Modarres, and Smythe (1995) where the sets of p order statistics are themselves random according to a uniform probability model; we shall refer to the selection of a random set of p order statistics as *random selection*. When the set of p order statistics is fixed we shall call it a case of *fixed selection*. Analysis of random selection is only a smoothing average measure to understand the distributions associated with the algorithm. We shall thus average over all sets of p order statistics and we shall speak of the average of the averages of all the fixed cases (we shall call it the *grand average*), which is stamped with the general character of the individual fixed cases. We shall also speak of the average variance (*grand variance*); we shall even speak of the "average distribution." To use a parallel language, we shall sometimes refer to this average distribution as the *grand distribution*. The grand distribution is an averaging of all distributions of the individual fixed cases and is indicative of their classes of probability distributions stamped with the general character of the individual fixed cases.

We characterize the limiting grand distribution of MQS when all the sets of p order statistics are equally likely. Thus we assume double randomness: The keys follow the random permutation model and the sets of p order statistics follow an independent uniform distribution.

In the analysis of MQS, we shall see that the behavior at level p is an inductive convolution involving the distributions associated with the number of comparisons at all the previous levels $1, \ldots, p-1$. The case $p = 1$, Hoare's FIND algorithm, provides the basis for this induction with an infinitely divisible distribution for a suitably normed version of the number of comparisons. Therefore, we shall take up the discussion of FIND and its analysis first.

7.5.1 Hoare's FIND

Hoare's FIND algorithm is designed to find some predesignated order statistic, say the mth. Descriptively, FIND operates as follows. It is a programming function *Find* (Figure 7.7) that takes in the parameters ℓ and u identifying respectively the lower and upper limits of the sublist $A[\ell \mathinner{..} u]$ being considered and returns the actual value of the mth order statistic; the initial external call is, of course, **call** *Find*$(1, n)$. Within FIND, m and the list itself are accessed globally. At the stage when the search has been narrowed down to the sublist extending between positions ℓ and u, FIND first goes through the partitioning process by the call

$$\textbf{call } Partition(\ell, u, k)$$

```
function Find (ℓ, u: integer) : integer;
    local p: integer;
    begin
        if ℓ = u then return(A[ℓ])
        else begin
                call Partition(ℓ, u, p);
                if p = m then return(A[p])
                else if p > m then return(Find(ℓ, p − 1))
                    else return(Find(p + 1, u));
            end;
    end;
```

Figure 7.7. Hoare's FIND algorithm for finding the mth order statistic.

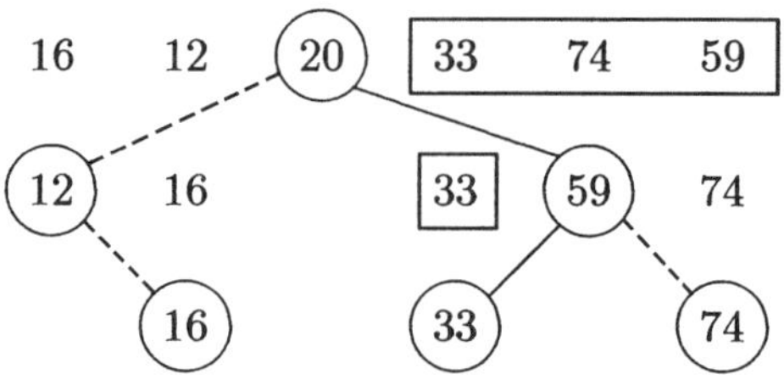

Figure 7.8. Operation of *Find* to find the 4th order statistic.

exactly as in QUICK SORT, moving the chosen pivot to its final position k (review the algorithm of Figure 7.2). If $k = m$, we are done; the element at position k is our mth order statistic. If $k > m$, the mth order statistic must be in the left sublist; we apply FIND recursively on the left sublist; the situation is handled by the call **call** *Find*$(\ell, k - 1)$; otherwise, the mth order statistic must be in the right sublist and it is now the $(m - k)$th smallest among the keys of the right sublist. This situation is handled by the call **call** *Find*$(k + 1, u)$. Figure 7.8 illustrates the operation of FIND on the running example of this chapter to find the 4th order statistic. In Figure 7.8 the boxed subarrays are subjected to the recursive operation; the unboxed subarrays are ignored. The effect on the associated tree is to prune every subtree corresponding to a truncated side. The edges pruned from the Quick Sort tree to obtain the tree associated with FIND are dashed in Figure 7.8; only the solid edges remain in FIND's tree.

As discussed in the previous paragraphs, for each fixed order statistic the limit distribution for the number of comparisons (suitably normalized) will depend on the particular order statistic chosen, and these may be difficult to obtain. As an average measure, we analyze FIND's grand distribution, when all order statistics are equally likely.

Let C_n be the number of comparisons that FIND makes between list elements when applied to a random list of size n (assimilated as usual by a random permutation of $\{1, \dots, n\}$) to find a randomly chosen order statistic.

The method of Rösler that we used for QUICK SORT can develop a functional relation for the limit law of the number of comparisons when suitably normed. For QUICK SORT we only implicitly characterized the limiting distribution by a distributional functional equation. Fortunately, for FIND where the algorithm is completely one sided, the distributional functional equation is solvable and we will be able to characterize the limiting distribution completely by its characteristic function.

The order statistic that will be searched for is first generated independently of the input from the discrete UNIFORM$[1 \mathinner{..} n]$ distribution. Let us call such a randomly generated value M_n. FIND then operates to locate the M_nth order statistic in some input list of size n and the ranks of the input are assimilated by a random permutation of $\{1, \dots, n\}$. Let P_n be the position to which the pivot moves. As the starting list is assimilated by a random permutation, $P_n \overset{\mathcal{D}}{=}$ UNIFORM$[1 \mathinner{..} n]$. The first partitioning stage makes $n - 1$ comparisons and we have the conditional behavior:

$$C_n = \begin{cases} n-1+C_{P_n-1}, & \text{if } M_n < P_n; \\ n-1, & \text{if } P_n = M_n; \\ n-1+C_{n-P_n}, & \text{if } M_n > P_n. \end{cases}$$

We thus have the distributional equation

$$C_n \overset{D}{=} n-1+C_{P_n-1}\mathbf{1}_{\{M_n<P_n\}}+\tilde{C}_{n-P_n}\mathbf{1}_{\{M_n>P_n\}}, \tag{7.10}$$

where for every j, $\tilde{C}_j \overset{D}{=} C_j$, and the two families $\{C_j\}$, $\{\tilde{C}_j\}$ are independent. Hence the expectation of C_n is

$$\mathbf{E}[C_n] \overset{D}{=} n-1+\mathbf{E}\big[C_{P_n-1}\mathbf{1}_{\{M_n<P_n\}}\big]+\mathbf{E}\big[\tilde{C}_{n-P_n}\mathbf{1}_{\{M_n>P_n\}}\big].$$

The last two additive terms are symmetric. So, by conditioning on P_n we find

$$\begin{aligned}\mathbf{E}[C_n] &= n-1+2\mathbf{E}\big[C_{P_n-1}\mathbf{1}_{\{M_n<P_n\}}\big] \\ &= n-1+2\sum_{k=1}^{n}\mathbf{E}\big[C_{P_n-1}\mathbf{1}_{\{M_n<P_n\}} \mid P_n=k\big]\,\mathbf{Prob}\{P_n=k\} \\ &= n-1+\frac{2}{n}\sum_{k=1}^{n}\mathbf{E}\big[C_{k-1}\mathbf{1}_{\{M_n<k\}}\big].\end{aligned}$$

Note here that even though C_{P_n-1} and $\mathbf{1}_{\{M_n<P_n\}}$ are dependent (via P_n), these random variables are conditionally independent (given $P_n=k$). So,

$$\mathbf{E}\big[C_{k-1}\mathbf{1}_{\{M_n<k\}}\big] = \mathbf{E}[C_{k-1}]\,\mathbf{E}\big[\mathbf{1}_{\{M_n<k\}}\big].$$

As the order statistic sought is randomly selected, $M_n \overset{D}{=} \text{UNIFORM}[1\,..\,n]$; the expectation of the indicator is

$$\begin{aligned}\mathbf{E}\big[\mathbf{1}_{\{M_n<k\}}\big] &= 0\times\mathbf{Prob}\{\mathbf{1}_{\{M_n<k\}}=0\}+1\times\mathbf{Prob}\{\mathbf{1}_{\{M_n<k\}}=1\} \\ &= \frac{k-1}{n}.\end{aligned}$$

We thus have

$$\mathbf{E}[C_n] = n-1+\frac{2}{n^2}\sum_{k=1}^{n}(k-1)\mathbf{E}[C_{k-1}].$$

We solve this recurrence by differencing. Let $a_n = \mathbf{E}[C_n]$. In terms of a_n the recurrence is

$$n^2 a_n = (n-1)n^2 + 2\sum_{k=1}^{n}(k-1)a_{k-1}.$$

A version of this recurrence for $n-1$ is

$$(n-1)^2 a_{n-1} = (n-2)(n-1)^2 + 2\sum_{k=1}^{n-1}(k-1)a_{k-1}.$$

Subtracting, we obtain the form

$$a_n = \frac{n^2-1}{n^2}a_{n-1} + \frac{(3n-2)(n-1)}{n^2},$$

to be solved with the initial condition $a_1 = \mathbf{E}[C_1] = 0$. This recurrence is easily linearized—set $x_n = na_n/(n+1)$, and write the recurrence in the form

$$\begin{aligned} x_n &= x_{n-1} + \frac{(3n-2)(n-1)}{n(n+1)} \\ &= x_{n-1} + 3 + \frac{2}{n} - \frac{10}{n+1}. \end{aligned}$$

Iterating the recurrence we obtain the following.

Theorem 7.5 *(Knuth, 1973). The average number of comparisons made by Hoare's FIND algorithm to find a randomly chosen order statistic in a random input of size n is*

$$\mathbf{E}[C_n] = 3n - 8H_n + 13 - \frac{8H_n}{n} \sim 3n.$$

The result presented in Knuth (1973) is for fixed selection where the jth (fixed) order statistic is sought. Knuth's analysis assumes Sedgewick's partitioning algorithm (Exercise 7.2), which takes $n+1$ data comparisons on a list of size n. The analysis can be mimicked for PARTITION, the partitioning algorithm used here, which makes $n-1$ data comparisons to partition a list of size n. The average number of comparisons in the fixed selection of the jth order statistic depends on j with asymptotic values ranging from $2n$ for extreme order statistics (the very small and the very large) to $(2 + 2\ln 2)n \approx 3.39n$ for the median. By assuming the choice of the order statistic is made randomly under PARTITION, then averaging Knuth's results we obtain the grand average presented in Theorem 7.5.

We focused here on a direct approach to the grand average because as mentioned before we shall develop a limit grand distribution for the case of random selection. The asymptotic $3n$ grand average of Theorem 7.5 will then be our centering factor. We also need a scale factor to norm the random variable. This is the asymptotic square root of the variance of C_n, which we shall take up next.

We can first determine the second moment of C_n by a method similar to the one we used for the grand average, though the computations become substantially more extensive. We start by squaring the distributional equation (7.10). The squares of the indicators $\mathbf{1}_{\{M_n<P_n\}}$ and $\mathbf{1}_{\{M_n>P_n\}}$ are of course the indicators themselves and their cross product is 0 as they indicate mutually exclusive events. We obtain the distributional equation

$$C_n^2 \stackrel{D}{=} (n-1)^2 + C_{P_n-1}^2 \mathbf{1}_{\{M_n<P_n\}} + \tilde{C}_{n-P_n}^2 \mathbf{1}_{\{M_n>P_n\}} \\ + 2(n-1)C_{P_n-1}\mathbf{1}_{\{M_n<P_n\}} + 2(n-1)\tilde{C}_{n-P_n}\mathbf{1}_{\{M_n>P_n\}}.$$

Once again, utilizing the symmetries $C_{P_n-1} \stackrel{D}{=} \tilde{C}_{n-P_n}$ and $\mathbf{1}_{\{M_n<P_n\}} \stackrel{D}{=} \mathbf{1}_{\{M_n>P_n\}}$, we have a recurrence for the second moment

$$\mathbf{E}[C_n^2] = (n-1)^2 + 2\mathbf{E}\big[C_{P_n-1}^2 \mathbf{1}_{\{M_n<P_n\}}\big] + 4(n-1)\mathbf{E}\big[C_{P_n-1}\mathbf{1}_{\{M_n<P_n\}}\big].$$

Proceeding as we did for the grand average,

$$\begin{aligned}\mathbf{E}[C_n^2] &= (n-1)^2 + 2\sum_{k=1}^{n} \mathbf{E}\big[C_{P_n-1}^2 \mathbf{1}_{\{M_n<P_n\}} \mid P_n = k\big]\,\mathbf{Prob}\{P_n = k\} \\ &\quad + 4(n-1)\sum_{k=1}^{n} \mathbf{E}\big[C_{P_n-1}\mathbf{1}_{\{M_n<P_n\}} \mid P_n = k\big]\,\mathbf{Prob}\{P_n = k\} \\ &= (n-1)^2 + \frac{2}{n}\sum_{k=1}^{n} \mathbf{E}\big[C_{k-1}^2 \mathbf{1}_{\{M_n<k\}}\big] + \frac{4(n-1)}{n}\sum_{k=1}^{n} \mathbf{E}\big[C_{k-1}\mathbf{1}_{\{M_n<k\}}\big] \\ &= (n-1)^2 + \frac{2}{n}\sum_{k=1}^{n} \mathbf{E}\big[C_{k-1}^2\big]\frac{k-1}{n} + \frac{4(n-1)}{n}\sum_{k=1}^{n} \mathbf{E}[C_{k-1}]\frac{k-1}{n}.\end{aligned}$$

Introducing the notation $b_n = \mathbf{E}[C_n^2]$, and again using the notation a_n for $\mathbf{E}[C_n]$, we have the recurrence

$$b_n = (n-1)^2 + \frac{2}{n^2}\sum_{k=1}^{n}(k-1)b_{k-1} + \frac{4(n-1)}{n^2}\sum_{k=1}^{n}(k-1)a_{k-1}.$$

Differencing two versions of this recurrence with indexes n and $n-1$, we obtain

$$b_n = \frac{n^2-1}{n^2}b_{n-1} + \frac{4(n-1)^3 + \nabla g(n)}{n^2},$$

where

$$g(n) = 4(n-1)\sum_{k=1}^{n}(k-1)a_{k-1},$$

and the averages a_{k-1} are given in Theorem 7.5. This recurrence is to be solved with the initial condition $b_1 = 0$. The recurrence is linearized by setting $y_n = nb_n/(n+1)$. In its linear iteratable form the recurrence is

$$y_n = y_{n-1} + \frac{4(n-1)^3 + \nabla g(n)}{n(n+1)}.$$

Unwinding the recurrence by direct iteration we obtain the second moment. Subtracting off the square of the grand average (Theorem 7.5) we obtain the following.

Theorem 7.6 *(Mahmoud, Modarres, and Smythe, 1995). The grand variance of the number of comparisons Hoare's FIND algorithm makes to find a randomly chosen order statistic in a random input of size n is*

$$\begin{aligned}\mathbf{Var}[C_n] &= n^2 - 10n - 16H_n^2 + 108H_n - 47 - 48H_n^{(2)} - 80\frac{H_n^2}{n} \\ &\qquad + 204\frac{H_n}{n} - 48\frac{H_n^{(2)}}{n} - 64\frac{H_n^2}{n^2} \\ &\sim n^2.\end{aligned}$$

Kirschenhofer and Prodinger (1998) compute the variance of the number of comparisons in fixed selection. For the jth (fixed) order statistic they find a variance of the form $v_j n^2$, where v_j are coefficients that depend on j. The grand variance of Theorem 7.6 for random selection is obtained by subtracting the square of the grand average from an "average second moment," which is the average of the second moments over all the fixed cases. But again, as in the grand average, we considered a direct approach to the grand variance of random selection (bypassing calculation of the fixed cases) because we shall consider the limiting grand distribution for the case of random selection; the square root of the grand variance is the appropriate scaling factor.

Having produced the asymptotic grand average and grand variance, we can begin to work toward the asymptotic distribution: The factor $3n$ will be used for centering (Theorem 7.5) and the factor n will be used for scaling (Theorem 7.6); introduce

$$Y_n = \frac{C_n - 3n}{n}.$$

We normalize (7.10) by writing it in the form

$$\begin{aligned}\frac{C_n - 3n}{n} &\stackrel{\mathcal{D}}{=} \frac{C_{P_n - 1} - 3(P_n - 1)}{P_n - 1} \times \frac{P_n - 1}{n}\mathbf{1}_{\{M_n < P_n\}} \\ &\qquad + \frac{\tilde{C}_{n-P_n} - 3(n - P_n)}{n - P_n} \times \frac{n - P_n}{n}\mathbf{1}_{\{M_n > P_n\}}\end{aligned}$$

$$+\frac{3(P_n - 1)}{n}\mathbf{1}_{\{M_n < P_n\}} + \frac{3(n - P_n)}{n}\mathbf{1}_{\{M_n > P_n\}}$$
$$+\frac{n-1}{n} - \frac{3n}{n}.$$

In terms of the normalized random variables, this is

$$Y_n \stackrel{D}{=} \mathbf{1}_{\{M_n < P_n\}}\frac{P_n - 1}{n}\big(Y_{P_n - 1} + 3\big) + \mathbf{1}_{\{M_n > P_n\}}\frac{n - P_n}{n}\big(\tilde{Y}_{n-P_n} + 3\big) - 2 - \frac{1}{n}, \tag{7.11}$$

where for each i, $\tilde{Y}_i \stackrel{D}{=} Y_i$ and the families $\{Y_i\}$, $\{\tilde{Y}_i\}$, $\{P_i\}$, and $\{M_i\}$ are mutually independent. This representation suggests a limiting functional equation as follows. We first note that both P_n/n and M_n/n asymptotically behave like two independent standard uniform random variables. Let U and W be independent UNIFORM(0,1) random variables. Then, as in Exercise 7.15,

$$\frac{P_n}{n} \xrightarrow{\mathcal{D}} U,$$
$$\frac{M_n}{n} \xrightarrow{\mathcal{D}} W,$$
$$\mathbf{1}_{\{M_n < P_n\}}\frac{P_n}{n} \xrightarrow{\mathcal{D}} \mathbf{1}_{\{W < U\}}U.$$

Now, if Y_n converges to a limit Y, then so will $Y_{P_n - 1}$ because $P_n \xrightarrow{\text{a.s.}} \infty$, and it is plausible to hypothesize that the combination $\mathbf{1}_{\{M_n < P_n\}}\frac{P_n - 1}{n}\big(Y_{P_n - 1} + 3\big)$ converges in distribution to $\mathbf{1}_{\{W<U\}}U(Y + 3)$. A symmetric argument applies to the second additive term in (7.11), and likewise it is plausible to hypothesize that $\mathbf{1}_{\{M_n > P_n\}}\frac{n - P_n}{n}\big(\tilde{Y}_{n-P_n} + 3\big) \xrightarrow{\mathcal{D}} \mathbf{1}_{\{W>U\}}(1 - U)(\tilde{Y} + 3)$, where $\tilde{Y} \stackrel{D}{=} Y$ and is independent of it. As in the analysis of QUICK SORT, the additive terms of (7.11) are dependent (via P_n). However, this dependence is weak enough to allow the following limit form to hold.

Theorem 7.7 *(Mahmoud, Modarres, and Smythe, 1995). There exists a limiting random variable Y such that*

$$Y_n \xrightarrow{\mathcal{D}} Y.$$

Let U and W be two independent UNIFORM*(0,1) random variables. The limit Y satisfies the distributional functional equation*

$$Y + 2 \stackrel{D}{=} \mathbf{1}_{\{W<U\}}U(Y + 3) + \mathbf{1}_{\{W>U\}}(1 - U)(\tilde{Y} + 3).$$

Proof. The proof of Theorem 7.7 goes along the main lines of that of Theorem 7.4. Having guessed a functional equation for the limit, we proceed by showing that the second-order Wasserstein distance between Y_n and Y converges to 0; and consequently $Y_n \xrightarrow{\mathcal{D}} Y$. To be able to compute the second-order Wasserstein distance we first seek a unified representation for Y_n and Y. As the pivot lands at a random position,

$$P_n \stackrel{D}{=} \lceil nU \rceil.$$

So, Y_n can be represented by the functional equation (see 7.11)

$$\begin{aligned} Y_n + 2 \stackrel{D}{=} \mathbf{1}_{\{W<U\}} &\frac{\lceil nU \rceil - 1}{n} \big(Y_{\lceil nU \rceil - 1} + 3\big) \\ &+ \mathbf{1}_{\{W>U\}} \frac{n - \lceil nU \rceil}{n} \big(\tilde{Y}_{n - \lceil nU \rceil} + 3\big) + O_p\Big(\frac{1}{n}\Big). \end{aligned}$$

The O_p term is $O(\frac{1}{n})$ in probability and contains terms like $-1/n$ and the correction terms changing the representation from P_n and M_n to U and W. An instance of this correction term is $(\mathbf{1}_{\{M_n<P_n\}} - \mathbf{1}_{\{W<U\}})(\lceil nU \rceil - 1)(Y_{P_n - 1} + 3)/n$. If Y_n has a limit, such a term converges to 0 in probability because the difference between the two indicators diminishes quickly down to 0. The last representation of Y_n expedites the computation of the second order Wasserstein distance between Y_n and Y. The technical development is similar to that in the proof of Theorem 7.4 (the details are left for Exercise 7.16). ■

An equivalent but simpler form for the functional equation in Theorem 7.7 is convenient and will generalize to MULTIPLE QUICK SELECT.

Theorem 7.8 *(Mahmoud, Modarres, and Smythe, 1995). The random variable $Y^* = Y + 2$ satisfies the functional equation*

$$Y^* \stackrel{D}{=} X(Y^* + 1),$$

where X and Y are independent with X having the density

$$f(x) = \begin{cases} 2x, & 0 < x < 1; \\ 0, & \textit{elsewhere.} \end{cases}$$

Proof. Let $Y^* = Y + 2$, and let $\phi_{Y^*}(t)$ be its characteristic function. According to Theorem 7.7, Y^* satisfies

$$Y^* \stackrel{D}{=} \mathbf{1}_{\{W<U\}} U(Y^* + 1) + \mathbf{1}_{\{W>U\}} (1 - U)(\tilde{Y}^* + 1),$$

where $\tilde{Y}^* \stackrel{D}{=} Y^*$, and is independent of it. We can condition on U and W to obtain a representation for the characteristic function. These two random variables are independent UNIFORM(0,1), and their joint distribution is uniform on the unit square. That is, their joint density is 1 in the unit square and 0 elsewhere. We have

$$\phi_{Y^*(t)} = \int_0^1 \int_0^1 \mathbf{E}\big[\exp\big\{it\big(\mathbf{1}_{\{W<U\}}U(Y^*+1) \\ + \mathbf{1}_{\{W>U\}}(1-U)(\tilde{Y}^*+1)\big\} \,|\, U=u, W=w\big]\,dw du.$$

By splitting the inner integral into two integrals, one from $w = 0$ to u and the other from $w = u$ to 1, we simplify the integration because in each of the two integrals only one of the indicators $\mathbf{1}_{\{u<w\}}$ and $\mathbf{1}_{\{u>w\}}$ is 1 and the other is 0. We obtain

$$\phi_{Y^*(t)} = \int_0^1 \int_0^u \mathbf{E}\big[e^{itu(Y^*+1)}\big]\,dw du + \int_0^1 \int_u^1 \mathbf{E}\big[e^{it(1-u)(Y^*+1)}\big]\,dw du \\ - \int_0^1 u\mathbf{E}\big[e^{itu(Y^*+1)}\big]\,du + \int_0^1 (1-u)\mathbf{E}\big[e^{it(1-u)(Y^*+1)}\big]\,du.$$

The last two integrals are identical; the change $v = 1 - u$ of the integration variable in the second integral renders it identical to the first. Hence

$$\phi_{Y^*(t)} = 2\int_0^1 ue^{itu}\phi_{Y^*}(tu)\,du.$$

On the other hand, by conditioning the random variable $X(Y^*+1)$ on $X = x$ we see that the characteristic function $\phi_{X(Y^*+1)}(t)$ of $X(Y^*+1)$ is

$$\phi_{X(Y^*+1)}(t) = \int_0^1 \mathbf{E}\big[e^{itX(Y^*+1)} \,|\, X = x\big] f(x)\,dx \\ = \int_0^1 2xe^{itx}\phi_{Y^*}(tx)\,dx.$$

Thus Y^* and $X(Y^*+1)$ have the same characteristic function, i.e., the same distribution. ■

The limit random variable Y, as we shall see shortly, belongs to the class of infinitely divisible random variables. A random variable S is called infinitely divisible if for every positive integer k, there exist independent identically distributed (i.i.d.) random variables $S_{1,k}, \ldots, S_{k,k}$ such that

$$Y \stackrel{\mathcal{D}}{=} S_{1,k} \oplus \cdots \oplus S_{k,k}. \tag{7.12}$$

For example, the standard normal variate $\mathcal{N}(0, 1)$ is infinitely divisible because linear combinations of independent normal variates are normal—the standard normal variate has the representations

$$
\begin{aligned}
\mathcal{N}(0,1) &= \mathcal{N}\Big(0,\frac{1}{2}\Big) \oplus \mathcal{N}\Big(0,\frac{1}{2}\Big)\\
&= \mathcal{N}\Big(0,\frac{1}{3}\Big) \oplus \mathcal{N}\Big(0,\frac{1}{3}\Big) \oplus \mathcal{N}\Big(0,\frac{1}{3}\Big)\\
&= \mathcal{N}\Big(0,\frac{1}{4}\Big) \oplus \mathcal{N}\Big(0,\frac{1}{4}\Big) \oplus \mathcal{N}\Big(0,\frac{1}{4}\Big) \oplus \mathcal{N}\Big(0,\frac{1}{4}\Big)\\
&\ \ \vdots
\end{aligned}
$$

When a random variable S has an infinitely divisible distribution, $\phi_S(t)$, the characteristic function of S is said to have a kth root for all integers $k \ge 1$, meaning that the kth root of $\phi_S(t)$ is a characteristic function, too, of some random variable. This is so because for any positive integer k, the random variable S has a representation as a convolution of k independent and identically distributed random variables as in (7.12)—if these random variables have the common characteristic function $\psi(t)$, then $\phi_S(t) = \psi^k(t)$. So, the kth root $\phi_S^{1/k}(t)$ is $\psi(t)$, and has a meaning as a characteristic function of some random variable. This property, true for infinitely divisible distributions, is not true for other characteristic functions.

We specify the limit Y in terms of its characteristic function.

Theorem 7.9 *(Mahmoud, Modarres, and Smythe 1995). FIND's limiting normalized number of comparisons Y is infinitely divisible with characteristic function*

$$
\phi_Y(t) = \exp\Big(2\int_0^1 \frac{e^{itx} - 1 - itx}{x}\,dx\Big).
$$

Proof. Consider the functional equation of Theorem 7.8 for $Y^* = Y + 2$. By conditioning on X, we obtain an equality for the characteristic function ϕ_{Y^*} of Y^* in the integral form

$$
\begin{aligned}
\phi_{Y^*}(t) &= \int_0^1 \mathbf{E}\big[e^{itx(Y^*+1)} \mid X = x\big] f(x)\,dx\\
&= \int_0^1 2xe^{itx}\phi_{Y^*}(tx)\,dx.
\end{aligned}
$$

The change of integration variable to $u = tx$ yields

$$
\phi_{Y^*}(t) = \int_0^t \frac{2ue^{iu}\phi_{Y^*}(u)}{t^2}\,du.
$$

The random variable $Y^* = Y + 2$ has a finite second moment. Hence, its characteristic function is at least twice differentiable. Taking the first derivative with respect to t of the last equality gives the first-order differential equation:

$$t^2 \frac{d\phi_{Y^*}(t)}{dt} = 2t\big(e^{it} - 1\big)\phi_{Y^*}(t),$$

with the solution

$$\phi_{Y^*}(t) = \exp\Big(2\int_0^t \frac{e^{iu} - 1}{u}\, du\Big).$$

Reverting to $Y = Y^* - 2$, we find the characteristic function

$$\begin{aligned}\phi_Y(t) &= e^{-2it}\phi_{Y^*}(t) \\ &= \Big(e^{-2i\int_0^t du}\Big)\exp\Big(2\int_0^t \frac{e^{iu} - 1}{u}\, du\Big).\end{aligned}$$

The statement of the theorem is obtained after setting the integration variable $u = tx$. The characteristic function $\phi_Y(t)$ has Kolmogorov's canonical form of an infinitely divisible distribution with finite variance. ∎

As discussed when infinitely divisible distributions were introduced (see the text following (7.12)), $\phi_Y(t)$ has a kth root for any positive integer k. Exercise 7.10 elicits the special meaning of the square root.

7.5.2 MULTIPLE QUICK SELECT

The techniques developed for the analysis of FIND are extendible to MULTIPLE QUICK SELECT and we are now ready to tackle the more general problem of the selection of a number of order statistics and its analysis. The multiple selection of order statistics has numerous applications in statistical inference where it is often desired to construct statistics based on a few order statistics (typically five or less) as discussed in Section 1.2.

Chambers (1971) suggested the use of a variant of Hoare's FIND algorithm to locate multiple order statistics. We shall refer to the adaptation of QUICK SORT to find several order statistics as MULTIPLE QUICK SELECT (MQS).

Suppose $1 \le j_1 < \cdots < j_p \le n$ is a collection of indexes specifying p order statistics required for an application. (For convenience, we introduce $j_0 = 0$ and $j_{p+1} = n + 1$.) To find $X_{(j_1)}, \ldots, X_{(j_p)}$, we need not completely sort the given sample. MQS first goes through a partitioning stage just like that of QUICK SORT. If k, the pivot's final position, happens to coincide with one of the indexes, say j_i, MQS announces the pivot (now moved to $A[k]$) as the j_ith order statistic and continues to operate recursively on $A[1 \mathinner{..} k-1]$ to find $X_{(1)}, \ldots, X_{(j_i - 1)}$, and on $A[k+1 \mathinner{..} n]$ to find $X_{(j_i+1)}, \ldots, X_{(p)}$. If instead $j_i < k < j_{i+1}$, for some $0 \le i \le p$, MQS operates recursively on $A[1 \mathinner{..} k-1]$ to find $X_{(1)}, \ldots, X_{(j_i)}$, and on $A[k+1 \mathinner{..} n]$ to find $X_{(j_i+1)}, \ldots, X_{(p)}$. We think of the algorithm as a hierarchy of levels (at level i the algorithm looks for i order statistics). The algorithm begins at level p, and recursively moves down to level 0. At level 0 no order statistics are sought and the

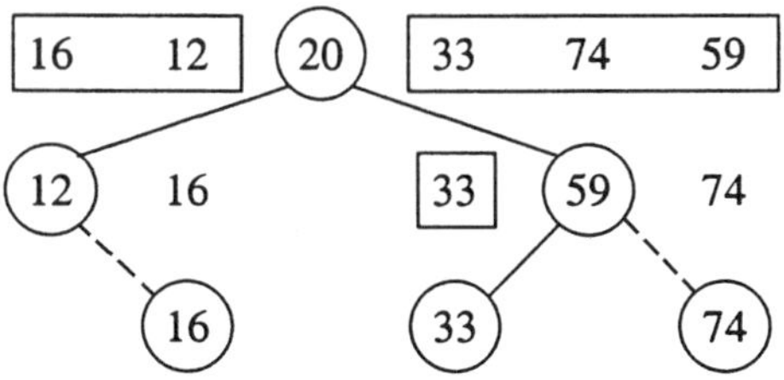

Figure 7.9. Operation of MQS to find the 1st, 4th, and 5th order statistics.

algorithm terminates its recursion. Only segments containing desired order statistics are searched and the rest are truncated. Figure 7.9 shows the operation of MQS on the running example of this chapter (with $n = 6$), to find the first, fourth, and fifth order statistics and the tree associated with it. Only segments containing order statistics are boxed (they are recursively searched). The associated tree, to be called the *MQS tree*, is derived by pruning the Quick Sort tree, removing from it every node labeled with a key not used as a pivot in the search. In Figure 7.9 the edges of the MQS tree are solid, the rest of the Quick Sort tree edges that are not in the MQS tree are dashed. When MQS is done finding the order statistics j_r and j_{r+1}, all data with ranks between these two numbers are moved to positions lying between the final positions of these two order statistics. We can thus find groups of data with ranks unspecified but known to be between two bounding order statistics. This is useful for many applications. For example, the α-trimmed mean of $X_1, \dots, X_n$, is a statistic that throws out the upper and lower α proportion of the (sorted) data ($0 < \alpha < 1$), deeming them too extremal to be representative of the nature of the data. These outliers may exist because of errors or inaccuracies in measurements. The α-trimmed mean is given by

$$\frac{1}{n - 2\lfloor \alpha n \rfloor} \sum_{i=\lfloor \alpha n \rfloor + 1}^{n - \lfloor \alpha n \rfloor} X_{(i)}.$$

We may use MQS to identify the two bounding order statistics $X_{\lfloor \alpha n \rfloor}$ and $X_{n - \lfloor \alpha n \rfloor + 1}$. In the process, MQS will place all intermediate order statistics in the array between positions $(\lfloor \alpha n \rfloor + 1), \dots, (n - \lfloor \alpha n \rfloor)$, not necessarily in sorted order. For the computation of the α-trimmed mean we need not sort these intermediate order statistics; we are only seeking their average. The only information required for this is their sum and their number.

The formal algorithm is given in Figure 7.10. The procedure MQS globally accesses the data array $A[1 \mathinner{..} n]$ and another array $OS[1..p]$ storing the p sorted ranks of the order statistics sought. The procedure takes in ℓ and u, the lower and upper delimiters of a stretch of the data array to be searched, and b and t, the top and bottom delimiters of a stretch of OS; at a general stage with these parameters, the procedure looks for the order statistics with ranks stored in the stretch $OS[b \mathinner{..} t]$ within $A[\ell \mathinner{..} u]$. The outside call is

$$\textbf{call } MQS(1, n, 1, p);$$

```
procedure MQS (ℓ, u, b, t: integer);
    local k, q: integer;
    begin
        if b ≤ t then
            begin
                call Partition(ℓ, u, k);
                q ← call Search(k, b, t);
                if q = b − 1 then call MQS(k + 1, u, b, t)
                else if OS[q] = k then
                        begin
                            print(k, A[k]);
                            call MQS(ℓ, k − 1, b, q − 1);
                            call MQS(k + 1, u, q + 1, t);
                        end;
                    else
                        begin
                            call MQS(ℓ, k − 1, b, q);
                            call MQS(k, u, q + 1, t);
                        end;
            end;
    end;
```

Figure 7.10. MULTIPLE QUICK SELECT.

The algorithm invokes PARTITION first. Upon receipt of k, the final position of the pivot, MQS searches for k, using the function $Search$, among $OS[b\,..\,t]$, and returns into q the largest rank in $OS[b\,..\,t]$ that is less than or equal to k. According as whether k is among the ranks $OS[b\,..\,t]$ or not, the algorithm either announces finding an order statistic or not, then moves down to lower levels, searching for fewer order statistics (stored in the appropriate segments of OS) within the appropriate segments of A. Two special cases may arise when q falls at the beginning or the end of $OS[b\,..\,t]$. In the special case that k is less than all the ranks stored in $OS[b\,..\,t]$, $Search$ returns $b - 1$, an index value outside the stretch $OS[b\,..\,t]$. The next stage in the algorithm becomes one sided and looks for $OS[b\,..\,t]$, all greater than k, among $A[k+1\,..\,u]$. This special case is handled by a conditional check. In the special case when k is not found in $OS[b\,..\,t]$ and $q = t$, when all the ranks in $OS[b\,..\,t]$ are less than k, the next stage is again one-sided and is covered by the call **call** $MQS(\ell, k - 1, b, q)$.

When $\ell > u$ the stretch $A[\ell\,..\,u]$ does not exist. However, this condition need not be checked, because $b \le t$ implies that there still are order statistics to be found and a stretch of A of length $u - \ell + 1 \ge t - b + 1$ must exist; the condition $\ell \le u$ is guaranteed when $b \le t$.

For each fixed set of p order statistics one should anticipate a limit distribution, depending on the particular set chosen, for a suitably centered and scaled version of the random number of comparisons (possibly the norming factors are different for different sets of p order statistics). As an average measure, we analyze the "average distribution," that is, we characterize the limiting distribution of MQS when all the sets of p order statistics are equally likely.

In MQS the occasional truncation of one side of the recursion is bound to improve the focused searching process over complete sorting by QUICK SORT. Indeed, the running average time is asymptotic to only $c_{j_1,\dots,j_p}n$, for fixed p, as $n \to \infty$. Of course, the constant $c_{j_1,\dots,j_p}$ depends on the particular set $\{j_1, \dots, j_p\}$ of order statistics being sought. Prodinger (1995) specifies these constants explicitly. Averaged over all possible sets of p order statistics (assuming all $\binom{n}{p}$ sets of p order statistics are equally likely) the grand average (average of averages) is seen to be asymptotically linear in n, as we shall prove in the next theorem. We shall take a direct shortcut to the grand average of random selection (bypassing fixed case calculation) because we shall develop the grand distribution.

The technique for finding the asymptotic grand average is based on the relationship of the MQS tree to the Quick Sort tree. Let $C_n^{(p)}$ be the number of comparisons made by MQS on a random input of size n to find a random set of p order statistics. If we sort the input by QUICK SORT, we would have an underlying Quick Sort tree. But if we only search for a selected collection of p order statistics, several nodes and branches are pruned. If a key is chosen as a pivot at some stage in MQS, it is compared with all the keys in the subtree that it fathers. For example, the first pivot (the label of the root of both the Quick Sort tree and the MQS tree) is compared against $n-1$ other keys. In MQS only subarrays containing order statistics are searched recursively. In the Quick Sort tree only nodes belonging to the MQS tree are the roots of subtrees containing desired order statistics and are compared against the keys in the subtrees they father; the rest are pruned. Let $S_j^{(n)}$ be the number of descendants of the node whose rank is j. Let

$$I_j^{(n)} = \begin{cases} 1, & \text{if key ranked } j \text{ is an ancestor of at least} \\ & \text{one of the } p \text{ order statistics;} \\ 0, & \text{otherwise.} \end{cases}$$

(In what follows we suppress the superscript in both $S_j^{(n)}$ and $I_j^{(n)}$). Then

$$C_n^{(p)} = \sum_{j=1}^{n} S_j I_j,$$

whose expectation is

$$\begin{aligned} \mathbf{E}[C_n^{(p)}] &= \sum_{j=1}^{n} \mathbf{E}[S_j I_j] \\ &= \sum_{j=1}^{n} \sum_{k=0}^{n-1} \mathbf{E}[S_j I_j \mid S_j = k]\, \mathbf{Prob}\{S_j = k\}. \end{aligned}$$

The value of the conditional expectation involved can be argued as follows:

$$\mathbf{E}\big[I_j S_j \mid S_j = k\big] = 0 \times \mathbf{Prob}\{I_j = 0 \mid S_j = k\} + k \times \mathbf{Prob}\{I_j = 1 \mid S_j = k\}.$$

As the p order statistics are chosen at random, the event that the key ranked j is not an ancestor of any of the required p order statistics, given there are k descendants of the key ranked j, amounts to choosing the p order statistics from among all the other nodes of the tree that are not in the subtree rooted at the key ranked j. Given that condition, there are $n - k - 1$ nodes outside the subtree rooted at k and any p of them re equally likely to be chosen for the random selection process. So, we have the following representation for $\mathbf{E}[C_n^{(p)}]$:

$$\begin{aligned}\mathbf{E}\big[C_n^{(p)}\big] &= \sum_{j=1}^{n}\sum_{k=0}^{n-1} k\left[1 - \frac{\binom{n-k-1}{p}}{\binom{n}{p}}\right]\mathbf{Prob}\{S_j = k\}\\ &= \sum_{j=1}^{n}\sum_{k=0}^{n-1} k\,\mathbf{Prob}\{S_j = k\} - \sum_{j=1}^{n}\sum_{k=0}^{n-1}\frac{\binom{n-k-1}{p}}{\binom{n}{p}}\,k\,\mathbf{Prob}\{S_j = k\}.\end{aligned}$$

The first double sum is reducible to a familiar term:

$$\sum_{j=1}^{n}\sum_{k=0}^{n-1} k\,\mathbf{Prob}\{S_j = k\} = \sum_{j=1}^{n}\mathbf{E}[S_j] = \mathbf{E}\big[C_n^{(n)}\big];$$

to see this, consider the sum of the number of descendants of a node taken over every node in the Quick Sort tree. This is the number of comparisons QUICK SORT would take if we wanted to sort the input instead. (QUICK SORT can be viewed as the special case $p = n$ of MQS when order statistics of *all* ranks are to be found; in this special case all the indicators I_j are 1.) The double sum is also an equivalent form for the internal path length of the QUICK SORT tree (a node at depth d is counted in this sum d times, once for every ancestor); the average number of ancestors is the same as the average number of descendants (Exercise 1.6.6). Hence, this double sum is $\mathbf{E}[C_n^{(n)}]$, the average number of comparisons of QUICK SORT on a random input of size n and is given by Theorem 7.1. Therefore

$$\mathbf{E}\big[C_n^{(p)}\big] = \mathbf{E}\big[C_n^{(n)}\big] - \sum_{j=1}^{n}\sum_{k=0}^{n-1}\frac{\binom{n-k-1}{p}}{\binom{n}{p}}\,k\,\mathbf{Prob}\{S_j = k\}. \tag{7.13}$$

It only remains to asymptotically analyze this exact expression. The asymptotic analysis is facilitated by Pochhammer's symbol for the rising factorial:

$$\langle x \rangle_p = x(x+1)\dots(x+p-1) = \sum_{r=1}^{p} \begin{bmatrix} p \\ r \end{bmatrix} x^r, \tag{7.14}$$

where $\left[{p \atop r} \right]$ is the rth signless Stirling number of the first kind. We shall also utilize the first derivative with respect to x of this identity:

$$\langle x \rangle_p \sum_{j=0}^{p-1} \frac{1}{x+j} = \sum_{r=1}^{p} r \begin{bmatrix} p \\ r \end{bmatrix} x^{r-1}. \tag{7.15}$$

In the Pochhammer notation (7.13) can be written as

$$\begin{aligned} \mathbf{E}\big[C_n^{(p)}\big] &= \mathbf{E}\big[C_n^{(n)}\big] - \frac{p!}{\langle n-p+1 \rangle_p} \sum_{j=1}^{n} \sum_{k=0}^{n-1} \frac{\langle n-p-k \rangle_p}{p!} \times k\, \mathbf{Prob}\{S_j = k\} \\ &= \mathbf{E}\big[C_n^{(n)}\big] - \sum_{j=1}^{n} \sum_{k=0}^{n-1} \Bigg[\sum_{r=1}^{p} \frac{\left[{p \atop r} \right] (n-p-k)^r}{\langle n-p+1 \rangle_p} \Bigg] \times k\, \mathbf{Prob}\{S_j = k\}. \end{aligned} \tag{7.16}$$

Theorem 7.10 *(Prodinger 1995; Lent and Mahmoud, 1996a). The average number of data comparisons for MULTIPLE QUICK SELECT to locate a random set of p order statistics in a random input of size n is*

$$\mathbf{E}\big[C_n^{(p)}\big] = (2H_p + 1)n - 8p \ln n + O(1),$$

when p is fixed as $n \to \infty$.

Proof. In the exact expression (7.16) for the grand average, expand the terms $(n - p - k)^r$ by the binomial theorem:

$$\begin{aligned} \mathbf{E}\big[C_n^{(p)}\big] &= \mathbf{E}\big[C_n^{(n)}\big] - \sum_{j=1}^{n} \sum_{k=0}^{n-1} k \sum_{r=1}^{p} \frac{\left[{p \atop r} \right]}{\langle n-p+1 \rangle_p} \\ &\qquad \times \Bigg[\sum_{t=0}^{r} (-1)^t \binom{r}{t} k^t (n-p)^{r-t} \Bigg] \mathbf{Prob}\{S_j = k\} \\ &= \mathbf{E}\big[C_n^{(n)}\big] - \sum_{j=1}^{n} \sum_{r=1}^{p} \begin{bmatrix} p \\ r \end{bmatrix} \sum_{t=0}^{r} (-1)^t \binom{r}{t} \frac{(n-p)^{r-t}}{\langle n-p+1 \rangle_p} \mathbf{E}\big[S_j^{t+1}\big]. \end{aligned}$$

Isolating the terms containing the first and second moments of S_j, we obtain:

$$\mathbf{E}\big[C_n^{(p)}\big] = \mathbf{E}\big[C_n^{(n)}\big] - \sum_{j=1}^{n} \sum_{r=1}^{p} \begin{bmatrix} p \\ r \end{bmatrix} \frac{(n-p)^r}{\langle n-p+1 \rangle_p} \mathbf{E}[S_j] \tag{7.17}$$

$$+\sum_{j=1}^{n}\sum_{r=1}^{p}\left[{p \atop r}\right]\frac{r(n-p)^{r-1}}{\langle n-p+1\rangle_p}\mathbf{E}[S_j^2] \tag{7.18}$$

$$+\sum_{j=1}^{n}\sum_{r=1}^{p}\left[{p \atop r}\right]\sum_{t=2}^{r}(-1)^{t+1}\binom{r}{t}$$

$$\times\frac{(n-p)^{r-t}}{\langle n-p+1\rangle_p}\mathbf{E}[S_j^{t+1}]. \tag{7.19}$$

First, using (7.14)

$$\begin{aligned}\sum_{j=1}^{n}\sum_{r=1}^{p}\left[{p \atop r}\right]\frac{(n-p)^{r}}{\langle n-p+1\rangle_p}\mathbf{E}[S_j] &= \frac{\sum_{r=1}^{p}\left[{p \atop r}\right](n-p)^r}{\langle n-p+1\rangle_p}\sum_{j=1}^{n}\mathbf{E}[S_j]\\ &= \frac{\langle n-p\rangle_p}{\langle n-p+1\rangle_p}\mathbf{E}[C_n^{(n)}]\\ &= \left(1-\frac{p}{n}\right)\mathbf{E}[C_n^{(n)}]\\ &= \mathbf{E}[C_n^{(n)}]-2pH_n+O(1);\end{aligned} \tag{7.20}$$

the last asymptotic relation is obtained by an application of Theorem 7.1. For terms involving the second moment, we use the identity

$$\sum_{j=1}^{n}\mathbf{E}[S_j^2]=3n^2+17n-10(n+1)H_n,$$

the derivation of which is relegated to Exercise 7.17. By way of the identity (7.15) we find

$$\begin{aligned}\sum_{j=1}^{n}\sum_{r=1}^{p}\frac{\left[{p \atop r}\right]r(n-p)^{r-1}}{\langle n-p+1\rangle_p}\mathbf{E}[S_j^2] &= \left(\sum_{j=1}^{n}\mathbf{E}[S_j^2]\right)\frac{\langle n-p\rangle_p}{\langle n-p+1\rangle_p}\sum_{i=0}^{p-1}\frac{1}{n-p+i}\\ &= \left(3n^2+17n-10(n+1)H_n\right)\\ &\qquad\times\left(1-\frac{p}{n}\right)\sum_{i=0}^{p-1}\left[1+O\left(\frac{1}{n}\right)\right]\frac{1}{n}\\ &= \frac{1}{n}\left[3n^2-10(n+1)H_n+17n\right]\\ &\qquad\times\left(1-\frac{p}{n}\right)\left[p+O\left(\frac{1}{n}\right)\right]\\ &= 3pn-10pH_n+O(1).\end{aligned} \tag{7.21}$$

Rearranging the terms involving the higher moments, we have

$$\sum_{r=1}^{p} \left[{p \atop r} \right] \frac{(n-p)^r}{\langle n-p+1 \rangle_p} \sum_{t=2}^{r} (-1)^{t+1} \binom{r}{t} \Big(1 - \frac{p}{n}\Big)^{-t} \sum_{j=1}^{n} \frac{\mathbf{E}[S_j^{t+1}]}{n^t}.$$

To simplify this, we use the asymptotic approximation

$$\Big(1 - \frac{p}{n}\Big)^{-t} = 1 + O\Big(\frac{1}{n}\Big).$$

We can write the result of Exercise 7.17 in the form

$$\sum_{j=1}^{n} \frac{\mathbf{E}[S_j^{t+1}]}{n^t} = n\Big(1 + \frac{2}{t}\Big) + O(1).$$

When we substitute these into expression (7.19), the sum on t becomes (for fixed $r \ge 1$)

$$\begin{aligned}
&\sum_{t=2}^{r} (-1)^{t+1} \binom{r}{t} \Big[n\Big(1 + \frac{2}{t}\Big) + O(1) \Big] \\
&\quad = n\Big[\sum_{t=0}^{r} (-1)^{t+1} \binom{r}{t} + 1 - r \Big] + n\Big[2 \sum_{t=1}^{r} \frac{(-1)^{t+1}}{t} \binom{r}{t} - 2r \Big] + O(1) \\
&\quad = n(1 + 2H_r - 3r) + O(1),
\end{aligned}$$

where we have used the binomial theorem and the identity

$$\sum_{t=1}^{r} \frac{(-1)^{t+1}}{t} \binom{r}{t} = H_r,$$

which in turn follows from the binomial theorem:

$$\begin{aligned}
\sum_{t=1}^{r} \frac{(-1)^{t+1}}{t} \binom{r}{t} &= \sum_{t=1}^{r} \int_0^1 (-1)^{t+1} \binom{r}{t} x^{t-1} \, dx \\
&= -\int_0^1 \frac{1}{x} \sum_{t=1}^{r} \binom{r}{t} (-x)^t \, dx \\
&= -\int_0^1 \frac{(1-x)^r - 1}{x} \, dx \\
&= \int_0^1 \frac{v^r - 1}{v - 1} \, dv \\
&= \int_0^1 (1 + v + v^2 + \cdots + v^{r-1}) \, dv.
\end{aligned}$$

Next we observe that

$$\frac{(n-p)^p}{\langle n-p+1\rangle_p} = 1 + O\Big(\frac{1}{n}\Big),$$

and that for each $r < p$,

$$\begin{bmatrix} p \\ r \end{bmatrix} \frac{(n-p)^r}{\langle n-p+1\rangle_p}(2H_r + 1 - 3r)n = O(1).$$

Then we may rewrite (7.19) as

$$\frac{(n-p)^p}{\langle n-p+1\rangle_p}(2H_p+1-3p)n + \sum_{r=1}^{p-1}\begin{bmatrix} p \\ r \end{bmatrix}\frac{(n-p)^r}{\langle n-p+1\rangle_p}(2H_r+1-3r)n + O(1),$$

which reduces to

$$(2H_p + 1 - 3p)n + O(1). \qquad (7.22)$$

Substituting (7.20), (7.21), and (7.22) for (7.17), (7.18), and (7.19) respectively gives the required result. ■

Theorem 7.10 develops the average number of *data* comparisons for MQS. Additional comparisons are needed for the search after each partition step for the position of the pivot of the step in the relevant portion of the array OS. Each of these comparisons are made between integer indexes and may generally be faster than one comparison of a pair of data. As discussed in Exercise 7.19, collectively over all the stages of the algorithm, the average number of these additional index comparisons is $O(1)$ under any reasonable search algorithm and can only affect the lower-order term of the average speed of MQS, or the rate of convergence to that average.

Let the outcome of the random process for determining the order statistics be $V_1, \ldots, V_p$; these order statistics are determined independently of the random input. Let P_n be the landing position of the first pivot. For a random permutation

$$P_n \overset{D}{=} \text{UNIFORM}[1 \,..\, n].$$

Right after the first partition step, for some $1 \le r \le p$, either the pivot (now moved to position P_n) splits the array into two segments: $A[1 \,..\, P_n - 1]$ containing $V_1, \ldots, V_r$, and $A[P_n \,..\, n]$ containing $V_{r+1}, \ldots, V_p$, or $P_n = V_r$, in which case $A[P_n]$ is one of the desired order statistics, $A[1 \,..\, P_n - 1]$ contains $V_1, \ldots, V_{r-1}$, and $A[P_n + 1 \,..\, n]$ contains $V_{r+1}, \ldots, V_r$. Under the hypothesis of random subarrays, we have a distributional equation:

$$C_n^{(p)} \overset{D}{=} (n-1) + \sum_{r=0}^{p}\Big(C_{P_n-1}^{(r)} + \tilde{C}_{n-P_n}^{(p-r)}\Big)I_n^{(r)} + \sum_{r=1}^{p}\Big(\hat{C}_{P_n-1}^{(r-1)} + \check{C}_{n-P_n}^{(p-r)}\Big)\tilde{I}_n^{(r)}, \qquad (7.23)$$

where for every j and k, $\tilde{C}_k^{(j)} \overset{D}{=} \hat{C}_k^{(j)} \overset{D}{=} \check{C}_k^{(j)} \overset{D}{=} C_k^{(j)}$ and the families $\{C_k^{(j)}\}$, $\{\tilde{C}_k^{(j)}\}$, $\{\hat{C}_k^{(j)}\}$, $\{\check{C}_k^{(j)}\}$, $0 \le j \le k < \infty$, are independent; the random variable $I_n^{(r)}$ is the indicator of the event $V_r < P_n < V_{r+1}$, and $\tilde{I}_n^{(r)}$ is the indicator of the event $V_r = P_n$. Only one of the indicators assumes the value 1, as they indicate mutually exclusive events. So, one indicator will pick up the two correct terms for the partitioning while all the other terms are filtered out of the picture.

The asymptotic grand average $(2H_p + 1)n$ in Theorem 7.10 suggests the use of this factor for centering. We use n as a scale factor since it is the order of the standard deviation for the case $p = 1$ (see Theorem 7.6), and we anticipate it to be the order of the variance for higher p as well; we shall see that indeed n is the exact order of the standard deviation for all fixed p. We normalize the random variable $C_n^{(p)}$ by introducing

$$Y_n^{(p)} = \frac{C_n^{(p)} - (2H_p + 1)n}{n}$$

(note that $Y_n^{(0)} = -1$). In normalized form the basic recurrence (7.23) is

$$\begin{aligned} Y_n^{(p)} + 2H_p \overset{D}{=} -\frac{1}{n} &+ \sum_{r=0}^{p}\Big\{\Big(Y_{P_n-1}^{(r)} + 2H_r + 1\Big)\frac{P_n - 1}{n} \\ &+ \Big(\tilde{Y}_{n-P_n}^{(p-r)} + 2H_{p-r} + 1\Big)\frac{n - P_n}{n}\Big\} I_n^{(r)} \\ &+ \sum_{r=1}^{p}\Big\{\Big(\hat{Y}_{P_n-1}^{(r-1)} + 2H_{r-1} + 1\Big)\frac{P_n - 1}{n} \\ &+ \Big(\check{Y}_{n-P_n}^{(p-r)} + 2H_{p-r} + 1\Big)\frac{n - P_n}{n}\Big\} \tilde{I}_n^{(r)}, \end{aligned}$$

where for every j and k, $\tilde{Y}_k^{(j)} \overset{D}{=} \hat{Y}_k^{(j)} \overset{D}{=} \check{Y}_k^{(j)} \overset{D}{=} Y_k^{(j)}$, $0 \le j \le k < \infty$, and the families $\{Y_k^{(j)}\}$, $\{\tilde{Y}_k^{(j)}\}$, $\{\hat{Y}_k^{(j)}\}$, $\{\check{Y}_k^{(j)}\}$ are independent.

The event $P_n = V_r$ occurs with probability $1/n$ in a random permutation. Thus $\tilde{I}_n^{(r)}$ converges to 0 in probability. And so, if, for every fixed p, $Y_n^{(p)}$ converges in distribution to a random variable, the sum

$$\sum_{r=0}^{p}\Big\{\Big(\hat{Y}_{P_n-1}^{(r-1)} + 2H_{r-1} + 1\Big)\frac{P_n - 1}{n} + \Big(\check{Y}_{n-P_n}^{(p-r)} + 2H_{p-r} + 1\Big)\frac{n - P_n}{n}\Big\} \tilde{I}_n^{(r)}$$

will be $o_P(1)$ (i.e., $o(1)$ in probability) and subsequently will play no role in the limiting distribution. We can therefore write the essentials of the recurrence in the form

$$Y_n^{(p)} + 2H_p \stackrel{D}{=} \sum_{r=0}^{p} \Big\{ \Big(Y_{P_n - 1}^{(r)} + 2H_r + 1 \Big) \frac{P_n - 1}{n} + \Big(\tilde{Y}_{n-P_n}^{(p-r)} + 2H_{p-r} + 1 \Big) \frac{n - P_n}{n} \Big\} I_n^{(r)} + o_P(1). \tag{7.24}$$

We shall show a little later that $Y_n^{(p)}$ converges in distribution to a limiting random variable, to be called $Y^{(p)}$. (FIND's limiting random variable Y is $Y_n^{(1)}$ in this notation.) The next theorem will then follow.

Theorem 7.11 *(Mahmoud and Smythe, 1998). For each $p \ge 1$, the limiting normalized number of comparisons $Y^{(p)}$ satisfies the characteristic equation*

$$Y^{(p)} + 2H_p \stackrel{D}{=} \sum_{r=0}^{\lfloor p/2 \rfloor} J_r \Big\{ X_r^{(p)} \big(Y^{(r)} + 2H_r + 1 \big) + (1 - X_r^{(p)}) \big(\tilde{Y}^{(p-r)} + 2H_{p-r} + 1 \big) \Big\},$$

where, for any r, $\tilde{Y}^{(r)} \stackrel{D}{=} Y^{(r)}$, $X_r^{(p)}$ is a random variable with density $f_r^{(p)}(x) = (p+1)\binom{p}{r} x^r (1-x)^{p-r}$, for $0 \le x \le 1$, and $J_0, J_1, \ldots, J_{\lfloor p/2 \rfloor}$ are the components of a $\lfloor p/2 \rfloor$-component vector $\mathbf{J}_p$ of Bernoulli random variables with joint distribution

$$\mathbf{Prob}\{\mathbf{J}_p^T = (0, \ldots, 0, 1, 0, \ldots, 0)\} = \frac{2}{p+1},$$

when the 1 appears at the rth position, for any $1 \le r < \lfloor p/2 \rfloor$, and

$$\mathbf{Prob}\{\mathbf{J}_p^T = (0, 0, \ldots, 0, 1)\} = \begin{cases} 1/(p+1), & \text{if } p \text{ is even;} \\ 2/(p+1), & \text{if } p \text{ is odd.} \end{cases}$$

Furthermore the families $\{X_r^{(p)}\}$, $\{Y^{(r)}\}$, and $\{\tilde{Y}^{(r)}\}$ are independent and the random variables J_ℓ are independent of $X_r^{(p)}$, $Y^{(j)}$, and $\tilde{Y}^{(k)}$ for all ℓ, r, p, j and k.

Proof. Consider the characteristic function of $Y_n^{(p)} + 2H_p$. Upon conditioning on the given ranks $V_1 = j_1, \ldots, V_p = j_p$; $P_n = m$ and using the various independence and identical distribution assumptions, the recurrence (7.24) yields

$$\phi_{Y_n^{(p)} + 2H_p}(t) = \frac{1}{n\binom{n}{p}} \sum_{m=1}^{n} \sum_{r=0}^{p} \sum_{j_1 < \ldots < j_r < m < j_{r+1} < \ldots < j_p} \phi_{Y_{m-1}^{(r)} + 2H_r + 1}\Big(\frac{m-1}{n} t \Big) \times \phi_{Y_{n-m}^{(p-r)} + 2H_{p-r} + 1}\Big(\frac{n-m}{n} t \Big) + o(1)$$

$$= \frac{1}{n\binom{n}{p}} \sum_{m=1}^{n} \sum_{r=0}^{p} \binom{m-1}{r} \binom{n-m}{p-r} \phi_{Y_{m-1}^{(r)}+2H_r+1}\Big(\frac{m-1}{n}t\Big)$$

$$\times\, \phi_{Y_{n-m}^{(p-r)}+2H_{p-r}+1}\Big(\frac{n-m}{n}t\Big) + o(1). \qquad (7.25)$$

Assuming the limit $Y^{(p)}$ exists, as will be proved later on, we can write the limit of the last equation in the form of a convolution:

$$\phi_{Y^{(p)}+2H_p}(t) = \frac{1}{p+1} \sum_{r=0}^{p} \int_0^1 \Big\{\phi_{Y^{(r)}+2H_r+1}(tx)$$

$$\times\, \phi_{Y^{(p-r)}+2H_{p-r}+1}\big(t(1-x)\big)\Big\} f_r^{(p)}(x)\,dx. \qquad (7.26)$$

For p odd, the sum in (7.26) can be split into two sums, one with the index r running from 0 to $\lfloor p/2 \rfloor$, the other from $\lceil p/2 \rceil$ to p. The change of index from r to $p-r$ in the second sum yields two identical summations. Splitting the sum into two identical sums is a convenient way of identifying the correct probabilities of the vector $\mathbf{J}_p$. The range $\lceil p/2 \rceil, \dots, p$ of the index r does not introduce new random variables; each of these terms coincides with one from the range $0, \dots, \lfloor p/2 \rfloor$, only doubling its probability. One gets

$$\phi_{Y^{(p)}+2H_p}(t) = \frac{2}{p+1} \sum_{r=0}^{\lfloor p/2 \rfloor} \int_0^1 \phi_{Y^{(r)}+2H_r+1}(tx)$$

$$\times\, \phi_{Y^{(p-r)}+2H_{p-r}+1}\big(t(1-x)\big) f_r^{(p)}(x)\,dx.$$

The right-hand side is the characteristic function of

$$\sum_{r=0}^{\lfloor p/2 \rfloor} J_r \Big\{X_r^{(p)}\big(Y^{(r)} + 2H_r + 1\big) + \big(1 - X_r^{(p)}\big)\big(\tilde{Y}^{(p-r)} + 2H_{p-r} + 1\big)\Big\}. \qquad (7.27)$$

A similar argument applies to the case p even. The only difference is that we cannot split the sum into two exact halves; a middle term that does not repeat appears. That is, each term in the range $0, \dots, \frac{1}{2}p - 1$ is doubled by a matching term from the range $\frac{1}{2}p + 1, \dots, p$, and a middle term corresponding to $r = p/2$ appears by itself. So, the right-hand side of (7.26) is the characteristic function of a random variable of the form (7.27), the only difference for even p being that the probabilities of the Bernoulli random variables are different; for $r = 1, \dots, \frac{1}{2}p - 1$:

$$J_r = \begin{cases} 1, & \text{with probability } 2/(p+1); \\ 0, & \text{with probability } 1 - 2/(p+1), \end{cases}$$

whereas

$$J_{p/2} = \begin{cases} 1, & \text{with probability } 1/(p+1); \\ 0, & \text{with probability } 1 - 1/(p+1). \end{cases}$$

Thus for p odd or even we get the same characteristic distributional equation, but with different weights assigned to the Bernoulli random variables involved, as stated. ■

The special case $p = 1$ is characterized by the distributional equation

$$Y^{(1)} + 2 \stackrel{D}{=} X_0^{(1)}(Y^{(1)} + 3),$$

with $X_0^{(1)}$ having the density $2x$, for $0 \le x \le 1$; that is, $X_0^{(1)}$ is X of Theorem 7.8.

As in the case $p = 1$, the technique for convergence in distribution gives us convergence of the first two moments as well. So, $Y^{(p)}$ is square-integrable. Thus it has a finite second moment for which a recurrence relation can be formulated from Theorem 7.11. Let

$$v_p \stackrel{\text{def}}{=} \mathbf{Var}[Y^{(p)}] = \mathbf{E}[(Y^{(p)})^2].$$

We first square the characteristic equation of Theorem 7.11, then take expectation of both sides. Simplifications follow from the fact that $Y^{(r)}$ are all centered random variables (having mean 0, for all r), and that the components of $\mathbf{J}_p$ are mutually exclusive (all cross-products $J_i J_k \equiv 0$, when $i \ne k$), and from the various independence and equality in distribution assumptions.

For odd p,

$$\begin{aligned} v_p + 4H_p^2 = \frac{2}{p+1}\Bigg(& \mathbf{E}\Big[\big(1 - X_0^{(p)}\big)^2\big(v_p + (2H_p + 1)^2\big)\Big] \\ & + \sum_{r=1}^{p-1} \big(v_r + (2H_r + 1)^2\big)\mathbf{E}\big[\big(X_r^{(p)}\big)^2\big] \\ & + \sum_{r=1}^{\lfloor p/2 \rfloor} 2\mathbf{E}\big[X_r^{(p)}\big(1 - X_{p-r}^{(p)}\big)\big](2H_r + 1)(2H_{p-r} + 1)\Bigg). \end{aligned}$$

Write the last summation as

$$\begin{aligned} & \sum_{r=1}^{\lfloor p/2 \rfloor} \mathbf{E}\big[X_r^{(p)}\big(1 - X_{p-r}^{(p)}\big)\big](2H_r + 1)(2H_{p-r} + 1) \\ & \qquad + \sum_{r=1}^{\lfloor p/2 \rfloor} \mathbf{E}\big[X_r^{(p)}\big(1 - X_{p-r}^{(p)}\big)\big](2H_r + 1)(2H_{p-r} + 1). \end{aligned}$$

Replace $p - r$ by r in the second sum and use the symmetry

$$X_r^{(p)} \overset{D}{=} 1 - X_{p-r}^{(p)}$$

to write just one sum extending over the index r from 1 to p. This further simplifies the variance recurrence to

$$v_p = \frac{2}{(p+1)^2(p+2)} \sum_{r=1}^{p-1} (r+1)(r+2)v_r + g_p,$$

where

$$\begin{aligned} g_p = {} & \frac{2}{p+1}(2H_p + 1)^2 - 4\frac{p+3}{p+1}H_p^2 \\ & + \frac{2}{(p+1)^2(p+2)} \sum_{r=1}^{p-1} (r+1)(r+2)(2H_r + 1)^2 \\ & + \frac{2}{(p+1)^2(p+2)} \sum_{r=1}^{p-1} (r+1)(p-r+1)(2H_r + 1)(2H_{p-r} + 1). \end{aligned}$$

Even though the existence of a middle term in the case p even changes the probabilities of $\mathbf{J}_p$, similar work shows that the last recurrence is a valid form for v_p, for even p, too. This recurrence can be linearized by first differencing a version with $p - 1$ replacing p to get

$$(p+1)^2(p+2)v_p - p^2(p+1)v_{p-1} = 2p(p+1)v_{p-1} + \nabla\{(p+1)^2(p+2)g_p\},$$

which simplifies to the linear recurrence

$$(p+1)v_p = pv_{p-1} + \frac{\nabla\{(p+1)^2(p+2)g_p\}}{(p+1)(p+2)}.$$

This form can be iterated, yielding

$$v_p = \frac{2}{p+1} + \sum_{j=2}^{p} \frac{\nabla\{(j+1)^2(j+2)g_j\}}{(j+1)(j+2)}, \tag{7.28}$$

where we used the boundary condition $v_1 = 1$, from Theorem 7.8. The sums involved in g_p vary in complexity. Several combinatorial identities leading to the required results are given in Exercise 7.18.

Theorem 7.12 *(Mahmoud and Smythe, 1998; Panholzer and Prodinger, 1998).*

$$v_p = 7 - 4H_{p+1}^{(2)} - \frac{4(p^2 + 3p + 5)}{3(p+1)^2(p+2)}.$$

Theorem 7.12 gives the asymptotic variance of the normalized number of comparisons of MQS, for p fixed. So,

$$\mathbf{Var}[C_n^{(p)}] \sim v_p n^2.$$

The assumption that p is fixed is crucial throughout. For example, the centering factor of Theorem 7.10 was obtained under this assumption and the asymptotic approximation of the binomial coefficients to make the transition from (7.25) to (7.26) uses this assumption and is not valid otherwise. Curiously, even though this analysis is done under the assumption of fixed p, the result coincides with a special case in which p is not fixed. When $p = n$, MQS finds *all* order statistics, i.e., sorts the input. Namely, MQS becomes QUICK SORT and if $p = n \to \infty$, the form of v_p in Theorem 7.12 converges to $7 - 2\pi^2/3$, QUICK SORT's variance of the limiting normalized number of comparisons (see Theorem 7.2). The exact variance computation of Panholzer and Prodinger (1998) for all fixed selection cases sheds light on why this is the case.

By an argument that generalizes that of the case $p = 1$, we shall prove that $Y_n^{(p)} \xrightarrow{\mathcal{D}} Y^{(p)}$. We shall then give an implicit characterization of the random variable $Y^{(p)}$ as the solution of a first-order differential equation.

We shall need additional notation and a little preparation to prove convergence in distribution. As in the random selection of one order statistic ($p = 1$), the proof is based on convergence in the Wasserstein metric space.

To be able to compute the Wasserstein second-order distance, we seek a unified representation of $Y_n^{(p)}$ and $Y^{(p)}$. Let $U, U_1, \ldots, U_p$ be independent UNIFORM(0,1) random variables, and let $U_{(1)}, \ldots, U_{(p)}$ be the order statistics of $U_1, \ldots, U_p$. We assume this family of random variables to be independent of all the other families that appeared so far. For convenience we also introduce $U_0 \equiv 0$ and $U_{p+1} \equiv 1$. Introduce the indicators

$$K_r^{(p)} = \mathbf{1}_{\{U_{(0)} < \ldots < U_{(r)} < U < U_{(r+1)} < \ldots < U_{(p+1)}\}},$$

which have a distribution function differing only by $O(1/n)$ from the distribution function of $I_n^{(p)}$. Also recall that

$$P_n \stackrel{\mathcal{D}}{=} \lceil nU \rceil.$$

An equivalent representation of $Y_n^{(p)}$ is

$$\begin{aligned} Y_n^{(p)} + 2H_p \stackrel{\mathcal{D}}{=} \sum_{r=0}^{p} \Big\{ & \Big(Y_{\lceil nU \rceil - 1}^{(r)} + 2H_r + 1 \Big) \frac{\lceil nU \rceil - 1}{n} \\ & + \Big(\tilde{Y}_{n - \lceil nU \rceil}^{(p-r)} + 2H_{p-r} + 1 \Big) \frac{n - \lceil nU \rceil}{n} \Big\} K_r^{(p)} \\ & + o_P(1). \end{aligned}$$

Lemma 7.1

$$Y^{(p)} + 2H_p \overset{\mathcal{D}}{=} \sum_{r=0}^{p} \{ (Y^{(r)} + 2H_r + 1)U$$
$$+ (\tilde{Y}^{(p-r)} + 2H_{p-r} + 1)(1 - U) \} K_r^{(p)}.$$

Proof. Let $\psi^{(p)}(t)$ denote the distribution function of the random variable on the right-hand side in the lemma. Condition on $U = u$, $U_1 = u_1, \ldots, U_n = u_n$. Only one of the indicators $K_r^{(p)}$ assumes the value 1; the rest are 0. Hence, summing over all permutations $i_1, \ldots, i_p$ of $\{1, \ldots, p\}$, we have

$$\psi^{(p)}(t) = \sum_{i_1, i_2, \ldots, i_p} \sum_{r=0}^{p} \int\int_{u_{i_1} < \ldots < u_{i_r} < u < u_{i_{r+1}} < \ldots < u_{i_p}} \cdots \int \mathbf{E}\Big[\exp\Big\{it\Big\{(Y^{(r)} + 2H_r + 1)u$$
$$+ (\tilde{Y}^{(p-r)} + 2H_{p-r} + 1)(1 - u)\Big\}\Big\}\Big]\, du\, du_1 \ldots du_p$$
$$= p! \sum_{r=0}^{p} \int\int_{u_1 < \ldots < u_r < u < u_{r+1} < \ldots < u_p} \cdots \int \mathbf{E}\Big[\exp\Big\{it\Big\{(Y^{(r)} + 2H_r + 1)u$$
$$+ (\tilde{Y}^{(p-r)} + 2H_{p-r} + 1)(1 - u)\Big\}\Big\}\Big]\, du\, du_1 \ldots du_p.$$

Notice that, for any integrable function $\mathcal{H}(u)$,

$$\int\int_{u_1 < \ldots < u_r < u < u_{r+1} < \ldots < u_p} \cdots \int \mathcal{H}(u)\, du\, du_1 \ldots du_p = \frac{1}{(p+1)!} \int_0^1 \mathcal{H}(u) f_r^{(p)}(u)\, du,$$

reconstructing the densities $f_r^{(p)}(u)$ in the representation; comparing this with (7.26) the function $\psi^{(p)}(t)$ coincides with $\phi_{Y^{(p)} + 2H_p}(t)$. ■

By an induction similar to that of the proof of Theorem 7.4 it can be shown that the Wasserstein distance between the distribution functions of $Y_n^{(p)}$ and $Y^{(p)}$ converges to 0, as $n \to \infty$. The induction here is a double induction. We assume the statement to be true for all n for all $p < \tilde{p}$; we then use this hypothesis to show the assertion true for all n at $p = \tilde{p}$. The details are left as an exercise.

As discussed at the beginning of this section, it follows from this convergence in the Wasserstein metric space that

$$Y_n^{(p)} \xrightarrow{\mathcal{D}} Y^{(p)},$$

and

$$\mathbf{E}\big[(Y_n^{(p)})^2\big] \to \mathbf{E}\big[(Y^{(p)})^2\big].$$

Having shown that a limiting distribution exists, we turn to characterizing it. In what follows we shall denote $\phi_{Y^{(p)}+2H_p}(t)$ by $R_p(t)$. With this notation, Lemma 7.1 enables us to write a recurrence for this characteristic function:

$$R_p(t) = e^{it} \sum_{k=0}^{\lfloor p/2 \rfloor} \mathcal{P}_k \int_0^1 R_k(tx) R_{p-k}\big(t(1-x)\big) f_k^{(p)}(x)\, dx, \tag{7.29}$$

where $\mathcal{P}_k$ are probabilities defined as

$$\mathcal{P}_k = \frac{2}{p+1}, \qquad k = 1, \ldots, \left\lfloor \frac{p}{2} \right\rfloor - 1,$$

and

$$\mathcal{P}_{\lfloor p/2 \rfloor} = \begin{cases} 1/(p+1), & \text{if } p \text{ is even;} \\ 2/(p+1), & \text{if } p \text{ is odd.} \end{cases}$$

The recurrence (7.29) gives us an inductive characterization of the distributions of $Y^{(p)}$; the infinitely divisible distribution whose characteristic function is

$$\exp\Big\{2 \int_0^t \Big(\frac{e^{iu} - 1 - iu}{u}\Big)\, du\Big\} \tag{7.30}$$

is at the basis of this induction.

Assume by induction on p that $R_1(t), \ldots, R_{p-1}(t)$ have already been determined. Let

$$G_p(t) = e^{it} \sum_{k=1}^{\lfloor p/2 \rfloor} \mathcal{P}_k \int_0^1 R_k(tx) R_{p-k}\big(t(1-x)\big) f_k^{(p)}(x)\, dx,$$

which is now assumed known. Then the recurrence (7.29) can be written as

$$\begin{aligned} R_p(t) &= \frac{2e^{it}}{p+1} \int_0^1 e^{-itx} R_p(t(1-x)) f_0^{(p)}(x)\, dx + G_p(t) \\ &= \frac{2}{t^{p+1}} \int_0^t u^p e^{iu} R_p(u)\, du + G_p(t). \end{aligned}$$

The limit $Y^{(p)}$ has finite second moment, whence $R_p(t)$ is at least twice differentiable; taking derivatives once with respect to t, we arrive at the differential equation

$$R'_p(t) + \frac{p+1-2e^{it}}{t} R_p(t) = \frac{(p+1)G_p(t) + G'_p(t)}{t}.$$

In principle, this first-order differential equation can be solved by well-known techniques and $R_p(t)$ will be expressed in terms of convolution integrals of the complicated function (7.30). These may be manipulated numerically to determine the shape of the distribution.

This equation may not admit an explicit solution because of the complex nature of the function G_p. However, bounds may be developed on $R_P(t)$ for large t and one can prove that $R_p(t)$ is integrable. Therefore the limit $Y^{(p)}$ possesses a continuous density.

EXERCISES

7.1 What adaptation does the text's PARTITION and QUICK SORT need to handle data files with possible replication of entries?

7.2 (Sedgewick, 1980) A popular partitioning algorithm is based on the following idea. Set up two indexes F (forward) and B (backward). Choose a pivot from the array A. Keep advancing F as long as you see array elements $A[F]$ less than the pivot. Then come down from the other end by retracting B until you find an element $A[B]$ less than the pivot. At this stage both $A[F]$ and $A[B]$ are on the "wrong side" of the pivot's correct position. Swap $A[F]$ and $A[B]$ and continue the process on $A[F+1\,..\,B-1]$ until F and B cross.

(a) Implement Sedgewick's partition algorithm to partition $A[1\,..\,n]$. (Hint: Introduce a dummy $+\infty$ at the end of the array as a sentinel.)

(b) What value serves as sentinel, when QUICK SORT (with Sedgewick's partition algorithm) handles an arbitrary stretch $A[\ell\,..\,u]$, $1 \le \ell < u \le n$?

(c) How many comparisons does Sedgewick's partitioning algorithm make to partition $A[1\,..\,n]$?

(d) How many swaps does Sedgewick's partitioning algorithm require in the best case, worst case, and on average? Determine the central limit tendency of the number of swaps.

7.3 Construct an input permutation of $\{1, \dots, n\}$ to drive QUICK SORT to make $O(n^{3/2})$ comparisons.

7.4 Assume the sample space used for the analysis of QUICK SORT has $n!$ equally likely input permutations. On this sample space, interpret $C_0, C_1, \dots, C_{n-1}$, which appear in the recurrence equations for C_n.

7.5 (Rösler, 1991) By taking the variance of both sides of the distributional equality (7.7), and using the independence assumptions for this equation, find the leading term in the variance of the number of comparisons of QUICK SORT.

7.6 Calculate the second-order Wasserstein distance between

(a) The distribution functions of two identically distributed random variables.

(b) The distribution functions of U = UNIFORM(0,1) and $W_n = U(1 + 1/n)$. Conclude that $W_n \xrightarrow{\mathcal{D}} U$.

7.7 What constant minimizes the second-order Wasserstein distance between a degenerate distribution function and that of a random variable?

7.8 (Mahmoud, Modarres, and Smythe, 1995) Assuming the text's PARTITION algorithm, prove that the mean and variance of Q_n, the number of comparisons in Hoare's FIND for locating the minimum (or the maximum, by symmetry) in a list of size n, are given by

$$\mathbf{E}\big[Q_n\big] = 2n - 2H_n$$
$$\sim 2n, \qquad \text{as } n \to \infty,$$

and

$$\mathbf{Var}\big[Q_n\big] = \frac{n(n-9)}{2} + 8H_n - 4H_n^{(2)}$$
$$\sim \frac{1}{2}n^2, \qquad \text{as } n \to \infty.$$

7.9 (Mahmoud, Modarres, and Smythe, 1995) Let Q_n be FIND's number of comparisons for locating the minimum (or the maximum, by symmetry). Show that

$$\frac{Q_n}{n} \xrightarrow{\mathcal{D}} Q$$

where $Q \overset{D}{=} UQ + 1$, and U is a UNIFORM(0, 1) random variable that is independent of Q.

7.10 (Mahmoud, Modarres, and Smythe, 1995) Demonstrate that the square root of the characteristic function of FIND's limiting number of comparisons Y for random selection is the characteristic function of FIND's limiting number of comparisons Q for the fixed selection of the minimum (or the maximum, by symmetry); see the previous exercise for the definition of Q.

7.11 (Mahmoud, Modarres, and Smythe, 1995) Let Y be the limiting normalized number of comparisons of FIND. Show that the support of Y is unbounded, that is, the probability that Y belongs to a bounded set of the real line is 0. (Hint: Argue that if Y were to have a bounded support, $\mathbf{Var}[Y]$ would have to be 0.)

7.12 (Mahmoud and Smythe, 1998) Let Y be the limiting normalized number of comparisons of FIND. Show that the distribution function of Y is absolutely continuous. (Hint: Show that $\phi_Y(t)$, the characteristic function of Y is asymptotic to t^{-2}, hence integrable, as $t \to \infty$.)

7.13 (Mahmoud, Modarres, and Smythe, 1995) Develop the following large deviation inequality for the upper tail of the distribution of the limiting random variable Y, FIND's normalized number of comparisons: For $y > 0$,

$$\mathbf{Prob}\{Y > y\} \le \Big(\frac{e}{1+y}\Big)^{1+y}.$$

(Hint: Use a Chernoff-like technique—start with Chebyshev's inequality in its exponential form and minimize the exponent.)

7.14 (Mahmoud, Modarres, and Smythe, 1995) Develop the following large deviation inequality for the lower tail of the distribution of the limiting random variable Y, FIND's normalized number of comparisons: For $y > 0$,

$$\mathbf{Prob}\{Y < -y\} \le \exp\Big(-\frac{y^2}{2}\Big);$$

see the hint to the previous exercise.

7.15 Suppose P_n and M_n are two independent families of random variables with the nth member of either family being distributed like UNIFORM$[1 \mathinner{..} n]$. These families arise in the analysis of random selection, for example, P_n may be the landing position of the pivot and M_n may be a randomly selected order statistic to be located by FIND. Let U and W be two independent UNIFORM(0, 1) random variables. By taking limits of the distribution functions of P_n/n and M_n/n and $\mathbf{1}_{\{M_n<P_n\}}P_n/n$, show that these random variables respectively converge to U, W and $\mathbf{1}_{\{W<U\}}U$.

7.16 Compute the second-order Wasserstein distance between $Y_n^{(p)}$, the normalized number of comparisons MQS makes in locating a randomly selected order statistic, and the limiting random variable $Y^{(p)}$.

7.17 (Lent and Mahmoud, 1996a) For the analysis of the average number of comparisons of MQS to find a random set of p order statistics in a random file of size n we need to develop sums of the form $\sum_{j=1}^{n} \mathbf{E}[S_j^t]$, where (simplifying notation) S_j is the number of descendants of node j in a random binary search tree.

(a) From the first moment of S_j found in Exercise 1.6.5 derive the identity

$$\sum_{j=1}^{n} \mathbf{E}[S_j] = 2(n+1)H_n - 4n.$$

(b) From the second moment of S_j found in Exercise 1.6.5 derive the identity

$$\sum_{j=1}^{n} \mathbf{E}[S_j^2] = 3n^2 + 17n - 10(n+1)H_n.$$

(c) For all moments of S_j develop the asymptotic relation

$$\mathbf{E}[S_j^m] = \Big(\frac{1}{m-1} - \frac{2}{m-2}\Big)\big[(n-j)^{m-1} + j^{m-1}\big] + \frac{m}{m-2} n^{m-1} + O\big(n^{m-2}\big).$$

For moments higher than the second develop the asymptotic relation

$$\sum_{j=1}^{n} \mathbf{E}[S_j^m] = \frac{m+1}{m-1} n^m + O\big(n^{m-1}\big).$$

7.18 Simplify (7.28) by reducing the combinatorial sums involving g_j by proving the following identities:

(a)

$$\sum_{r=1}^{p-1} (r+1)(r+2)(2H_r+1)^2 = \Big(\frac{4}{3}p^2 + 4p + \frac{8}{3}\Big) p H_{p-1}^2 + \Big(\frac{4}{9}p^3 + \frac{4}{3}p^2 + \frac{8}{9}p + \frac{8}{3}\Big) H_{p-1} + \frac{5}{27}p^3 + \frac{11}{9}p^2 + \frac{100}{27}p - \frac{46}{9}.$$

(b) Starting with

$$\sum_{r=1}^{p-1} (r+1)(p-r+1)(2H_r+1)(2H_{p-r}+1) = \sum_{r=1}^{p-1} (r+1)(p-r+1) + 2\sum_{r=1}^{p-1} (r+1)(p-r+1)H_r + 2\sum_{r=1}^{p-1} (r+1)(p-r+1)H_{p-r} + 4\sum_{r=1}^{p-1} (r+1)(p-r+1)H_r H_{p-r},$$

show that the first sum is

$$\frac{1}{6}p^3 + p^2 - \frac{1}{6}p - 1;$$

show that the second and third sums are both equal to:

$$\Big(\frac{1}{6}p^2 + p + \frac{5}{6}\Big) p H_{p-1} - \frac{5}{36}p^3 - \frac{2}{3}p^2 - \frac{1}{36}p + \frac{5}{6}.$$

Let

$$a_p = \sum_{r=1}^{p-1} (r+1)(p-r+1) H_r H_{p-r}$$

and show that the generating function

$$A(z) = \sum_{p=0}^{\infty} a_p z^p$$

is the square of some elementary function of z. Extract coefficients from this elementary function to prove that

$$\begin{aligned} a_p = {} & \frac{1}{6}(p+1)(p+2)(p+3)\big(H_p^2 - H_p^{(2)}\big) \\ & + \frac{1}{108}(37p^2 + 186p + 209)p \\ & - \frac{1}{18}(5p^2 + 30p + 37)pH_p. \end{aligned}$$

7.19 What is the average number of partition steps in MQS when it runs on a random input of n keys to find a random set of order statistics of size p? What does that tell us about the number of index comparisons performed within the array OS of predesignated order statistics?

8

Sample Sort

The term *sample sorts* refers to a class of algorithms that are based on QUICK SORT where a sampling stage precedes application of QUICK SORT. The pathological cases of QUICK SORT, with $\Omega(n^2)$ behavior, arise when the splitting pivot does not split the given list near its middle at most recursive levels, as, for example, in the case of an already sorted list of numbers. The idea in sample sort algorithms is to avoid such pathological splitting by taking a sample from the list and using it to produce a situation favorable to middle splitting with probability higher than that of the chance of a single pivot to split near the middle. This preprocessing stage helps QUICK SORT reach its speed potential.

Some sample sort variations take a small sample. The median of the small sample is employed as the splitting pivot to speed up QUICK SORT. The sampling is then reapplied at each level of recursion. Another approach to sampling as an aid to speeding up QUICK SORT uses a large sample once at the top level of recursion. The elements of the sample are then inserted in their proper places, producing a large number of random segments, each handled by standard QUICK SORT. In either case, QUICK SORT's behavior is improved but the algorithm becomes more complex.

8.1 THE SMALL SAMPLE ALGORITHM

In the small sample approach to sorting $n \to \infty$ elements, a fixed-size sample is chosen. The median of the sample is then used as a pivot. Recursively, at each level the sampling is reapplied.

Median selection is not always the easiest task. We saw instances of median-finding algorithms based on adaptations of some standard sorting algorithms. The point of a small sample approach is to avoid complicating sample sort by burdening the median finding stage. If the sample size is $2k + 1$ for some small fixed k, simple median finding algorithms may be used. (Standard QUICK SORT, discussed extensively in Chapter 7, is itself the special case $k = 0$.) In the practicable choice $k = 1$, we have a sample of size 3 at each level of recursion. This special version of sample sort is commonly referred to as MEDIAN OF THREE QUICK SORT. The sample can be taken from any three positions. Under the random permutation model the joint distribution of data ranks at any three positions are the same; it does not really matter which three positions to take as the sample. For consistency in implementation, at any level of recursion we may take our sample, say, from the first three positions of the stretch of the host array under consideration. The process continues recursively until arrays of size two or less are reached, when it is no longer possible to sample three elements. The arrangement of arrays of size two is resolved by a simple **if** statement; smaller arrays are of course already sorted.

One can find the median of three items by an algorithm that uses only **if** statements. For example, Figure 1.6 provides a complete sorting algorithm for three items using only **if** statements. Finding the median of three distinct items by comparisons is equivalent to sorting the three elements (see Lemma 1.1). This simple, yet worst-case optimal, algorithm for median selection from among three items either makes two comparisons (with probability 1/3) or three comparisons (with probability 2/3); the theoretic lower bound of $\lceil \lg 3! \rceil = 3$ is attained (Theorem 1.1). That is, the random number of comparisons, B, for median selection among three items is given by

$$B = 2 + \text{BERNOULLI}\Big(\frac{2}{3}\Big),$$

whose average is $\mathbf{E}[B] = 8/3$.

Algorithms based on arithmetic operations are also available for finding the median of three elements (see Exercise 1.9.4).

For a meaningful contrast with standard QUICK SORT, we analyze MEDIAN-OF-THREE QUICK SORT under a comparison-based median selection algorithm that does not introduce other operations like arithmetic operations, for example, that are harder to evaluate relative to comparisons. We shall use the worst-case optimal algorithm of Figure 1.6 to assess the kind of improvement one gets by sampling over standard QUICK SORT.

Let C_n be the number of comparisons needed by MEDIAN-OF-THREE QUICK SORT to sort a random input. The basic distributional equation is essentially the same as that of the standard QUICK SORT, see (7.1). The difference is that at each level of recursion an additional step for choosing the pivot from the sample is required. Once a pivot is chosen, it can be compared to the other $n-1$ elements of the array by PARTITION, QUICK SORT's partitioning algorithm. Two subarrays are created, to which the sampling, median-of-three finding, and the partitioning process are repeated recursively. Let P_n be the position of the pivot (which is the median of a sample); then (for $n \ge 4$)

$$C_n \overset{\mathcal{D}}{=} C_{P_n - 1} + \tilde{C}_{n-P_n} + n - 1 + B, \tag{8.1}$$

where $\tilde{C}_n \overset{\mathcal{D}}{=} C_n$ and the families $\{C_n\}_{n=1}^{\infty}$, $\{\tilde{C}_n\}_{n=1}^{\infty}$ and $\{B\}$ are independent. No longer is P_n uniformly distributed as in standard QUICK SORT; it has a biased distribution humping at the middle, and that is where MEDIAN-OF-THREE outperforms QUICK SORT.

Under the random permutation model, all $\binom{n}{3}$ triples of ranks are equally likely to appear in our sample. For a sample of size three to be favorable to the event $\{P_n = p\}$, the median of the sample must be of rank p; one of the other two ranks in the sample must be smaller (can be chosen in $p - 1$ ways), and the other must be larger (can be chosen in $n - p$ ways). The probability distribution for the pivot position is

$$\mathbf{Prob}\{P_n = p\} = \frac{(p-1)(n-p)}{\binom{n}{3}}. \tag{8.2}$$

Taking the expectation of (8.1),

$$\begin{aligned} \mathbf{E}[C_n] &= \mathbf{E}[C_{P_n-1}] + \mathbf{E}[\tilde{C}_{n-P_n}] + n - 1 + \mathbf{E}[B] \\ &= 2\sum_{p=1}^{n} \mathbf{E}[C_{P_n-1} \mid P_n = p]\frac{(p-1)(n-p)}{\binom{n}{3}} + n - 1 + \frac{8}{3} \\ &= \frac{2}{\binom{n}{3}} \sum_{p=1}^{n} \mathbf{E}[C_{p-1}](p-1)(n-p) + n + \frac{5}{3}, \end{aligned} \tag{8.3}$$

where we used the symmetry $P_n - 1 \overset{D}{=} n - P_n$. The biased probability distribution of the pivot's landing position gives us this recurrence that does not directly telescope as in the case of standard QUICK SORT. This recurrence calls for a different technique. We shall tackle it by generating functions. Let

$$C(z) = \sum_{n=0}^{\infty} \mathbf{E}[C_n]z^n.$$

Multiplying the recurrence (8.3) throughout by $\binom{n}{3}z^{n-3}$ and summing over the range of validity of the recurrence, we obtain

$$\begin{aligned} \sum_{n=4}^{\infty} \binom{n}{3}\mathbf{E}[C_n]z^{n-3} &= 2\sum_{n=4}^{\infty} z^{n-3} \sum_{p=1}^{n} (p-1)(n-p)\mathbf{E}[C_{p-1}]z^{n-3} \\ &\quad + \sum_{n=4}^{\infty} \frac{3n+5}{3}\binom{n}{3}z^{n-3}, \end{aligned}$$

or

$$\begin{aligned} &\sum_{n=4}^{\infty} n(n-1)(n-2)\mathbf{E}[C_n]z^{n-3} \\ &\qquad = 12\sum_{n=4}^{\infty}\sum_{p=1}^{n} \big(\mathbf{E}[C_{p-1}](p-1)z^{p-2}\big)\big((n-p)z^{n-p-1}\big) \end{aligned}$$

$$+\frac{1}{3}\sum_{n=4}^{\infty} n(n-1)(n-2)(3n+3+2)z^{n-3}$$

$$= 12\Big(\sum_{k=1}^{\infty} \mathbf{E}[C_{k-1}](k-1)z^{k-2}\Big)\Big(\sum_{j=0}^{\infty}(j+1)z^j\Big)$$

$$+\sum_{k=4}^{\infty}(n+1)n(n-1)(n-2)z^{n-3}$$

$$+\frac{2}{3}\sum_{k=4}^{\infty} n(n-1)(n-2)z^{n-3}$$

$$= 12\Big(\sum_{k=1}^{\infty} \mathbf{E}[C_{k-1}](k-1)z^{k-2}\Big)\Big(\sum_{j=0}^{\infty}(j+1)z^j\Big)$$

$$+\frac{d^4}{dz^4}\sum_{k=4}^{\infty} z^{n+1} + \frac{2}{3}\Big(\frac{d^3}{dz^3}\sum_{k=4}^{\infty} z^n\Big).$$

In differential form

$$C'''(z) - 6\mathbf{E}[C_3] = \frac{12C'(z)}{(1-z)^2} + \Big(\frac{24}{(1-z)^5} - 24\Big) + \Big(\frac{4}{(1-z)^4} - 4\Big).$$

Observing the initial conditions $\mathbf{E}[C_0] = \mathbf{E}[C_1] = 0$, $\mathbf{E}[C_2] = 1$, and $\mathbf{E}[C_3] = \mathbf{E}[B] = \frac{8}{3}$, we obtain the Euler differential equation

$$C'''(z) = \frac{12C'(z)}{(1-z)^2} + \frac{24}{(1-z)^5} + \frac{4}{(1-z)^4} - 12,$$

with boundary conditions $C(0) = C'(0) = 0$, $C''(0) = 2$. The homogeneous part,

$$C'''(z) - \frac{12C'(z)}{(1-z)^2} = 0,$$

has $(1-z)^{-\lambda}$ as solution for every λ satisfying the characteristic equation

$$\lambda(\lambda+1)(\lambda+2) - 12\lambda = 0.$$

The solutions of the characteristic equation are 2, 0, and -5. The homogeneous solution is therefore

$$\frac{K_1}{(1-z)^2} + K_2 + K_3(1-z)^5,$$

where K_1, K_2, K_3 are three constants to be determined from the boundary conditions. To develop the particular solution, we use a well-known operator method. Introduce the operator

$$\mathcal{L} = \frac{d^3}{dz^3} - \frac{12}{(1-z)^2} \times \frac{d}{dz}.$$

A term $a/(1-z)^\theta$ on the right-hand side of the functional equation

$$\mathcal{L}\{C(z)\} = \frac{a}{(1-z)^\theta}$$

gives rise to the solution $K(1-z)^{-\lambda}$ when

$$\mathcal{L}\{K(1-z)^{-\lambda}\} = \frac{K\lambda(\lambda+1)(\lambda+2) - 12K\lambda}{(1-z)^{\lambda+3}} = \frac{a}{(1-z)^\theta}.$$

This is possible only if $\lambda = \theta - 3$, $K = a/[\lambda(\lambda+1)(\lambda+2) - 12\lambda] = a/[(\theta-3)(\theta-2)(\theta-1) - 12(\theta-3)]$, and θ is not a root of the equation

$$h(\theta) \stackrel{\text{def}}{=} (\theta-3)(\theta-2)(\theta-1) - 12(\theta-3) = 0, \tag{8.4}$$

i.e., when θ is not 5, 3, or -2. Hence, every term $a/(1-z)^\theta$ on the right-hand side with $h(\theta) \neq 0$ contributes to the solution the particular integral

$$\frac{a}{h(\theta)(1-z)^{\theta-3}}.$$

In our equation we have two terms matching this form: the term $\frac{4}{(1-z)^4}$, with $\theta = 4$ and $a = 4$, and the term -12, with $\theta = 0$ and $a = -12$. The corresponding solutions are respectively $-\frac{2}{3}(1-z)^{-1}$, and $-\frac{2}{5}(1-z)^3$.

When θ is a root of (8.4), we can again start with the formal requirement of a solution with symbolic θ:

$$\mathcal{L}\left\{\frac{a}{h(\theta)(1-z)^{\theta-3}}\right\} = \frac{a}{(1-z)^\theta},$$

and take derivatives with respect to θ to obtain the form

$$\frac{\partial}{\partial\theta}\mathcal{L}\left\{\frac{a}{(1-z)^{\theta-3}}\right\} = \frac{ah(\theta)}{(1-z)^\theta}\ln\left(\frac{1}{1-z}\right) + \frac{ah'(\theta)}{(1-z)^\theta}.$$

At a root of $h(\theta) = 0$, the last formal representation suggests a particular integral:

$$\mathcal{L}\left\{\frac{a}{(1-z)^{\theta-3}h'(\theta)}\ln\left(\frac{1}{1-z}\right)\right\} = \frac{a}{(1-z)^\theta}.$$

In our case, the only term involving a root of $h(\theta) = 0$ is $24/(1-z)^5$, which according to the argument just presented gives the particular integral $12/[7(1-z)^2]\ln[1/(1-z)]$.

Collecting contributions of homogeneous and particular solutions and matching the initial conditions to determine the constants K_1, K_2 and K_3 in the homogeneous solution, the complete solution to the differential equation is obtained:

$$C(z) = \frac{12}{7(1-z)^2}\ln\Big(\frac{1}{1-z}\Big) - \frac{207}{245(1-z)^2} - \frac{2}{3(1-z)} + \frac{82(1-z)^5}{735} - \frac{2(1-z)^3}{5} + \frac{9}{5}.$$

Extracting coefficients, we find the required averages.

Theorem 8.1 *(Sedgewick, 1980). The average number of comparisons for MEDIAN-OF-THREE QUICK SORT to sort $n \ge 6$ random keys is*

$$\mathbf{E}[C_n] = \frac{12}{7} n H_n - \frac{627}{245} n + \frac{12}{7} H_n - \frac{1111}{735} \sim \frac{12}{7} n \ln n, \qquad \textit{as } n \to \infty.$$

The nature of the distributional equation for MEDIAN-OF-THREE QUICK SORT is the same as that of standard QUICK SORT. Orders of magnitude of both mean and variance are the same for both algorithms, only differing in the coefficients of their respective leading terms. The same methods that were used in Chapter 7 can be used again to establish the existence of a limit C for the centered and scaled random variable

$$C_n^* \stackrel{\text{def}}{=} \frac{C_n - \frac{12}{7} n \ln n}{n}.$$

Subtracting off the asymptotic mean and scaling by n, the order of the standard deviation (as we did in QUICK SORT) gives us the analogous distributional equation satisfied by the limit.

Theorem 8.2 *(Rösler, 1991). Let C_n^* be the centered and scaled number of comparisons made by MEDIAN-OF-THREE QUICK SORT to sort n random keys. Then C_n^* converges in distribution to a limiting random variable C satisfying the equation*

$$C \stackrel{D}{=} LC + (1-L)\tilde{C} + g(L),$$

where $\tilde{C} \stackrel{D}{=} C$, and C, $\tilde{C}$ and L are independent, with L having the density

$$f_L(x) = 6x(1-x), \qquad \textit{for } 0 \le x \le 1,$$

and

$$g(u) = 1 + \frac{12}{7} u \ln u + \frac{12}{7}(1-u)\ln(1-u).$$

Recall that for standard QUICK SORT the limit law for P_n/n is that of UNIFORM(0,1). For MEDIAN-OF-THREE QUICK SORT it humps at the middle. The random variable L is the limit of P_n/n.

A full expansion of $\mathbf{Var}[C_n]$ may be possibly obtained at a great effort by the methods of Kirschenhofer, Prodinger and Martínez (1997). Such a formula may turn out to be several pages long. The leading term in that formula can be found directly by taking the variance of both sides of the distributional equation of the limit law. From the symmetry of L and $1 - L$, and the identical distribution of C and $\tilde{C}$, and from the various independencies,

$$\mathbf{Var}[C] = \mathbf{E}[C^2] = 2\mathbf{E}[C^2]\mathbf{E}[L^2] + \mathbf{E}\big[g^2(L)\big].$$

This yields

$$\mathbf{E}[C^2] = \frac{\mathbf{E}[g^2(L)]}{1 - 2\mathbf{E}[L^2]} = \frac{\int_0^1 g^2(u) f_L(u)\,du}{1 - 2\int_0^1 u^2 f_L(u)\,du} = \frac{485}{98} - \frac{24}{49}\pi^2.$$

We conclude this section with a few remarks about the performance of MEDIAN-OF-THREE QUICK SORT. Asymptotically MEDIAN-OF-THREE is slightly faster than plain QUICK SORT, and is also more concentrated. Its asymptotic average of $\frac{12}{7}n \ln n$ improves over standard QUICK SORT's $2n \ln n$ asymptotic average to sort n elements, as $n \to \infty$. The improvement is about %14, and one would expect greater savings for larger samples, i.e., for MEDIAN-OF-$(2k+1)$ QUICK SORT for $k > 1$. The methods of analysis that were used for MEDIAN-OF-THREE remain valid and versatile tools for MEDIAN-OF-$(2k+1)$ QUICK SORT. The differential equation for the generating function of averages will be of higher order, but it still is an Euler differential equation and can be handled by the operator method discussed earlier in the section.

Some versions of MEDIAN-OF-THREE QUICK SORT, like the Australian version, try to take advantage of the comparisons made during the median-finding stage. (Obvious generalizations to similar savings in MEDIAN-OF-$(2k+1)$ QUICK SORT are also possible improving variations on the same theme.) Sorting a file of n keys, the relative ranking of two elements is already known with respect to the pivot after median selection. The Australian version uses this partial information to reduce the $n - 1$ comparisons needed by PARTITION to only $n - 3$ as two pivot comparisons against data are unnecessary. It is shown in Kirschenhofer, Prodinger and Martínez (1997) that this reduction affects only lower-order terms in the analysis.

8.2 THE LARGE SAMPLE ALGORITHM

The choice of a large sample is perhaps only interesting from a theoretical viewpoint. With a properly chosen sample size QUICK SORT can be optimized to approach asymptotically the theoretic bound of $\lceil \lg n! \rceil \sim n \lg n$ on average. Let us suppose the sample size is $s = s(n)$, a yet-to-be chosen function of n in a way that optimizes the sampling approach to QUICK SORT. The algorithm proceeds in stages: A random sample is chosen, and in a practicable implementation the sample is inserted among the other data, partitioning them into segments. To facilitate the insertion of the sample, the sample itself is sorted prior to its insertion.

Each segment is then sorted. Assume, as usual, that the data reside in a host array $A[1 \mathinner{..} n]$. Any s positions of data following the random permutation model form a random subset of size s in the sense that the ranks of the s items are equally likely to be any of the $\binom{n}{s}$ possible choices of s ranks. So, we might as well take the first s positions residing in the segment $A[1 \mathinner{..} s]$ as our sample.

Three stages are present in the algorithm:

(a) The sorting of the sample itself—assume this stage makes X_n comparisons.
(b) The insertion of the sample—assume this stage makes Y_n comparisons.
(c) Sorting the subfiles—assume this stage makes Z_n comparisons.

Let the total number of comparisons made by such a large-sample algorithm be C_n. That is,

$$C_n = X_n + Y_n + Z_n.$$

Of course, the character of all these random variables will depend also on s. However, we shall choose s as a suitable function of n, and there is no need to use a heavier notation with s as a second subscript.

As we are interested in lowering the number of comparisons of QUICK SORT to ideally approach the theoretic lower bound, we can use any good sorting algorithm for stage (a). Such an algorithm should have the correct asymptotic order of magnitude, but not necessarily the correct coefficient. For example, we shall assume in the following paragraphs that QUICK SORT itself is chosen for this purpose. QUICK SORT makes about

$$2s \ln s \tag{8.5}$$

to sort a large sample of size s, which is $2 \ln 2 \approx 1.386$ as much as an optimal algorithm would. A correct order, not necessarily having the asymptotic equivalent of an optimal algorithm, may be sufficient for optimization because after all we are applying it to a sample, which is necessarily of a size that is an order of magnitude below n. The analysis below shows that indeed QUICK SORT for the sample insertion still leads to an optimized algorithm that we shall refer to as the large sample algorithm. The sample size will be assumed to be $s = o(n)$, but still grows to infinity as $n \to \infty$.

For the insertion stage, we may repeatedly use PARTITION, a building block of QUICK SORT (the algorithm of Figure 7.2)—this algorithm makes n comparisons to insert a pivot among n other elements. We can of course insert the sample elements in whichever order. To speed up the process on average, we shall try to split files in the middle. We first insert the sample's median, followed by its two quartiles, then octiles that are not among the preceding insertions, and so on. We shall call this scheme the *quantile scheme*. This way, for example, on average, when the two quartiles are inserted none of them falls in an unduly long subfile, which is to be contrasted with a case like inserting the sample according to its order (smallest key in the sample, second smallest, third smallest, etc.). The contrast is exemplified in Exercise 8.2.

Lemma 8.1 *(Frazer and McKellar, 1970). Following the quantile scheme,*

$$\mathbf{E}[Y_n] \sim (n - s) \lg s.$$

Proof. Following the quantile scheme, the median of the sample will be inserted first with $n - s$ comparisons. This creates two subfiles of sizes N_1 and N_2, totaling to $N_1 + N_2 = n - s$. In turn, the insertion of the lower quartile in N_1 elements will make N_1 comparisons and the insertion of the upper quartile in N_2 elements will make N_2 comparisons. Combined, the insertion of the lower and upper quartiles requires $n - s$ comparisons, the next four octiles combined will require $n - s$ comparisons, and so forth. That is,

$$\mathbf{E}[Y_n] = \underbrace{(n - s) + \cdots + (n - s)}_{(\lfloor \lg s \rfloor - 1) \text{ times}} + L_{n,s},$$

where $L_{n,s}$ is a leftover quantity that corresponds to the insertion of the last $s - (2^{\lfloor \lg s \rfloor} - 1)$ pivots. Clearly, $L_{n,s} \le n - s$. ■

Suppose the sorted sample keys are $K_{(1)} \le K_{(2)} \ldots \le K_{(s)}$. After the insertion stage, there are N_1 unsorted elements that are below $K_{(1)}$ and appear in $A[1 \,..\, N_1]$, there are N_2 elements that are between $K_{(1)}$ and $K_{(2)}$ and appear in $A[N_1 + 2 \,..\, N_1 + N_2 + 1]$, and so on. At the tail of the file there are N_{s+1} unsorted keys that are larger than $K_{(s)}$ and appear in $A[N_1 + \ldots + N_s + s \,..\, n]$ as illustrated in Figure 8.1. Of course, the N_j's are random and dependent as they must satisfy

$$N_1 + \cdots + N_{s+1} = n - s,$$

the size of nonsample elements. QUICK SORT is applied to each of these $s + 1$ subarrays. We shall assume that QUICK SORT requires Q_i comparison to sort an array of size i. On the ith file QUICK SORT performs Q_{N_i} comparisons. Therefore Z_n, the number of comparisons to sort these $s + 1$ subarrays, is

$$Z_n \overset{\mathcal{D}}{=} Q^{(1)}_{N_1} + Q^{(2)}_{N_2} + \cdots + Q^{(s+1)}_{N_{s+1}},$$

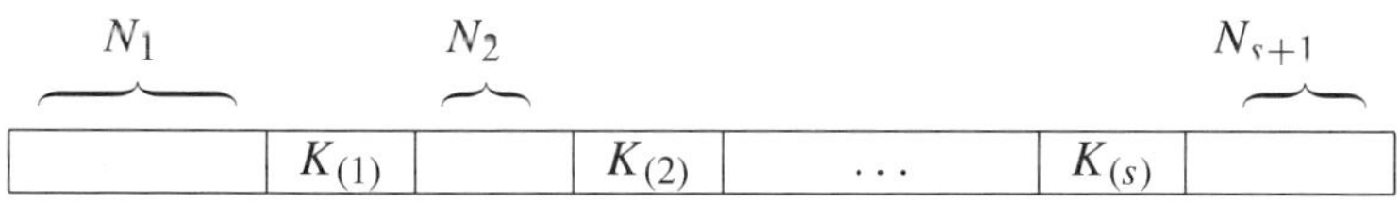

Figure 8.1. Partitioning by a large sample.

where, for each i and j, the random variables $Q_i^{(j)} \stackrel{D}{=} Q_i$, and for any two different indexes j and k, the random variables $Q_\mu^{(j)}$ and $Q_\nu^{(k)}$ are independent for all μ and ν by the independence of the action of the sorting algorithms in the subarrays.

Lemma 8.2 *(Frazer and McKellar, 1970).*

$$\mathbf{E}[Z_n] = 2(n+1)(H_{n+1} - H_{s+1}) - 2(n-s).$$

Proof. QUICK SORT is used in the subarrays, giving an expected total number of comparisons equal to

$$\mathbf{E}[Z_n] = \mathbf{E}[Q_{N_1}^{(1)}] + \cdots + \mathbf{E}[Q_{N_{s+1}}^{(s+1)}] = (s+1)\mathbf{E}[Q_{N_1}],$$

by symmetry of the sizes of the $(s+1)$ subarrays. Then, by conditioning on the share of the first subarray,

$$\mathbf{E}[Z_n] = (s+1)\sum_{j=0}^{n-s} \mathbf{E}[Q_j]\,\mathbf{Prob}\{N_1 = j\}.$$

According to the random permutation model, all $\binom{n}{s}$ choices of ranks are equally likely to appear in the sample. We can easily count the number of those choices that are favorable to the event $\{N_1 = j\}$. For this event to happen, the key with rank $j+1$ must appear in the sample, and no key with smaller rank should appear in the sample (if a key with a lower rank appears, N_1 would have to be less than j). We must choose the key ranked $j+1$ in the sample and the remaining $s-1$ keys in the sample must come from a set of keys with ranks $j+2, \ldots, n$, which can happen in $\binom{n-j-1}{s-1}$ ways. That is, the probability in question is $\binom{n-j-1}{s-1}/\binom{n}{s}$. For $\mathbf{E}[Q_j]$ we have an exact expression from Hoare's theorem (Theorem 7.1). The issue now is only a matter of combinatorial reduction of

$$\begin{aligned}\mathbf{E}[Z_n] &= \frac{(s+1)}{\binom{n}{s}} \sum_{j=0}^{n-s} \mathbf{E}[Q_j]\binom{n-j-1}{s-1} \\ &= \frac{(s+1)}{\binom{n}{s}} \sum_{j=0}^{n-s} [2(j+1)H_j - 4j]\binom{n-j-1}{s-1}.\end{aligned}$$

This formula collapses to the exact value of the lemma—this follows from straightforward combinatorial reductions and the following nontrivial combinatorial identity suggested by Knuth (1973):

$$\sum_{k=0}^{n} \binom{k}{m} \frac{1}{n+1-k} = \binom{n+1}{m} (H_{n+1} - H_m).$$

This latter identity can be proved by induction as in Frazer and McKellar (1970). ∎

The average number of comparisons for each of the three stages has been determined in (8.5) and Lemmas 8.1 and 8.2. Putting the three averages together, we have an asymptotic expression for the large-sample algorithm.

Theorem 8.3 *(Frazer and McKellar, 1970). Let C_n be the number of comparisons the large-sample algorithm ($s \to \infty$) makes to sort a random input of size n. This random variable has the asymptotic average value*

$$\mathbf{E}[C_n] \sim 2s \ln s + (n - s) \lg s + 2(n + 1)(H_{n+1} - H_{s+1}) - 2(n - s).$$

We shall next appeal to a heuristic argument to optimize the sample size. It will turn out that as simple as it is the heuristic gives a result that cannot be improved asymptotically on grounds of the universal lower bound result on comparison-based algorithms (Theorem 1.1). The heuristic, based on approximating the functions involved, gives a value of s that drives the algorithm asymptotically down to $n \lg n$ comparisons. Had we worked with the exact expression or sharper asymptotic estimates, we could not have improved beyond the asymptotic equivalent $n \lg n$; perhaps we could have only improved the lower-order terms (the rate of convergence to the theoretic limit $\lceil \lg n! \rceil \sim n \lg n$).

An important consequence of Theorem 8.3 is that it enables us to choose an optimal s.

Corollary 8.1 *(Frazer and McKellar, 1970). When the large-sample algorithm is applied to a large number of keys $n \to \infty$, s that satisfies the equation*

$$s \ln s = n$$

is an asymptotically optimal choice of the sample size as it asymptotically minimizes the average number of comparisons over the three stages of the algorithm.

Proof. Consider an asymptotic equivalent of the expression of Theorem 8.3 when harmonic numbers are replaced by their approximate value, when both n and s tend to infinity:

$$g(s) \stackrel{\text{def}}{=} 2s \ln s + (n - s) \lg s + 2n(\ln n - \ln s) - 2(n - s).$$

Treat this approximate expression for some large n as a function of s. To minimize $g(s)$, set its derivative with respect to s equal to zero:

$$g'(s) = 4 + 2\ln s - \frac{\ln s}{\ln 2} + \frac{n - s}{s \ln 2} - \frac{2n}{s} = 0.$$

For large n, the solution of this equation is asymptotically equivalent to that of

$$2\ln s - \frac{\ln s}{\ln 2} + \frac{n}{s\ln 2} - \frac{2n}{s} = 0.$$

Collecting common factors, the latter equation is

$$s \ln s = n.$$

Now, plugging the chosen s into the asymptotic expression $g(s)$ of the average, we obtain

$$\begin{aligned}
\mathbf{E}[C_n] &\sim 2s\ln s + (s\ln s - s)\lg s + 2s(\ln s)\big[\ln(s\ln s) - \ln s\big] - 2s\ln s + 2s \\
&\sim (s\ln s - s)\lg s + 2s(\ln s)(\ln s + \ln\ln s - \ln s) + 2s \\
&\sim s(\ln s)\lg s \\
&\sim n\lg s.
\end{aligned}$$

However, because $s\ln s \sim n$, we also have $\lg s \sim \lg n$. The large sample algorithm has the asymptotic equivalent $n \lg n$, which is the lower bound on sorting by comparisons. ■

The large sample algorithm "barely" approaches optimality—in the proof of Corollary 8.1 we worked only with leading-order terms. The lower-order terms are almost comparable to the leading order for any practical n. If we implement this algorithm, the rate of approaching the asymptotic optimality will be extremely slow and n will have to be astronomically large for the leading terms both to stand out among all lower-order terms, and to stand out compared with the large overhead incurred by the various stages of the algorithm, especially the second stage, which in practice may require delicate handling with large overhead.

EXERCISES

8.1 Let P_n be the landing position of the pivot in MEDIAN-OF-THREE QUICK SORT. Show that $P_n/n \xrightarrow{\mathcal{D}} L$, where L is a random variable with density $6x(x-1)$, for $x \le 0 \le 1$.

8.2 In a large sample approach, the sample may be inserted within the rest of the data according to several schemes. In the text we discussed the quantile scheme. By contrast, suppose we insert the sample sequentially—for instance, in a given array $A[1\,..\,n]$ of size n, we may first sort $A[1\,..\,s]$ and use its s elements as our sample, to be inserted according to the sequence $A[1], A[2], \ldots, A[s]$. What is the number of comparisons required for the sample insertion by the usual PARTITION procedure (the algorithm of Figure 7.2)? How does this sequential scheme compare to the quantile scheme?

9

Heap Sort

One can sort data by manipulating a suitable data structure. For example, one can build a binary search tree and then sort the data by traversing the tree in a manner that recursively visits the left subtree, then the root node, then the right subtree. Each node of such a tree will have to carry two pointers. Yet another elegant data structure, the heap, gives rise to a nice sorting routine. The heap is essentially a pointer-free complete binary tree. The heap can be implemented within an array. By fixing a convention of where the children of a node are in the array, one can move freely up and down the tree by following that convention. In 1964 Williams described a top-down heap construction whose worst case for n items is $\Theta(n \ln n)$. Williams derived a sorting algorithm from the heap. Later that year, Floyd gave a bottom-up heap construction algorithm that builds the heap in $O(n)$ for any input of size n.

Roughly speaking, HEAP SORT works by creating a heap for n keys, then manipulating the root and restoring a heap of $n - 1$ elements, and so on. The restoration is facilitated by the fact that one does not restructure the heap from scratch, but from the two subtrees of an existing heap, which have ample information about a large partial order in the heap. One then encounters a sequence of heaps of diminishing size.

HEAP SORT is efficient because:

(a) Each of the n keys is made to travel at most the height of a heap of size at most n. The heaps encountered in the process have height that is at most $\lg n$, thus achieving $O(n \ln n)$ sorting for all inputs of size n.

(b) The method works in situ.

The method definitely destroys the randomness after the first stage, which makes probabilistic analysis rather convoluted.

9.1 THE HEAP

First, let us introduce the conceptual heap. The heap is a complete binary tree, labeled such that every node carries a label larger than the labels of all its descendants. Figure 9.1 shows a conceptual heap.

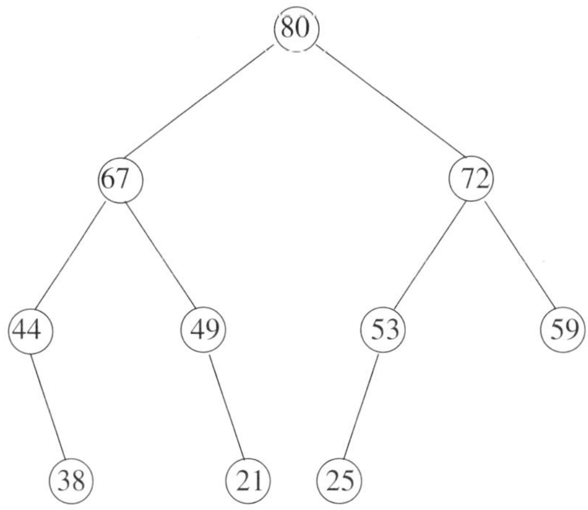

Figure 9.1. A conceptual heap.

HEAP SORT is designed to work in situ on a data array, say $A[1\,..\,n]$. Toward a pointerless implementation of this tree, we adopt the convention that the children of a node stored in $A[i]$ (if they both exist) reside at $A[2i]$ and $A[2i+1]$. To avoid gaps in the data representation, we also put an additional restriction on the admissible shapes of complete binary trees, always favoring left children—if a node has only one child, we choose to attach it as the left child. In other words, among the many possible complete binary trees on n nodes, we choose the heap to be the one with all the nodes at the highest level packed as far to the left as possible (see Figure 9.2).

For the rest of this chapter the term *heap* will refer to labeled trees meeting the required definition of the conceptual heap, and having the triangular shape of a complete binary tree with all the nodes at the highest level appearing as far to the left as possible. This choice guarantees that the data of the heap will appear contiguously in the first n position of the array, and that, for any "node" i, if $2i < n$, then $A[2i]$ contains a key from the heap (the left child of node i), and $A[2i+1]$ contains a key from the heap (the right child of node i); if $2i = n$, node i has only a left child at $A[2i]$.

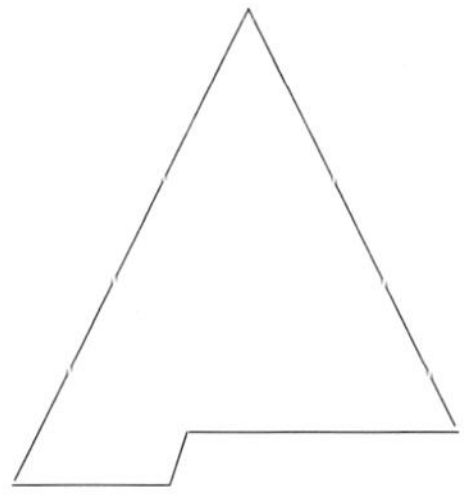

Figure 9.2. The admissible complete tree shape for a heap.

In HEAP SORT the heap construction is only a preprocessing stage followed by the data manipulation stage that actually sorts. Several good algorithms to construct a heap are known. To organize n keys into a heap is a process that operates in $\Theta(n)$ according to these various algorithms. So, the more clever ones can only improve the constant coefficient hidden in the $\Theta(n)$ expression. In some other applications where the heap is meant for its own sake the differentiation between heap construction algorithms may be of paramount importance. However, in a sorting application, as we know from the theoretical lower bound (Theorem 1.1), ordering n keys by comparisons within a heap will require $\Omega(n \ln n)$; an intricate but efficient heap construction algorithm will only improve the lower-order terms in the efficiency of HEAP SORT. Therefore we shall not dwell for too long on the choice of a subtle heap construction algorithm. We present below Floyd's algorithm, one of the earliest and most intuitive approaches to heap construction. It is the speed of repeated reconstruction of a heap after deleting its root that can make a noticeable difference in the efficiency of HEAP SORT.

Starting out with a (random) set of data in the array $A[1 \,..\, n]$, Floyd's algorithm works in a bottom-up fashion. It constructs the "shrubs" at the bottom of the heap (heaps of two levels). After all these bottom subheaps are obtained, their roots are combined via parent nodes to form larger subheaps on three levels, and so on.

For instance, to arrange the 10 keys

1	2	3	4	5	6	7	8	9	10
53	44	59	21	25	38	67	80	49	72

we start with the last index (10 in this case), and decide to combine the array entry (henceforth referred to interchangeably as a node) at the 5th position with its child at position 10 into a complete tree on two nodes. We see that $A[10] > A[5]$. To satisfy the heap property we must exchange these two keys to obtain the first subheap in our construction as shown in Figure 9.3(a). (In Figure 9.3 data are encircled and the number outside a node is its index in the array.) We then retract the index of the parent by one position to produce 4 as a new parent position, whose two children are at positions $2 \times 4 = 8$ and $2 \times 4 + 1 = 9$. We exchange the data as necessary to

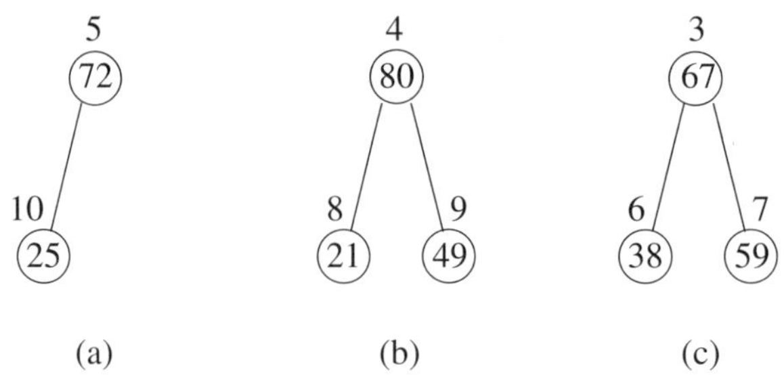

Figure 9.3. A round of Floyd's algorithm constructing subtrees of height 1.

arrange $A[4]$ and its two children $A[8]$ and $A[9]$ as the subheap of Figure 9.3(b), that is, put the largest of the three elements in the root of a complete tree of order 3 by exchanging 21 and 80. Position 3 is the next parent; $A[3]$ and its two children $A[6]$ and $A[7]$ are arranged into the subheap of Figure 9.3(c). The status of the array now is

1	2	3	4	5	6	7	8	9	10
53	44	67	80	72	38	59	21	49	25

Going down to position 2, its two children at positions 4 and 5 are already engaged as roots of subheaps; all we need do is combine these subheaps with the root at 2, to obtain a subheap of three levels. Regarding position 2 and all its descendents before the rearrangement as a complete tree we have the tree of Figure 9.4(a), which we want to reorganize as a heap. We observe here that only the root of the tree of Figure 9.4(a) violates the heap labeling property. The two subtrees are already heaps and we need only adjust the structure. One of the two subheaps has a root ($A[4] = 80$) larger than $A[2] = 44$ and larger than the root of the other subheap; we promote 80 to be the root of the subheap rooted at $A[2]$, and position 4 is a candidate recipient of 44. Yet, 44 there would violate the heap property as the right child $A[9] = 49 > 44$; we promote 49 to be at $A[4]$, and now use its position to store 44, obtaining the heap of Figure 9.4(b).

We are down to the last parent. The array now is:

1	2	3	4	5	6	7	8	9	10
53	80	67	49	72	38	59	21	44	25

and the last task is to combine the left heap rooted at position 2 (Figure 9.4(b)) with the right heap rooted at position 3 (Figure 9.3(c)) with $A[1] = 53$ into a heap of size 10. As we did before, we promote appropriate nodes: 80 goes down to position $A[1]$ and 72 goes down to $A[2]$, vacating $A[5]$ for 53 (note that 25 should not move down

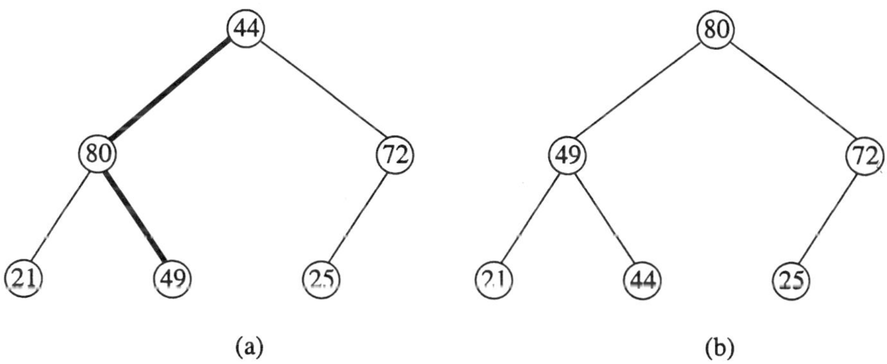

Figure 9.4. (a) Two subtrees (heaps) of two levels each and a root. (b) Rearrangement into a heap of three levels. The darkend edges are the path of maximal sons.

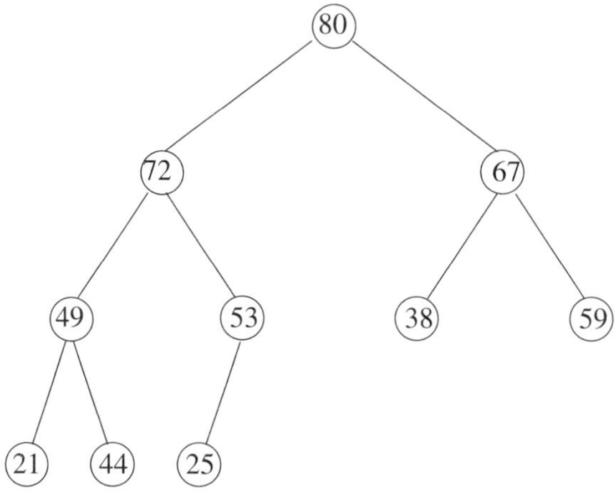

Figure 9.5. The final heap.

as it is less than 53)—the present example is not an instance of worst-case behavior; the root does not travel at every heap reconstruction all the way along a path of maximal sons to replace a terminal node. Finally we obtain the heap of Figure 9.5.

The general idea is to work backwards with subheap roots at positions $i = \lfloor n/2 \rfloor, \ldots, 2, 1$. For general $i < \lfloor n/2 \rfloor$, we combine two subheaps rooted at $A[2i]$ and $A[2i+1]$ with the root at $A[i]$, which may possibly violate the heap property. In the algorithm implementation we save $A[i]$ in a variable, say key. We trace a *path of maximal sons*, always choosing the larger of the two children for a comparison against key (we use j as the index of the larger child). An instance of the path of maximal sons in a heap is shown by the darkened edges in the tree of Figure 9.4(a). If the chosen child is larger than key, we promote it in the tree, and apply the operation recursively in the subheap rooted at the chosen child. We repeat these promotions until either we reach a node whose two children are smaller than key, or the doubling of the position j to find children throws us out of bounds. In either case, that last position is the place for key. It is clear that the operation of combining two subheaps with a root node, possibly out of the heap property, to rebuild a heap is repeatedly needed at various parent positions. The algorithm of Figure 9.6 is the core of Floyd's algorithm written in the form of a procedure that receives a parent at position i, and the size s of the heap to be constructed. The procedure assumes that the subtree rooted at $A[2i]$ and $A[2i+1]$ are heaps, and it returns the array with the complete subtree rooted at i as a heap. (As usual, the array A is accessed globally.)

The rest of Floyd's algorithm is to use the heap rebuilding procedure repeatedly at the roots $i = \lfloor \frac{n}{2} \rfloor, \ldots, 2, 1$ as in the code:

for $i \leftarrow \lfloor \frac{n}{2} \rfloor$ **downto** 1 **do**
    call $RebuildHeap(i, n)$;

```
procedure RebuildHeap(i, s : integer);
    local j, key: integer;
    begin
        key ← A[i];
        j ← 2i;
        while j ≤ s do
            begin
                if (j < s) and (A[j] < A[j + 1]) then
                    j ← j + 1;
                    if key ≥ A[j] then
                        begin
                            A[⌊j/2⌋] ← key;
                            return;
                        end;
                A[⌊j/2⌋] ← A[j];
                j ← 2j;
            end;
        A[⌊j/2⌋] ← key;
    end;
```

Figure 9.6. The core of Floyd's heap construction algorithm.

9.2 SORTING VIA A HEAP

Once a heap is available, sorting is a relatively easy matter. The heap construction guarantees that its root is the largest datum. We can simply take it out and claim it as the largest key. We do that by placing it at the last position by exchanging it with $A[n]$; the largest key is now at its correct position in the array. What do we have then in $A[1 \mathinner{..} n-1]$? We simply have a complete tree whose right and left subtrees are heaps; a small element came from position n and is now sitting at the root of the whole tree. We need to adjust a labeled complete binary tree structure on $n-1$ keys whose two subtrees are heaps, and its root may be in violation of making the entire aggregate $A[1 \mathinner{..} n-1]$ a heap. No new algorithm is required for this adjustment; the procedural part of Floyd's algorithm (Figure 9.6) is designed to do just that, if invoked with the right parameters (root at 1 and size $n-1$). After restoring a heap in $A[1 \mathinner{..} n-1]$ we can extract the root; exchange it with $A[n-1]$, and the second largest key is now at its correct position in the array. We then repeat the process on $A[1 \mathinner{..} n-2]$, etc. The root of the complete tree to be adjusted is always 1, but at the ith stage, the size of the subtree is $n-i+1$, as in the algorithm in Figure 9.7.

An easy-to-establish upper bound on the number of data moves can be found. The iteration dealing with a complete tree of size i looks for an insertion position in a path starting at the root. The largest number of comparisons that the root key needs to travel to its appropriate position in the heap is the height of the (extended) complete tree on i nodes, which is $\lceil \lg(i+1) \rceil$. In the worst case, insertion at the end

```
for i ← n downto 2 do
    begin
        call swap(A[1], A[i]);
        call RebuildHeap(1, i − 1);
    end;
```

Figure 9.7. HEAP SORT.

of such a path will occur at each iteration, giving

$$\sum_{i=2}^{n} \lceil \lg(i+1) \rceil = \sum_{i=2}^{n} (\lg i + O(1)) = \lg n! + O(n).$$

The term $\lg n!$ is the same quantity that appeared in the lower bound on the number of comparisons of a worst-case optimal sort. It was analyzed by Stirling's approximation and we found its leading term to be $n \lg n + O(n)$. So, the number of moves during repeated calls to *RebuildHeap* in the algorithm of Figure 9.7 cannot exceed $n \lg n + O(n)$. In fact, the number of moves during the whole sorting routine does not exceed $n \lg n + O(n)$, because the only additional moves come from the initial execution of Floyd's algorithm to obtain the first heap of size n. This execution takes $\Theta(n)$ time, as can be seen from the following argument. The terminal nodes of the heap are at distance $d_n = \lfloor \lg n \rfloor$ from the root. The first round of the bottom-up construction builds trees of two levels. Recall that this is done in Floyd's algorithm by sliding large keys toward the root. In the worst case every node that is a parent of a terminal node makes a move; there are at most $2^{d_n - 1}$ nodes in the layer before last. The second round that combines the two-level shrubs with roots to obtain three-level shrubs can move any of the $2^{d_n - 2}$ nodes at distance $d_n - 2$ two levels to become terminals, etc. The number of moves is bounded above by

$$\begin{aligned} 1 \times 2^{d_n - 1} + 2 \times 2^{d_n - 2} + 3 \times 2^{d_n - 3} + \cdots + d_n \times 1 &= 2^{d_n + 1} - d_n - 2 \\ &< 2^{1 + \lg n} \\ &< 2n. \end{aligned}$$

What about comparisons? In Floyd's algorithm moving a key involves two comparisons: one to find the larger of the two children, and one more to compare it with the key we are relocating. In the initial phase of heap construction the worst-case number of data comparisons is therefore at most $4n$. The worst-case number of data comparisons of the entire sort is asymptotic to $2n \lg n$; not quite the theoretical lower bound of $n \lg n$.

This gap invited researchers to come up with improvements. Floyd (1964) himself suggested in general terms that data comparisons be avoided till the last possible minute by an algorithm that goes all the way along a path of maximal sons without comparing the root to any of its members (Figure 9.4(a) illustrates in darkened edges a path of maximal sons). Once the terminal node of the path of maximal sons is found we reverse the direction and back up (halving the index by integer divisions) along the same path but this time comparing the key being inserted with the nodes on this path, sliding them toward the leaves (the opposite action of Floyd's algorithm) till the insertion position is found. The idea was followed through in the work of Carlsson (1987) and Wegener (1993), who established that at most $\frac{3}{2} n \lg n$ comparisons are needed for any input of size n to this algorithm.

Can we do better? An elegant idea of Gonnet and Munro (1986) saves the day. One may observe that at a global level Floyd's algorithm searches for a key in a tree. The method of search chosen in Floyd's algorithm is essentially a linear search along a path of maximal sons of length $\lg n + O(1)$. Gonnet and Munro (1986) suggested that a careful implementation of the algorithm (one that remembers the indexes of the path of maximal sons) can improve the overall cost. The labels of that path are in decreasing order. Unsuccessful binary search for a key in a sorted list of size i takes $\lg i + O(1)$ comparisons regardless of the destination gap that the new key belongs to; one can search the sorted sublist on the path of maximal sons (of length $\lg i + O(1)$) using binary search with $\lg \lg i + O(1)$ comparisons. The structure of this sorting algorithm is similar to the HEAP SORT we have already presented, except that a complete path of maximal sons is found (a root-to-terminal node path), the indexes of the path are stored (in an array, say), then binary search is performed on the subarray corresponding to the nodes of the path of maximal sons to insert the root key. When the heap's size is i the number of comparisons made along the path of maximal sons is the length of the path, $\lg i + O(1)$, then binary search takes at most $\lg \lg i + O(1)$. The overall number of comparisons is bounded above by

$$\sum_{i=2}^{n} \big(\lg i + \lg \lg i + O(1)\big) = n \lg n + o(n \ln n).$$

HEAP SORT based on binary search is a worst-case optimal comparison-based sorting algorithm. The space needed to remember the sequence of indexes on the path of maximal sons needs to hold at most $1 + \lg n$ indexes. So, this HEAP SORT can be considered an in-situ sorting algorithm. Schaffer and Sedgewick (1993) have shown that the standard version has $n \lg n$, asymptotic average, too.

EXERCISES

9.1 In the text the data arrangement

53 44 59 21 25 38 67 80 49 72

was used for illustration. Find a permutation of these data that is worst possible for Floyd's algorithm on 10 keys.

9.2 Starting with random data (a random permutation of $\{1, \ldots, n\}$) Floyd's algorithm constructs a heap, then HEAP SORT exchanges the root $A[1]$ with $A[n]$. After this exchange, is the array $A[1 \mathinner{..} n - 1]$ a random permutation of $\{1, \ldots, n - 1\}$?

9.3 How many heaps are possible for n distinct keys? (Hint: Write a recurrence relation based on the structure of the subtrees.)

10

Merge Sort

We have seen how the divide-and-conquer paradigm leads to the efficient sorting algorithm QUICK SORT, which becomes particularly good when the data file is split into two parts of about equal size at most recursive levels. For the abundance of instances of data where this happens, the algorithm possesses $\Theta(n \ln n)$ behavior. In the fewer instances when the recursion does not repeatedly split the file nearly equally, the behavior may deteriorate to $\Theta(n^2)$.

MERGE SORT is designed to take advantage of the idea that equal splitting in the divide-and-conquer paradigm gives good performance. Merging two sorted lists into one larger list is a very straightforward process and suits several common types of data structures like arrays and linked lists. MERGE SORT is a divide-and-conquer algorithm, recursive in its standard form, designed especially to observe that the segment of the file being sorted is split into two "halves" of sizes that are as close as they can be.[1] If at some stage the subfile considered is of length n, MERGE SORT splits it in a top-down fashion into two files of sizes $\lceil n/2 \rceil$ and $\lfloor n/2 \rfloor$, then each subfile is sorted recursively. This splitting is continued recursively until small files of size 1 are reached; these are already sorted. The algorithm then enters a phase of merging sorted sublists working from the ground up—merging the sorted small clusters into bigger and bigger clusters until the whole file is sorted.

A bottom-up version avoiding recursion will also be discussed. This bottom-up approach does not lead to any noticeable difference as the analysis will show. Both versions of the algorithm are $O(n \ln n)$ for all inputs. The only criticism that MERGE SORT algorithms may receive is that they are usually not in-situ. To sort n items in a data structure, a secondary data structure of the same size must be allocated in most standard merging algorithms.

10.1 MERGING SORTED LISTS

There are several ways to merge two *sorted* lists of data with sizes m and n. Without loss of generality we shall assume that $m \le n$, for we can always rename the variables denoting the sizes. We shall call a merging algorithm MERGE. Since MERGE

[1]When a file of size n is split in two parts of sizes $\lceil n/2 \rceil$ and $\lfloor n/2 \rfloor$, the two parts are of the exact same sizes when n is even, or differ in size by 1 when n is odd. The proportion of file keys in either part is asymptotic to $1/2$, and we shall liberally call the two parts of the split the two *halves* even when their sizes are not equal.

SORT is designed to optimize the splitting of data in the middle of a list, order data in each half, then merge the resulting two sorted halves, we shall later use MERGE on two sorted lists of sizes differing by at most 1. So, we shall focus in our presentation on merging algorithms that deal with two lists of sizes differing by at most 1. Moreover, for the purpose of sorting by merging, MERGE SORT will need to merge two lists residing in two different portions of the same host data structure. Any general MERGE procedure can be easily specialized to the needs of MERGE SORT.

We shall present the intuitive LINEAR MERGE algorithm, a rather simplistic, yet efficient merging algorithm. Other MERGE algorithms may be devised to pursue goals other than simplicity. For example, a MERGE algorithm may be sought to achieve merging in-situ.

As simple as it is, we shall see that LINEAR MERGE is indeed worst-case optimal for the purpose of MERGE SORT. Therefore in later sections we shall restrict the presentation of MERGE SORT and its analysis to a variation that uses LINEAR MERGE.

By way of contrast, we shall briefly sketch two other merging algorithms, BINARY MERGE and the HWANG-LIN algorithm. At first sight, they may appear to provide efficiency over LINEAR MERGE. A closer look shows, however, that neither has any real advantage over LINEAR MERGE in conjunction with MERGE SORT. They may offer an advantage in unbalanced merging (the merging of two lists with substantially different sizes), a situation that MERGE SORT avoids by design; MERGE SORT divides a given list into two sorted halves before it merges them.

To argue that the simplistic LINEAR MERGE is indeed a good merging algorithm for the relevant probability models, let us take up the theoretical issue of absolute bounds on MERGE. Introduce the variable S_{mn}, the number of comparisons sufficient to always merge two sorted lists of sizes m and n. This is the usual min-max definition for worst-case optimality considerations—S_{mn} is the number of comparisons the best MERGE algorithm would make on its worst-case pair of sorted lists of sizes m and n.

Let us first consider general bounds on merging to see if merging algorithms other than LINEAR MERGE would be worthwhile. We shall shortly see in Subsection 10.1.1 that LINEAR MERGE takes at most $m + n - 1$ comparisons on any two sorted lists of the sizes considered; this provides us with an obvious upper bound. A not-too-sharp lower bound follows from a decision tree argument similar to the one we used to find the theoretical lower bound on comparison-based sorting (by contrast, fortunately there the bound was sharp). Suppose the elements of the first sorted list are

$$X_1 < X_2 < \dots < X_m,$$

and those of the second sorted list are

$$Y_1 < Y_2 < \dots < Y_n.$$

The goal of an optimal MERGE algorithm is to determine the total order in the data set $\{X_1, \dots, X_m\} \cup \{Y_1, \dots, Y_n\}$. We assume the entire data collection consists of distinct elements (a stipulation that matches the random permutation model when all the elements of the whole data set are sampled from a continuous distribution). Construct a decision tree to model the action of the optimal MERGE algorithm in the usual way, with a node labeled by the query $X_i \overset{?}{\le} Y_j$ if it corresponds in the optimal MERGE algorithm to a comparison between X_i and Y_j. At the top of the tree, we must find leaves labeled with various total orders. For example, one possible input data set has all X's less than all Y's. Our optimal MERGE algorithm must ask a sufficient number of questions climbing along a path in the decision tree terminating at a leaf labeled

$$X_1 \ X_2 \ \dots \ X_m \ Y_1 \ Y_2 \ \dots \ Y_n.$$

To count the number of different orders in an input of X's and Y's such that the X's are sorted among themselves and the Y's are sorted among themselves, let us start with unconstrained permutations. There are $(m + n)!$ permutations of $m + n$ distinct keys. All $m!$ permutations of the X's within will give us the same sorted list of X's, and all $n!$ permutations of Y's will give us the same sorted list of Y's. In other words, there are $(n+m)!/(m!\,n!) = \binom{n+m}{m}$ total orders that respect the known partial order.

Our algorithm must be prepared to discover every single total order. There are $\binom{m+n}{m}$ leaves in the decision tree. A binary tree with this many leaves has height at least $\left\lceil \lg \binom{m+n}{m} \right\rceil$ (see Proposition 1.1), giving us a lower bound on the number of comparisons that must be made to discover the total order in our data set. Putting the two bounds together, we see that

$$\left\lceil \lg \binom{m+n}{m} \right\rceil \le S_{mn} \le m + n - 1.$$

MERGE SORT's strategy splits lists at the middle as discussed. We are interested particularly in bounds for the case where the two lists differ in size by at most one. By Stirling's approximation for factorials in the case where m and n are replaced with $\lfloor n/2 \rfloor$ and $\lceil n/2 \rceil$, we obtain the lower bound

$$S_{\lfloor n/2 \rfloor, \lceil n/2 \rceil} \ge n - \frac{1}{2} \lg n + O(1).$$

10.1.1 LINEAR MERGE

LINEAR MERGE is one of the simplest ways of merging two sorted lists. It thinks in terms of laying the two sorted lists down as two horizontal strings of data and aligning their left edges. LINEAR MERGE works in two stages—the *comparison stage* followed by the *tail transfer stage*. LINEAR MERGE begins the comparison stage by comparing the first element in each list (the leftmost in each), then transfers the smaller of the two to a new list. Each of the two numbers compared is the smallest in its list because each list is assumed sorted. Their minimum is therefore the absolute minimum in the entire collection of data (the set formed by the union of the two sets of data). The second datum in the list from which the element has been transferred is now promoted to be the first in that list (smallest among what is left in that list). This is the key to look at next; we compare it with the smallest element of the other list and this will be the second time this latter element is involved in a comparison. We transfer the smaller of the two, which must be the second smallest in the data set, to the recipient list. We continue in this fashion until one of the lists is exhausted. What is left in the other list is a "tail" collection of elements (possibly only one) in increasing order from left to right and all the elements in this tail are larger than all elements in the recipient list. The tail transfer stage is then executed by moving or copying the elements of the tail in order, starting with their smallest (leftmost) until the tail is exhausted. Transferring an element does not involve any comparisons and is implemented by operations like copying or assignments, which are generally cheaper than comparisons. Figure 10.1 illustrates the merging of two lists with the arrows indicating the next element to go out to the recipient list and Figure 10.2 gives the formal code of an implementation *LinearMerge* of the merging algorithm adapted to the specific needs of MERGE SORT. The implementation *LinearMerge*

Progress of the Recipient List		Progress of the Two Lists
–	← (first list)	12 17 18 13 16 19 23 25
12	← (second list)	17 18 13 16 19 23 25
12 13	← (second list)	17 18 16 19 23 25
12 13 16	← (first list)	17 18 19 23 25
12 13 16 17	← (first list)	18 19 23 25
12 13 16 17 18		– 19 23 25
12 13 16 17 18 19 23 25		– –

Figure 10.1. Merging two sorted lists by LINEAR MERGE.

```
procedure LinearMerge(ℓ, u: integer);
    local i, j, k, mid: integer;
        C: array[1 .. n] of integer;
    begin
        k ← 1;
        mid ← ⌊(ℓ+u)/2⌋;
        i ← ℓ;
        j ← mid + 1;
        while (i ≤ mid) and (j ≤ u) do
            begin
                if A[i] < A[j] then
                    begin
                        C[k] ← A[i];
                        i ← i + 1;
                    end;
                else begin
                        C[k] ← A[j];
                        j ← j + 1;
                    end;
                k ← k + 1;
        end;
        {transfer the tail}
        while i ≤ mid do
            begin
                C[k] ← A[i];
                i ← i + 1;
                k ← k + 1;
            end;
        while j ≤ u do
            begin
                C[k] ← A[j];
                j ← j + 1;
                k ← k + 1;
            end;
        {transfer sorted segment back into A}
        for j ← 1 to k − 1 do
            A[ℓ + j − 1] ← C[j];
    end;
```

Figure 10.2. An algorithm for merging sorted lists.

assumes the two lists reside in the stretch $A[\ell \mathinner{..} u]$ of the host array $A[1 \mathinner{..} n]$: The first sorted list is in the stretch $A[\ell \mathinner{..} \lfloor(\ell + u)/2\rfloor]$, the second sorted list is in the stretch $A[\lfloor(\ell + u)/2\rfloor + 1 \mathinner{..} u]$.

In a practical implementation of the data by an array or a linked list, the promotion of an element to the top of a list is done simply by advancing an index or changing a pointer. In the case of arrays, two index variables are maintained to indicate the "tops" of the two lists. When an index is down to position j in one list of size s, we consider $A[1 \mathinner{..} j-1]$, that is, the elements up to position $j-1$, to have been processed and copied to the recipient list at the right places. The subarray $A[j \mathinner{..} s]$ is in sorted order.

We analyze M_{mn}, LINEAR MERGE's number of comparisons to merge the two lists, by way of L_{mn}, the number of "leftover" elements that are transferred from the tail of one of the two lists in the closing stage of the merging algorithm. Since every element in the recipient list up to the moment before transferring the tail corresponds to one comparison, we have

$$M_{mn} = m + n - L_{mn}.$$

One can construct many instances of two lists of sizes $m \le n$ with $L_{mn} = 1$, as well as many other instances with $L_{mn} = n$; see Exercises 10.1 and 10.2. The range of the number of comparisons of LINEAR MERGE is therefore

$$m \le M_{mn} \le m + n - 1$$

We are assuming the ranks of the input to MERGE SORT are random permutations. Dividing such an input stream of data into two parts of sizes m and n creates parts with the relative ranks within each part being a random permutation of its respective size. Ranked among themselves, the data ranks of the first part are a random permutation of $\{1, \dots, m\}$; ranked among themselves, the data ranks of the second part are a random permutation of $\{1, \dots, n-m\}$. No matter what ranks we have in each list, the two lists will be sorted before they are merged. All $(n+m)!$ permutations with the same set of first m absolute ranks (and consequently the same set of $n-m$ ranks in the second part), will give exactly the same two sorted lists before they are merged. It is a combination counting situation: All $\binom{n+m}{m}$ subsets of absolute ranks are equally likely to appear in the first list.

To have leftover of size at least $\ell \ge 1$, we can put the elements with ranks $m+n-\ell+1, \dots, m+n$ at the tail of either of the two lists. If we place them at the tail of the second list, we can choose any of m ranks among the remaining $m+n-\ell$ ranks to appear in the first list (in sorted order, of course); the remaining $n-\ell$ ranks are then added to the second list in sorted order in front of the tail. This construction can be done in $\binom{m+n-\ell}{m}$ ways; by a symmetric argument we can construct two sorted lists of sizes m and n in $\binom{m+n-\ell}{n}$ ways with the first list holding the largest ℓ elements at its tail. Therefore,

$$\mathbf{Prob}\{L_{mn} \ge \ell\} = \frac{\dbinom{m+n-\ell}{m} + \dbinom{m+n-\ell}{n}}{\dbinom{m+n}{m}}. \tag{10.1}$$

We can find the mean of L_{mn} from a classical identity involving the tail probabilities:

$$\mathbf{E}\big[L_{mn}\big] = \sum_{\ell=0}^{\infty} \mathbf{Prob}\{L_{mn} > \ell\}.$$

The mean value of the length of the leftover tail is therefore

$$\begin{aligned}
\mathbf{E}[L_{mn}] &= \sum_{\ell=1}^{\infty} \mathbf{Prob}\{L_{mn} \ge \ell\} \\
&= \sum_{\ell=1}^{m+n} \frac{\binom{m+n-\ell}{m} + \binom{m+n-\ell}{n}}{\binom{m+n}{m}} \\
&= \frac{1}{\binom{m+n}{m}} \left[\sum_{k=0}^{m+n-1} \binom{k}{m} + \sum_{k=0}^{m+n-1} \binom{k}{n} \right] \\
&= \frac{\binom{m+n}{m+1} + \binom{m+n}{n+1}}{\binom{m+n}{m}} \\
&= \frac{n}{m+1} + \frac{m}{n+1}.
\end{aligned}$$

Therefore, the number of comparisons to merge sorted lists (of random ranks) of sizes m and n by LINEAR MERGE is on average

$$\begin{aligned}
\mathbf{E}[M_{mn}] &= m + n - \mathbf{E}[L_{mn}] \\
&= m + n - \frac{n}{m+1} - \frac{m}{n+1}.
\end{aligned} \qquad (10.2)$$

By a similar calculation one finds the variance of M_{mn} from another classical identity for the second factorial moment. The calculation involves similar combinatorial reductions as those used for the mean:

$$\begin{aligned}
\mathbf{E}\big[L_{mn}(L_{mn} - 1)\big] &= 2 \sum_{\ell=0}^{\infty} \ell \, \mathbf{Prob}\{L_{mn} > \ell\} \\
&= \frac{2m(m-1)}{(n+1)(n+2)} + \frac{2n(n-1)}{(m+1)(m+2)}.
\end{aligned}$$

The variance of M_{mn} now follows from

$$\begin{aligned}
\mathbf{Var}[M_{mn}] &= \mathbf{Var}[m + n - L_{mn}] \\
&= \mathbf{Var}[L_{mn}] \\
&= \mathbf{E}\big[L_{mn}(L_{mn} - 1)\big] + \mathbf{E}[L_{mn}] - \mathbf{E}^2[L_{mn}] \\
&= \frac{2m(m-1)}{(n+1)(n+2)} + \frac{2n(n-1)}{(m+1)(m+2)} + \frac{n}{m+1} + \frac{m}{n+1} \\
&\qquad - \Big(\frac{n}{m+1} + \frac{m}{n+1}\Big)^2.
\end{aligned}$$

In a typical application with MERGE SORT, LINEAR MERGE is used to merge the two halves of a list of size n. The random variable that will arise in that context is $M_{\lceil n/2\rceil, \lfloor n/2\rfloor}$. Note that here $n - M_{\lceil n/2\rceil, \lfloor n/2\rfloor} = L_{\lceil n/2\rceil, \lfloor n/2\rfloor}$ is narrowly concentrated. For instance, from the above calculations

$$\mathbf{E}\big|n - M_{\lceil n/2\rceil, \lfloor n/2\rfloor}\big| = \mathbf{E}\big[L_{\lceil n/2\rceil, \lfloor n/2\rfloor}\big] < 2,$$

and by similar calculations we find

$$\mathbf{E}\big|n - M_{\lceil n/2\rceil, \lfloor n/2\rfloor}\big|^2 = \mathbf{E}\big[L^2_{\lceil n/2\rceil, \lfloor n/2\rfloor}\big] < 6,$$

and

$$\mathbf{E}\big|n - M_{\lceil n/2\rceil, \lfloor n/2\rfloor}\big|^3 < 26.$$

This concentration will help us later establish a Gaussian law for MERGE SORT under LINEAR MERGE.

10.1.2 BINARY MERGE

The BINARY MERGE algorithm attempts to take advantage of the efficiency of binary searching. Assuming again our two disjoint lists are $\mathcal{X} = (X_1, X_2, \ldots, X_m)$, and $\mathcal{Y} = (Y_1, Y_2, \cdots, Y_n)$, with

$$X_1 < X_2 < \cdots < X_m,$$

and

$$Y_1 < Y_2 < \cdots < Y_n,$$

and without loss of generality assuming $m \le n$, BINARY MERGE proceeds by first searching for X_m within $\mathcal{Y}$ using the standard BINARY SEARCH algorithm. When a position i is found such that $Y_i < X_m < Y_{i+1} < \cdots < Y_n$, the data stretch $X_m, Y_{i+1}, \ldots, Y_n$ is transferred to the sorted merged output. We then do the same with X_{m-1} and the remainder of the Y's. This time, a data segment, say $X_{m-1} < Y_j < \ldots < Y_i$, is transferred to the output, and can be appended in front of the first data segment, because we are already assured that all these elements are smaller than what is already out there from the first stage. The algorithm proceeds in this fashion until one of the two lists is exhausted. After one list is exhausted, the keys remaining in the other are smaller than everything already on the output list; the algorithm transfers that tail appending it at the beginning of the output list.

Boundary interpretations are important. For instance, as usual, we may define $Y_i < \ldots < Y_j$ as an empty subset of $\mathcal{Y}$ that does not exist if $i > j$. Put together with all minor details and boundary considerations, the algorithm is quite simple and can be easily coded.

This algorithm performs poorly in the min-max sense. The worst-case scenario is when $X_1 > Y_n$. In this case, the size of the $\mathcal{Y}$ list is never diminished, and at the ith stage BINARY MERGE performs BINARY SEARCH to locate a position for X_{m-i+1} among the full $\mathcal{Y}$ list; every insertion will fall at the bottom of $\mathcal{Y}$, requiring BINARY SEARCH to make at least $\lg n + O(1)$ comparisons. In the worst case, BINARY MERGE performs about

$$m \lg n$$

comparisons.

Specifically, if used with MERGE SORT on n keys ($m = \lfloor n/2 \rfloor$), BINARY MERGE will have a number of comparisons asymptotically equal to

$$\frac{1}{2} n \lg n,$$

which for large n is higher than $n - 1$, LINEAR MERGE's worst-case number of comparisons. LINEAR MERGE seems to offer more advantages when used by MERGE SORT.

10.1.3 The HWANG-LIN Merging Algorithm

The HWANG-LIN merging algorithm is basically a generalization of BINARY MERGE. It considers a general probe position as the starting point of attack instead of BINARY MERGE's middle point. Adding this liberty in choosing the splitting point, the HWANG-LIN algorithm then considers a point that optimizes the number of comparisons for uniform data.

Again, assume our two disjoint lists are $\mathcal{X} = (X_1, X_2, \ldots, X_m)$, and $\mathcal{Y} = (Y_1, Y_2, \ldots, Y_n)$, with

$$X_1 < X_2 < \cdots < X_m,$$

and

$$Y_1 < Y_2 < \cdots < Y_n,$$

and without loss of generality assume $m \le n$.

Let us consider the data aspects that make BINARY MERGE not as good as it promises to be. Had we started with a random key and a random list, BINARY

MERGE could not be excelled. However, our situation is not exactly that. Consider, for instance, the particular case of two equal size lists ($m = n$). BINARY MERGE begins with finding a position for X_m in the $\mathcal{Y}$ list. The key X_m is the largest among m keys sampled from a continuous distribution. But so is Y_m. These two extremal keys are equidistributed. On average, one expects X_m to fall close to Y_m in the $\mathcal{Y}$ list rather than the middle. Contrary to BINARY MERGE, the HWANG-LIN algorithm favors a position close to the tail of the $\mathcal{Y}$ list as a starting point of attack.

Consider uniform data: The $\mathcal{X}$ list consists of m randomly picked data points from the UNIFORM(0,1) distribution, and the $\mathcal{Y}$ list consists of n randomly picked data points from the UNIFORM(0,1) distribution. What point of attack is suitable as a start for the more general situation of two lists of arbitrary sizes? On average, the m points of $\mathcal{X}$ split the unit interval into $m + 1$ equal intervals, of length $1/(m + 1)$ each. The jth order statistic lies on average at $j/(m + 1)$, and so the largest key in the $\mathcal{X}$ list lies on average at $m/(m + 1)$. Similarly, the jth order statistic in the $\mathcal{Y}$ list lies on average at $j/(n + 1)$. The question is, For large list sizes (both m and n tend to ∞ in such a way that $n = o(m^2)$), which order statistic Y_j in the $\mathcal{Y}$ list is on average closest to X_m? We determine this by asymptotically solving the equation

$$\frac{j}{n+1} = \frac{m}{m+1}.$$

This gives us the position

$$\begin{aligned} j &= \frac{n+1}{1+m^{-1}} \\ &= (n+1)\Big(1 - \frac{1}{m} + O\Big(\frac{1}{m^2}\Big)\Big) \\ &= n - \frac{n}{m} + 1 + O\Big(\frac{n}{m^2}\Big) \\ &\approx n - \frac{n}{m} + 1. \end{aligned}$$

For instance, if $m = n \to \infty$, the approximate solution suggests n as the starting point of attack. With high probability X_m will be close to Y_m, whereas if $m = \frac{1}{2}n \to \infty$, the approximate solution suggests $n - 1$ as a starting point of attack.

Only for aesthetic reasons, the HWANG-LIN algorithm uses an exponential representation for the ratio n/m. As a power of 2, this ratio is

$$\frac{n}{m} = 2^{\alpha},$$

for the α that satisfies

$$\alpha = \lg \frac{n}{m}.$$

Of course, for algorithmic purposes, the "position" j must be an integer; the HWANG-LIN algorithm takes the position

$$j = n - 2^{\lfloor \alpha \rfloor} + 1.$$

The HWANG-LIN merging algorithm proceeds in stages as follows:

- Stage 1: Let $\alpha = \lfloor \lg(n/m) \rfloor$ and define $j = n - 2^{\alpha} + 1$.
- Stage 2: Compare X_m with Y_j. If $X_m < Y_j$, move the data segment $Y_j < \cdots < Y_n$ to the output list. What is left to be done is to merge the $\mathcal{X}$ list among the set $Y_1 < \cdots < Y_{j-1}$. If, on the other hand, $X_m > Y_j$, merge X_m into $Y_{j+1} < \cdots < Y_n$ by BINARY MERGE. Suppose the BINARY SEARCH stage of BINARY MERGE identifies position k, such that $Y_{k-1} < X_m < Y_k$. Move the data stretch $X_m < Y_k < \cdots < Y_n$ to the output list. We are left with the problem of merging $X_1 < \cdots < X_{m-1}$ into $Y_1 < \cdots < Y_{k-1}$.
- Stage 3: Redefine m as the size of the smaller of the two remaining lists and n as the size of the larger, call the members of the shorter list X's and the elements of the longer Y's, then go back to invoke Stage 1 recursively on the remaining lists.

The HWANG-LIN algorithm is designed with the unbalanced cases of BINARY MERGE in mind. It senses when BINARY MERGE is about to become inefficient and remedies that by reversing the roles of the two lists, so that big chunks from the longer list are moved to the output list. Otherwise, the HWANG-LIN algorithm simply lets BINARY MERGE run its course. The HWANG-LIN algorithm is thus effective when there is a large difference between the two list sizes (which explains the choice of α). For the purpose of MERGE SORT, we need only consider the cases $m = \lfloor n/2 \rfloor$. For these cases of two equal halves, $\alpha = 0$, and the HWANG-LIN algorithm's worst-case number of comparisons is identical to LINEAR MERGE's number of comparisons in the worst case.

10.2 THE MERGE SORT ALGORITHM

The MERGE SORT algorithm is a recursive bookkeeper. We shall discuss the algorithm in the context of arrays. Adaptation to linked lists is a straightforward exercise. Given the array $A[1\,..\,n]$, MERGE SORT divides it into two "halves," of sizes $\lceil n/2 \rceil$ and $\lfloor n/2 \rfloor$, which are sorted recursively by MERGE SORT. Subsequent levels of recursion further divide these two halves into smaller subarrays until arrays of size 1 are reached. Of course, arrays of size 1 are already sorted and nothing needs to be done for these. The algorithm then returns from the ground up building sorted segments of increasing sizes starting from the singletons.

The general step handles a stretch $A[\ell\,..\,u]$, extending between the limits ℓ and u, by dividing it at the "middle" position $\lfloor (\ell + u)/2 \rfloor$. The implementation *MergeSort*

```
procedure MergeSort (ℓ, u: integer);
    begin
        if ℓ < u then
            begin
                call Mergesort(ℓ, ⌊(ℓ+u)/2⌋);
                call MergeSort(⌊(ℓ+u)/2⌋ + 1, u);
                call LinearMerge(ℓ, u);
            end;
    end;
```

Figure 10.3. The MERGE SORT algorithm.

of Figure 10.3 assumes that the host array A is accessed globally and calls the merging scheme *LinearMerge* of Figure 10.2. A recursive chain of calls is begun by the external call

call $MergeSort(1, n)$;

Figure 10.4 illustrates the step-by-step effects of *MergeSort* on an input of size 7.

The recursive form *MergeSort* naturally gives rise to a recurrence equation for the number of comparisons it makes. Letting the number of comparisons that *MergeSort* makes to sort a random list of n keys be C_n, we have the recurrence

$$C_n \stackrel{D}{=} C_{\lceil n/2 \rceil} + \tilde{C}_{\lfloor n/2 \rfloor} + Y_n, \tag{10.3}$$

where, for each j, $\tilde{C}_j \stackrel{D}{=} C_j$, and the random variable Y_n is the random number of comparisons needed for merging two sorted lists of random ranks and of sizes $\lceil n/2 \rceil$ and $\lfloor n/2 \rfloor$; the families $\{C_j\}$, $\{\tilde{C}_j\}$, $\{Y_j\}$ are independent. The random variable Y_n is $M_{\lfloor n/2 \rfloor, \lceil n/2 \rceil}$ of the discussion of merging in Section 10.1.1.

Like HEAP SORT, the MERGE SORT algorithm is a particularly important algorithm from the theoretical point of view as it gives a benchmark for the worst case that other competitors must race against. For *any* input of size n, MERGE SORT's number of comparisons is upper bounded by $O(n \ln n)$, as will be proved in the next theorem. This order, being also the order of the worst-case number of comparisons for any sorting algorithm performing on n keys (Theorem 1.1), renders MERGE SORT worst-case optimal. In the next theorem we derive the worst-case number of comparisons for the particular implementation *MergeSort*. Other variants of MERGE SORT may slightly improve the lower-order terms, but of course not the overall worst-case order, as it cannot be improved.

The fractional part of a number x is denoted as usual by $\{x\}$, which is a sawtooth function of x. A typical recurrence for a property associated with MERGE SORT involves ceils and floors resulting from the possibility of partitioning into nonequal "halves." A general technique for solving these recurrences is discussed in Section 1.11.4 and will be applied to a variety of similar recurrences with ceils and floors, like those recurrences for the best case, average, and variance of the number of

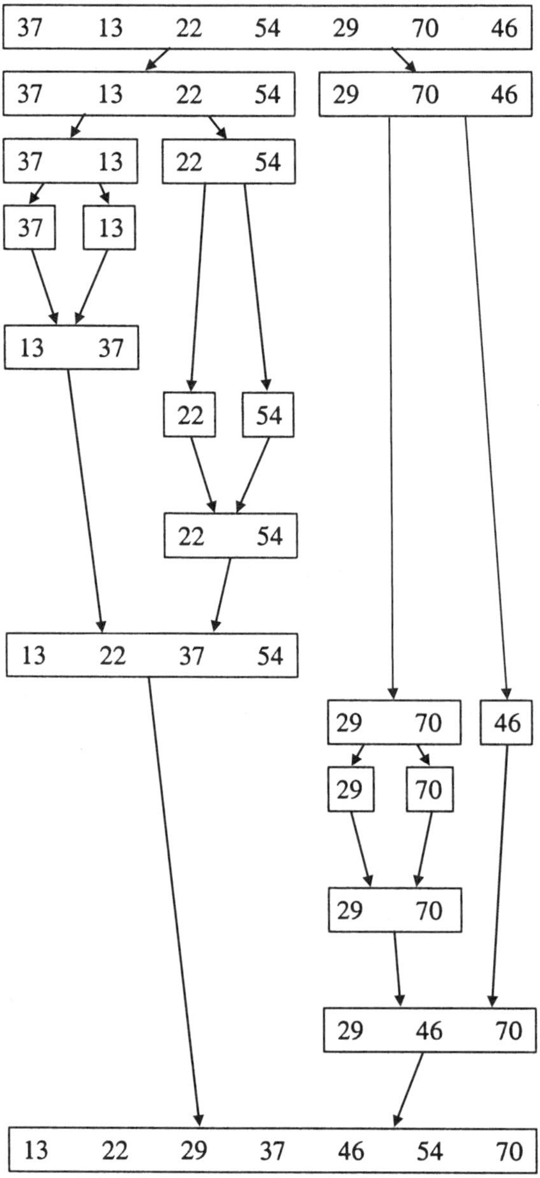

Figure 10.4. Operation of *MergeSort* on seven items of data.

comparisons of *MergeSort*. But first, we handle the worst case by more elementary standard methods.

Theorem 10.1 *Let W_n be the worst-case number of comparisons performed by MergeSort on any input of size n. Then*

$$\begin{aligned} W_n &= n\lceil \lg n \rceil - 2^{\lceil \lg n \rceil} + 1 \\ &= n \lg n + n\delta(\lg n) + 1, \end{aligned}$$

where δ is the periodic function

$$\delta(x) = 1 - \{x\} - 2^{1-\{x\}}.$$

Proof. To have *MergeSort* exercise its maximal number of comparisons on an input of size n, each side of the recursion is forced to perform its maximal number of comparisons for its respective size: The left recursion will perform $W_{\lceil n/2 \rceil}$ and the right recursion will perform $W_{\lfloor n/2 \rfloor}$ comparisons. Moreover, the MERGE stage will experience $n - 1$ comparisons, its maximal number of comparisons for merging the two halves, by a particular choice of interleaved ranks that go into each half (Exercise 10.2). Whence, the worst-case recurrence is similar to (10.3) when each term is replaced by its maximal possible value:

$$W_n = W_{\lceil n/2 \rceil} + W_{\lfloor n/2 \rfloor} + n - 1. \tag{10.4}$$

To solve this type of recurrence, we may first take a guess at what the answer is like without the clumsy ceils and floors. Let us consider $n = 2^k$ so that every level of recursion handles a list of length that is an exact power of 2, too. For this special form of n:

$$\begin{aligned} W_n &= n - 1 + 2W_{n/2} \\ &= n - 1 + 2\Big[\frac{n}{2} - 1 + 2W_{n/4}\Big] \\ &= (n-1) + (n-2) + 4\Big[\frac{n}{4} - 1 + 2W_{n/8}\Big] \\ &\ \vdots \\ &= (n-1) + (n-2) + (n-4) + \cdots + \big(n - 2^{k-1}\big) + 2^k W_{n/2^k}, \end{aligned}$$

with boundary condition $W_{n/2^k} = W_1 = 0$ (no comparisons are needed to sort one element). Thus

$$W_n = nk - 2^k + 1 = n \lg n - 2^{\lg n} + 1.$$

Fortunately this form, fudged by ceiling all $\lg n$ terms, gives a solution to the recurrence for general n, as well, as can be easily verified (see Exercise 10.4).

The second form in the theorem explicitly reflects the periodic fluctuations. This form follows directly for $n = 2^k$, in which case the fractional part of $\lg n$ is identically 0, and $\delta(\lg n) = -1$. For n not of the form 2^k, the substitution $\lceil \lg n \rceil = \lg n + 1 - \{\lg n\}$ gives the result. ■

The function $\{\lg n\}$ appears in the statement of the worst-case number of comparisons introducing periodic fluctuations in the worst-case behavior of *MergeSort*. These fluctuations are in the function $\delta(\lg n)$. The function $n\delta(\lg n)$ fluctuates between $-n$ and $-(0.915716564\ldots)n$, and the lower-order term is $O(n)$. Interpolating n to the positive real line x, and $\delta(\lg x)$ to be its real continuation $\delta(x)$, the graph of $\delta(x)$, as shown in Figure 10.5, is an oscillating function. So, the graph of $W(x)$, the real continuation of $W(n)$, takes the form of a rising function with the general shape of the function $x \lg x$ and small fluctuations about it. The relative magnitude of the fluctuations, compared to the nonperiodic rising function $x \lg x$, are not too pronounced, because the fluctuations appear in the lower-order linear term, and will not be seen unless $W(x)$ is drawn with a very magnified vertical scale.

Moving from the easier case $n = 2^k$, for some k, to the case of general n was a matter of a good guess here. This transition is a lot more complicated for the average-case recurrence, the variance, and several other MERGE SORT analyses. A more structured approach that involves no such guessing is developed in Section 1.11.4.

The general form (10.4) of the recurrence of the worst-case number of comparisons reappears in the analysis of several other characteristics of MERGE SORT's number of comparisons such as, for example, the average number of comparisons. The term $n - 1$ in (10.4) will be replaced by more complicated expressions for these other analyses. For example, the recurrence for the average number of comparisons

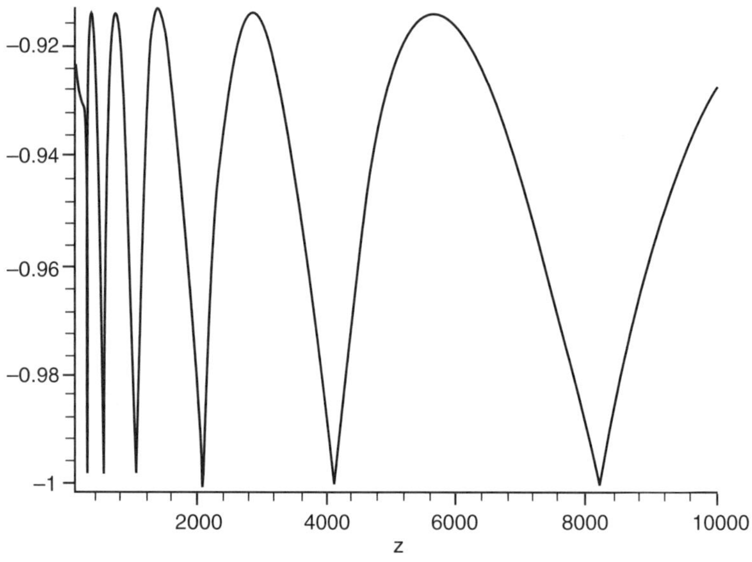

Figure 10.5. Oscillation in the worst-case behavior of MERGE SORT.

in the implementation *MergeSort* involves the same kind of ceils and floors in the recursive terms, and the nonrecursive term is $\mathbf{E}[Y_n]$ by specializing (10.2) to the specific situation of middle splitting. *MergeSort* splits the file to be sorted into two halves, and

$$Y_n = M_{\lfloor n/2 \rfloor, \lceil n/2 \rceil} = n - L_{\lfloor n/2 \rfloor, \lceil n/2 \rceil}, \tag{10.5}$$

with an average value

$$\mathbf{E}[Y_n] = \mathbf{E}\big[M_{\lfloor n/2 \rfloor, \lceil n/2 \rceil}\big] = n - \frac{\lceil n/2 \rceil}{\lfloor n/2 \rfloor + 1} - \frac{\lfloor n/2 \rfloor}{\lceil n/2 \rceil + 1}. \tag{10.6}$$

A technique for handling these recurrences by integral transform methods, due to Flajolet and Golin (1994), is discussed in Section 1.11.4. The technique provides a paradigm for asymptotically solving recurrences with ceils and floors of the general form

$$a_n = a_{\lceil n/2 \rceil} + a_{\lfloor n/2 \rfloor} + b_n, \tag{10.7}$$

where b_n is a nonrecursive term that depends on n.

Theorem 10.2 *(Flajolet and Golin, 1994). The average number of comparisons made by MergeSort to sort a random input of size n is*

$$\mathbf{E}[C_n] = n \lg n + n\eta(n) + O(1),$$

where η is a periodic function given by the Fourier expansion

$$\eta(u) = b_0 + \sum_{\substack{k=-\infty \\ k \neq 0}}^{\infty} b_k e^{2ik\pi u},$$

with

$$b_0 = \frac{1}{2} - \frac{1}{\ln 2} - \frac{2}{\ln 2} \sum_{m=1}^{\infty} \frac{1}{(m+1)(m+2)} \ln\Big(\frac{2m+1}{2m}\Big) \approx -1.24815204\ldots,$$

and

$$b_k = \frac{1 + \kappa(\chi_k)}{\chi_k(\chi_k + 1) \ln 2},$$

with $\chi_k = 2\pi i k / \ln 2$, and

$$\kappa(u) = \sum_{m=1}^{\infty} \frac{2}{(m+1)(m+2)} \Big(\frac{1}{(2m+1)^u} - \frac{1}{(2m)^u}\Big).$$

Proof. In the implementation *MergeSort*, the merging stage operates on two sorted lists of random ranks and of sizes $\lceil n/2 \rceil$ and $\lfloor n/2 \rfloor$. As discussed before, the average value of *LinearMerge*'s number of comparisons is

$$\mathbf{E}[Y_n] = n - \frac{\lceil n/2 \rceil}{\lfloor n/2 \rfloor + 1} - \frac{\lfloor n/2 \rfloor}{\lceil n/2 \rceil + 1}.$$

Then on average, *MergeSort* will make

$$\mathbf{E}[C_n] = \mathbf{E}\big[C_{\lceil n/2 \rceil}\big] + \mathbf{E}\big[C_{\lfloor n/2 \rfloor}\big] + n - \frac{\lceil n/2 \rceil}{\lfloor n/2 \rfloor + 1} - \frac{\lfloor n/2 \rfloor}{\lceil n/2 \rceil + 1}$$

comparisons. According to Theorem 1.6, under the initial condition $\mathbf{E}[C_1] = 0$, the last recurrence has the solution

$$\mathbf{E}[C_n] = \frac{n}{2\pi i} \int_{3-i\infty}^{3+i\infty} \frac{\eta(s)\, n^s}{s(s+1)(1-2^{-s})}\, ds,$$

with $\eta(s) = \sum_{n=1}^{\infty} \triangle \triangledown \mathbf{E}[Y_n]/n^s$. We have

$$\mathbf{E}[Y_{2n}] = 2n - \frac{n}{n+1} - \frac{n}{n+1} = 2n - 2 + \frac{2}{n+1},$$

and

$$\mathbf{E}[Y_{2n+1}] = 2n + 1 - \frac{n+1}{n+1} - \frac{n}{(n+1)+1} = 2n - 1 + \frac{2}{n+2}.$$

So,

$$\begin{aligned} \triangle \triangledown \mathbf{E}[Y_{2n}] &= \big(\mathbf{E}[Y_{2n+1}] - \mathbf{E}[Y_{2n})\big) - \big(\mathbf{E}[Y_{2n}] - \mathbf{E}[Y_{2n-1}]\big) \\ &= \mathbf{E}[Y_{2n+1}] - 2\mathbf{E}[Y_{2n}] + \mathbf{E}[Y_{2n-1}] \\ &= -\frac{2}{(n+1)(n+2)}. \end{aligned}$$

Similar algebraic manipulation for odd indexes also shows that

$$\triangle \triangledown \mathbf{E}[Y_{2n+1}] = \frac{2}{(n+1)(n+2)} = -\triangle \triangledown \mathbf{E}[Y_{2n}].$$

The Dirichlet transform of the second-order differences is

$$\begin{aligned} \eta(s) &= \frac{\triangle \triangledown \mathbf{E}[Y_1]}{1^s} + \frac{\triangle \triangledown \mathbf{E}[Y_2]}{2^s} + \frac{\triangle \triangledown \mathbf{E}[Y_3]}{3^s} + \cdots \\ &= \triangle \triangledown \mathbf{E}[Y_1] + \frac{\triangle \triangledown \mathbf{E}[Y_2]}{2^s} - \frac{\triangle \triangledown \mathbf{E}[Y_2]}{3^s} + \frac{\triangle \triangledown \mathbf{E}[Y_4]}{4^s} - \frac{\triangle \triangledown \mathbf{E}[Y_4]}{5^s} + \cdots \end{aligned}$$

$$= \mathbf{E}[Y_2] - 2\mathbf{E}[Y_1] + \mathbf{E}[Y_0] + \sum_{k=1}^{\infty}\Big(\frac{1}{(2k)^s} - \frac{1}{(2k+1)^s}\Big) \vartriangle \triangledown \mathbf{E}[Y_{2k}]$$

$$= 1 + \sum_{k=1}^{\infty}\Big(\frac{1}{(2k+1)^s} - \frac{1}{(2k)^s}\Big)\frac{2}{(k+1)(k+2)}.$$

Let $\psi(s)$ be defined as the sum in the last line. That is, $\eta(s) = 1 + \psi(s)$, and

$$\begin{aligned} \mathbf{E}[C_n] &= \frac{n}{2\pi i}\int_{3-i\infty}^{3+i\infty} \frac{\big(1+\psi(s)\big)n^s}{s(s+1)(1-2^{-s})}\, ds \\ &= \frac{n}{2\pi i}\int_{3-i\infty}^{3+i\infty} \frac{n^s}{s(s+1)(1-2^{-s})}\, ds + \frac{n}{2\pi i}\int_{3-i\infty}^{3+i\infty} \frac{\psi(s)\, n^s}{s(s+1)(1-2^{-s})}\, ds \\ &\stackrel{\text{def}}{=} \frac{n}{2\pi i}[I_1 + I_2]. \end{aligned}$$

As discussed in Subsection 1.11.4, both I_1 and I_2 are to be evaluated by residue computation. In the domain of convergence $\Re s > 1$, $\psi(s)$ is an entire function, with no singularities. The equation

$$1 - 2^{-s} = 0$$

has solutions at $\chi_k = 2\pi i k/\ln 2$, for $k = 0, \pm 1, \pm 2, \ldots$. The integrands of both I_1 and I_2 have poles at $\chi_{-1} = -1$, and χ_k, for $k = 0, \pm 1, \pm 2, \ldots$. The poles χ_{-1} and χ_k, $k = \pm 1, \pm 2, \ldots$, are all simple poles, whereas χ_0 is a double pole, and will give the dominant terms.

At its double pole χ_0 the integrand of I_1 has the residue

$$\lg n - \frac{1}{2} - \frac{1}{\ln 2},$$

and at χ_{-1} its residue is $1/n$. At the imaginary pole χ_k, $k \neq 0$, the integrand of I_1 has a residue

$$\frac{n^{\chi_k}}{\chi_k(\chi_k + 1)\, \ln 2}.$$

At its double pole χ_0, the integrand of I_2 has the residue

$$\frac{2}{\ln 2}\sum_{k=1}^{\infty} \frac{1}{(k+1)(k+2)} \ln\Big(\frac{2k+1}{2k}\Big).$$

At its simple pole χ_{-1}, the integrand of I_2 has a residue of $1/n$. At the imaginary pole χ_k, $k \neq 0$, the integrand of I_2 has a residue

$$\frac{n^{\chi_k}\psi(\xi_k)}{\chi_k(\chi_k+1)\ln 2}.$$

Collecting contributions of all poles, the theorem follows. ∎

Variance calculation is very similar in spirit. Taking the variance of the distributional equation (10.3),

$$\mathbf{Var}[C_n] = \mathbf{Var}[C_{\lceil n/2\rceil}] + \mathbf{Var}[C_{\lfloor n/2\rfloor}] + \mathbf{Var}[Y_n].$$

The variance operator was applied to each individual term on the right, as these three variables are independent. Application of Theorem 1.6 along the lines and residue calculation illustrated in the proof of Theorem 1.6 gives the variance as stated next. As usual, residue computations are more involved for the variance than for the mean; the details are left as an exercise.

Theorem 10.3 *(Flajolet and Golin, 1994). The variance of C_n, the number of comparisons made by MergeSort to sort a random input of size n, is asymptotically*

$$\mathbf{Var}[C_n] \sim n\,\theta(\lg n), \qquad as\ n \to \infty,$$

where θ is a periodic function given by the Fourier expansion

$$\theta(u) = \frac{1}{\ln 2}\sum_{k=-\infty}^{\infty} c_k e^{2\pi i k u},$$

and the coefficients c_k are given by

$$c_0 = \frac{1}{\ln 2}\sum_{m=1}^{\infty}\frac{2m(5m^2+10m+1)}{(m+1)(m+2)^2(m+3)^2}\ln\Big(1+\frac{1}{2m}\Big) \approx 0.345499568\ldots,$$

and for integer $k \ne 0$,

$$c_k = \frac{\zeta(\chi_k)}{\chi_k(\chi_k+1)\ln 2},$$

with $\chi_k = 2\pi i k/\ln 2$ and

$$\zeta(u) = \sum_{m=1}^{\infty}\frac{2m(5m^2+10m+1)}{(m+1)(m+2)^2(m+3)^2}\Big(\frac{1}{(2m+1)^u} - \frac{1}{(2m)^u}\Big).$$

The variance of *MergeSort*'s number of comparisons grows "almost" linearly; as the variance function rises, it goes through periodic fluctuations. The function $\theta(\lg n)$ fluctuates around the mean value $c_0 \approx 0.34549\ldots$. The fluctuations in the variance are more noticeable than in the average, as they occur in the variance's leading term.

10.3 DISTRIBUTIONS

A recursive computational formula for the exact distribution follows from the fundamental recurrence:

$$C_n \stackrel{D}{=} C_{\lceil n/2 \rceil} + \tilde{C}_{\lfloor n/2 \rfloor} + Y_n.$$

The recursive formula can be worked out in the usual inductive way to build the exact distribution inductively from the basis up. The three terms on the right-hand side in the fundamental recurrence are independent. Let $F_n(z)$ be the probability generating function of C_n, the number of comparisons made by *MergeSort*, and let $\xi_n(z)$ be the probability generating function of Y_n, the number of comparisons to merge the two sorted halves. Being a convolution, the right-hand side of the recurrence has a probability generating function equal to the product of the three probability generating functions of its three ingredients:

$$F_n(z) = F_{\lceil n/2 \rceil}(z) F_{\lfloor n/2 \rfloor}(z) \xi_n(z).$$

This recurrence can be unwound all the way back to simple cases. For example, for $n = 13$:

$$\begin{aligned} F_{13}(z) &= F_7(z) F_6(z) \xi_{13}(z) \\ &= \big[F_4(z) F_3(z) \xi_7(z)\big]\big[F_3^2(z) \xi_6(z)\big] \xi_{13}(z) \\ &\;\;\vdots \\ &= \xi_1^8(z) \xi_2^5(z) \xi_3^3(z) \xi_4(z) \xi_6(z) \xi_7(z) \xi_{13}(z). \end{aligned} \tag{10.8}$$

We have already implicitly computed the functions $\xi_n(z)$. In the context of *MergeSort*, where the splitting is done at the middle, the tail probability of the leftover in (10.1) becomes:

$$\textbf{Prob}\{L_{\lfloor n/2 \rfloor, \lceil n/2 \rceil} \ge k\} = \frac{\binom{n-k}{\lfloor n/2 \rfloor} + \binom{n-k}{\lceil n/2 \rceil}}{\binom{n}{\lfloor n/2 \rfloor}}$$

$$= \begin{cases} 2\dfrac{\binom{n-k}{n/2}}{\binom{n}{n/2}}, & \text{if } n \text{ is even;} \\[3ex] \dfrac{\binom{n-k+1}{(n+1)/2}}{\binom{n}{(n-1)/2}}, & \text{if } n \text{ is odd;} \end{cases}$$

the two binomial coefficients in the probability calculation for odd n were combined by Pascal's identity

$$\binom{m}{j} = \binom{m-1}{j} + \binom{m-1}{j-1}.$$

And so, for n even,

$$\begin{aligned}\mathbf{Prob}\{L_{\lfloor n/2\rfloor,\lceil n/2\rceil} = k\} &= 2\frac{\binom{n-k}{n/2} - \binom{n-k-1}{n/2}}{\binom{n}{n/2}} \\ &= 2\frac{\binom{n-k-1}{n/2-1}}{\binom{n}{n/2}},\end{aligned}$$

by Pascal's identity. For odd n, similar work yields

$$\mathbf{Prob}\{L_{\lfloor n/2\rfloor,\lceil n/2\rceil} = k\} = \frac{\binom{n-k}{(n-1)/2}}{\binom{n}{(n-1)/2}}.$$

Recall that $Y_n = n - L_{\lfloor n/2\rfloor,\lceil n/2\rceil}$. Therefore, Y_n then has the probability mass function

$$\mathbf{Prob}\{Y_n = k\} = \mathbf{Prob}\{L_{\lfloor n/2\rfloor,\lceil n/2\rceil} = n-k\},$$

and it follows that

$$\xi_n(z) = \begin{cases} \dfrac{2}{\binom{n}{n/2}} \displaystyle\sum_{k=1}^{n/2} \binom{n-k-1}{n/2-1} z^{n-k}, & \text{if } n \text{ is even;} \\ \dfrac{1}{\binom{n}{(n-1)/2}} \displaystyle\sum_{k=1}^{(n+1)/2} \binom{n-k}{(n-1)/2} z^{n-k} & \text{if } n \text{ is odd.} \end{cases}$$

For example,

$$\begin{aligned}\xi_1(z) &\equiv 1, \\ \xi_2(z) &= z, \\ \xi_3(z) &= \frac{1}{3}z + \frac{2}{3}z^2,\end{aligned}$$

$$\xi_4(z) = \frac{1}{3}z^2 + \frac{2}{3}z^3,$$

$$\xi_6(z) = \frac{1}{10}z^3 + \frac{3}{10}z^4 + + \frac{6}{10}z^5,$$

$$\xi_7(z) = \frac{1}{35}z^3 + \frac{4}{35}z^4 + \frac{10}{35}z^5 + \frac{20}{35}z^6,$$

$$\xi_{13}(z) = \frac{1}{1716}z^6 + \frac{7}{1716}z^7 + \frac{28}{1716}z^8 + \frac{84}{1716}z^9 + \frac{210}{1716}z^{10} + \frac{462}{1716}z^{11} + \frac{924}{1716}z^{12}.$$

The product (10.8) is:

$$F_{13}(z) = \frac{1}{18618600}(z^{22} + 22z^{23} + 241z^{24} + 1758z^{25} + 9604z^{26} + 41860z^{27} + 151542z^{28} + 467440z^{29} + 1242940z^{30} + 2844536z^{31} + 5522160z^{32} + 8842848z^{33} + 11172672z^{34} + 10367616z^{35} + 6209280z^{36} + 1774080z^{37}).$$

The exact mean of C_{13} is $F'_{13}(1) = 2029/60 = 33.81666666\ldots$, and the exact variance is $F''_{13}(1) + F'_{13}(1) - (F'_{13}(1))^2 = 10799/3600 = 2.99972222\ldots$. The probability mass function of C_{13} is represented by vertical lines in Figure 10.6, where the height of the line at k is $\mathbf{Prob}\{C_{13} = k\}$, for integer k. The same plot depicts the normal distribution $\mathcal{N}(2029/60, 10799/3600)$. The plot suggests that the exact probability distribution has a normal limit distribution and that the rate of convergence to this limit must be very fast because the exact distribution is already very close to the Gaussian distribution, even for a value of n as low as 13.

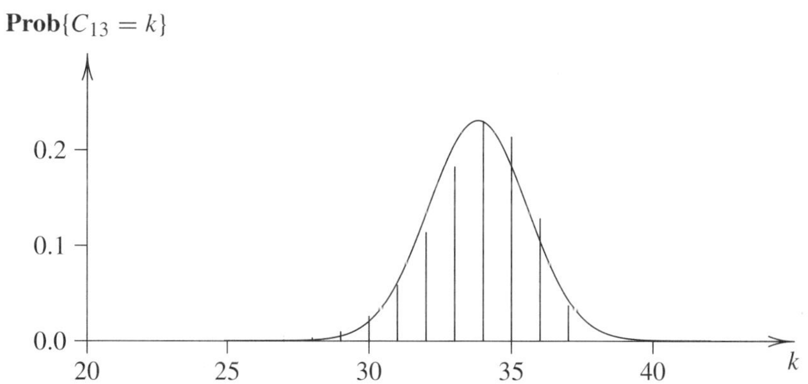

Figure 10.6. The exact probability distribution of C_{13} in comparison with a normal distribution.

To prove that the limiting distribution of the number of comparisons is normal, we may take the logarithm of the equation obtaining a recurrence of the general form (1.24):

$$\ln F_n(z) = \ln F_{\lceil n/2\rceil}(z) + \ln F_{\lfloor n/2\rfloor}(z) + \ln \xi_n(z).$$

For any fixed z, we can think of $\{\ln F_n(z)\}_{n=1}^{\infty}$, $\{\ln \xi_n(z)\}_{n=1}^{\infty}$ as sequences of numbers to which the paradigm of recurrences with floors and ceils may be applied. The computational difficulty in this route is that the poles of the Dirichlet transform involved will have locations depending on z.

The basic recurrence (10.3) permits verification of a condition like Lyapunov's, as was done by Flajolet and Golin. Tan (1993) checks Lindeberg's condition, which involves only the order of magnitude of the variance as given by Theorem 10.3. However, Tan's analysis is only along a sequence of exact powers of two: $n = 2^k$, as $k \to \infty$. Cramer (1997) uses convergence in the Zolotarev's metric of distances of third order, which remains simple so long as n is a power of 2, but then becomes complicated for general n.

Theorem 10.4 *(Flajolet and Golin, 1994; Cramer, 1997). Let C_n be the number of comparisons made by* $MergeSort$ *to sort a random file of n keys. Then*

$$\frac{C_n - n\lg n}{\sqrt{n\theta(\lg n)}} \xrightarrow{\mathcal{D}} \mathcal{N}(0, 1),$$

where $\theta(n)$ is the periodic function of Theorem 10.3.

Proof. Recall that C_n satisfies the basic recurrence

$$C_n \overset{D}{=} C_{\lceil n/2\rceil} + \tilde{C}_{\lfloor n/2\rfloor} + Y_n,$$

where Y_n is the number of comparisons needed to merge two sorted random lists of sizes $\lceil n/2\rceil$ and $\lfloor n/2\rfloor$ and whose exact probability distribution has been studied. Iterating the recurrence we have:

$$C_n \overset{D}{=} C_{\lceil \frac{1}{2}\lceil n/2\rceil\rceil} + \tilde{C}_{\lfloor \frac{1}{2}\lceil n/2\rceil\rfloor} + \check{C}_{\lceil \frac{1}{2}\lfloor n/2\rfloor\rceil} + \breve{C}_{\lfloor \frac{1}{2}\lfloor n/2\rfloor\rfloor} + \tilde{Y}_{\lceil n/2\rceil} + \check{Y}_{\lfloor n/2\rfloor} + Y_n,$$

where $\tilde{C}_i \overset{D}{=} \check{C}_i \overset{D}{=} \breve{C}_i \overset{D}{=} C_i$, $\tilde{Y}_j \overset{D}{=} \check{Y}_j \overset{D}{=} Y_j$, and the families $\{C_i\}$, $\{\tilde{C}_i\}$, $\{\check{C}_i\}$, $\{\breve{C}_i\}$ $\{Y_i\}$, $\{\tilde{Y}_i\}$, $\{\check{Y}_i\}$ are all independent. For every random variable C continue the recursion producing two Y's and a C with lower indexes. Continue this process till the indexes of the C's are reduced to 1, and use the boundary condition $C_1 \equiv 0$. This way we shall represent C_n as a sum of independent members of the $\{Y_i\}$ family (possibly some are repeats). For example,

$$\begin{aligned} C_5 &\stackrel{D}{=} C_3 + \tilde{C}_2 + Y_5 \\ &\stackrel{D}{=} (C_2 + C_1 + Y_3) + (\tilde{C}_1 + \check{C}_1 + Y_2) + Y_5 \\ &\stackrel{D}{=} (C_1 + \hat{C}_1 + \tilde{Y}_2) + 0 + Y_3 + 0 + 0 + Y_2 + Y_5 \\ &\stackrel{D}{=} Y_2 + \tilde{Y}_2 + Y_3 + Y_5. \end{aligned}$$

How many members of the $\{Y_i\}$ family enter this representation? By induction, C_n is the sum of $n-1$ independent members of the $\{Y_i\}$ family: If the induction hypothesis is true for values less than n, then $C_{\lceil n/2 \rceil}$ contributes $\lceil n/2 \rceil - 1$ random variables and $\tilde{C}_{\lfloor n/2 \rfloor}$ contributes $\lfloor n/2 \rfloor - 1$ random variables, and Y_n adds one more, that is, $\lceil n/2 \rceil - 1 + \lfloor n/2 \rfloor - 1 + 1 = n - 1$.

For each n, we have a representation of C_n in the form

$$C_n = X_1 + X_2 + \cdots + X_{n-1},$$

where $X_1, X_2, \ldots$ are independent and each one of them is a member of the $\{Y_i\}$ family. This representation as a sum of independent random variables is a perfect candidate for application of standard theorems. For instance, we can choose to verify Lyapunov's condition

$$\frac{\sum_{i=1}^{n-1} \mathbf{E}\big|X_i - \mathbf{E}[X_i]\big|^3}{\big(\mathbf{Var}[\sum_{i=1}^{n-1} X_i]\big)^2} \to 0,$$

as $n \to \infty$. Most of the verification has already been done: The denominator is $(\mathbf{Var}[C_n])^2$, which, by Theorem 10.3, is $\Theta(n^2)$. Each term in the numerator's sum is $O(1)$, which is a consequence of the concentrated nature of the $\{i - Y_i\}$ family as outlined next. By the triangle inequality, we have the term

$$\big|Y_i - \mathbf{E}[Y_i]\big| \le \big|Y_i - i\big| + \big|i - \mathbf{E}[Y_i]\big|.$$

So,

$$\begin{aligned} \mathbf{E}\big|Y_i - \mathbf{E}[Y_i]\big|^3 \le \mathbf{E}\big|Y_i - i\big|^3 &+ 3\mathbf{E}\big|Y_i - i\big|^2 \mathbf{E}\big|i - \mathbf{E}[Y_i]\big| \\ &+ 3\mathbf{E}\big|Y_i - i\big| \mathbf{E}\big|i - \mathbf{E}[Y_i]\big|^2 + \mathbf{E}\big|i - \mathbf{E}[Y_i]\big|^3. \end{aligned}$$

Each of these terms is uniformly bounded, as was demonstrated at the end of Section 10.1.1. As n goes to infinity, each term in the numerator's sum remains $O(1)$; the sum itself is $O(n)$. ■

10.4 BOTTOM-UP MERGE SORT

We present in this section a bottom-up approach to MERGE SORT that can be implemented nonrecursively. The only advantage of this method is that a special implementation of it can be applied to on-line files whose size is not known in advance. The efficiency of BOTTOM-UP MERGE SORT, as a review of the results will show, is only slightly worse than that of the top-down version discussed at length in the earlier sections of this chapter. Analysis methods founded in digital techniques were devised by Panny and Prodinger (1995). The problem also fits in the Flajolet-Golin paradigm, as will be sketched only for one result on the best-case performance of BOTTOM-UP MERGE SORT. The rest of the results can be proved similarly and will only be listed without proof to create a context for contrast with the top-down version.

The idea of BOTTOM-UP MERGE SORT is to go in rounds merging clusters of sizes that are, whenever possible, increasing powers of two, starting at the bottom from the singleton clusters and moving up to merge clusters of size 2, then clusters of size 4, and so on. If n is even, the first stage, the merging of singletons into pairs, will rearrange all the singletons, residing in $A[1\mathinner{..}n]$, into $n/2$ sorted pairs: $A[1\mathinner{..}2]$, $A[3\mathinner{..}4], \dots, A[n-1\mathinner{..}n]$. If n is odd, only the first $A[1\mathinner{..}n-1]$ elements are arranged into sorted pairs; the nth element is not merged with any other at this first round. Similarly, after j rounds of merging, the algorithm considers in the next round sorted clusters of size 2^j, merging them into larger sorted clusters of size 2^{j+1}. At the $(j+1)$st round the sorted clusters $A[1\mathinner{..}2^j]$ and $A[2^j+1\mathinner{..}2^{j+1}]$ are merged into the sorted cluster $A[1\mathinner{..}2^{j+1}]$, the two sorted clusters $A[2\times 2^j+1\mathinner{..}3\times 2^j]$ and $A[3\times 2^j+1\mathinner{..}4\times 2^j]$ are merged into $A[2\times 2^j+1\mathinner{..}4\times 2^j]$, and so on. This round takes care of $K_n^{(j)} \stackrel{\text{def}}{=} 2^{j+1}\lfloor n/2^{j+1}\rfloor$ data in "whole" clusters of size 2^j each. If the number $n - K_n^{(j)}$ of the remaining elements does not exceed 2^j, the size of a whole cluster for this round, this tail is left unaltered by this round. If this tail size exceeds 2^j, it is divided into two clusters: one whole in $A[K_n^{(j)}+1\mathinner{..}K_n^{(j)}+2^j]$ and one incomplete in $A[K_n^{(j)}+2^j+1\mathinner{..}n]$. The round is completed by one additional merge operation, combining them into a sorted stretch of data in $A[K_n^{(j)}+1\mathinner{..}n]$; we shall refer to this last merge as an *incomplete merge*. After $\lceil\lg n\rceil$ rounds are completed in this fashion the array A is completely sorted.

The algorithm has a recursive formulation that is essentially the same as the standard top-down version of Figure 10.3, only differing in the splitting policy. While the top-down version splits at the middle (position $\lceil n/2\rceil$), BOTTOM-UP MERGE SORT splits at $2^{\lceil\lg(n/2)\rceil}$, the first proper power of 2 past the middle point (note that the two policies coincide if n is an exact power of 2). For example, with $n = 23$, at the top level of the recursion, BOTTOM-UP MERGE SORT splits the array into two parts of sizes 16 and 7, whereas the standard top-down version splits the array into two halves of sizes 12 and 11.

The minor alteration of the standard recursive MERGE SORT to become BOTTOM-UP MERGE SORT is represented in the algorithm of Figure 10.7. Notice that a slightly modified MERGE algorithm is needed. The implementation *Merge* of MERGE that is used here assumes that the algorithm works with two adjacent sorted segments $A[i\mathinner{..}j]$ and $A[j+1\mathinner{..}k]$ and merges them together into one sorted stretch $A[i\mathinner{..}k]$. An implementation based on LINEAR MERGE will do the job efficiently.

```
procedure BottomUpMergeSort (ℓ, u: integer);
    local s: integer;
    begin
        if ℓ < u then
            begin
                s ← 2^⌈lg n/2⌉;
                call BottomUpMergeSort(ℓ, s);
                call BottomUpMergeSort(s + 1, u);
                call Merge(ℓ, s, u);
            end;
    end;
```

Figure 10.7. BOTTOM-UP MERGE SORT algorithm.

The splitting policy of BOTTOM-UP MERGE SORT admits a nonrecursive implementation, shown in Figure 10.8, because the boundaries of the clusters of increasing sizes are all known before going through any merging round. In this algorithm ℓ is the length of the typical segment of a round. It starts out at 1, so that pairs of singletons are merged. Then ℓ is doubled before the beginning of every new round to handle the sorted clusters of the increasing sizes 2, then 4, etc. The inner loop represents the operation of the ℓth round on whole clusters of length ℓ each. At the beginning of the ℓth round the two indexes p and q respectively demarcate the starting points of the first two clusters (of length ℓ each) for the round. After these two clusters are merged, the two indexes are shifted up by 2ℓ, to merge the following

```
procedure NonRecursiveMergeSort (n: integer);
    local p, q, ℓ: integer;
    begin
        ℓ ← 1;
        while ℓ < n do
            begin
                p ← 0;
                q ← ℓ;
                while q + ℓ ≤ n do
                    begin
                        call Merge(p + 1, p + ℓ, q + 1, q + ℓ);
                        p ← q + ℓ;
                        q ← p + ℓ;
                    end;
                if q < n then call Merge(p + 1, p + ℓ, q + 1, n);
                ℓ ← 2ℓ;
            end;
    end;
```

Figure 10.8. Nonrecursive BOTTOM-UP MERGE SORT algorithm.

two clusters of length ℓ each, and so forth. Only once in each round an incomplete merging operation may be performed, and only if certain conditions on the length of the tail are satisfied. The **if** statement between the two loops looks out for this special condition at the end of each round and completes the round by one additional incomplete merge, if necessary. Figure 10.9 illustrates the operation of nonrecursive BOTTOM-UP MERGE SORT on the same input file used for the illustration of the top-down version version (Figure 10.4).

Let $(b_k b_{k-1} \dots b_1 b_0)_2$ be the binary expansion of n, with b_k being the most significant 1 in the expansion, thus $k = \lfloor \lg n \rfloor$. At the jth round, BOTTOM-UP MERGE SORT performs $\lfloor n/2^{j+1} \rfloor$ merging operations on pairs of whole clusters of size 2^j each, and when n is not an exact power of two, the algorithm may perform an additional incomplete merge on the tail only if its size $n - K_n^{(j)}$ exceeds the size of clusters of this round—that is, if and only if n **mod** $2^{j+1} > 2^j$. So, the incomplete merge takes place if and only if n **mod** $2^{j+1} > 2^j$, a condition captured if and only if the product $b_j (b_{j-1} b_{j-2} \dots b_1 b_0)_2 > 0$.

As we argued for the worst-case of the top-down version of MERGE SORT, the best case for BOTTOM-UP MERGE SORT round occurs when every merge process encounters the best instances possible for all complete merges, as well as for the incomplete merge, when it exists. At the first round ($j = 0$), the algorithm performs $\lfloor n/2 \rfloor$ merges on pairs of singletons (one comparison for each pair). In the best case, at the second round as the algorithm merges pairs of clusters (of size 2 each), it will encounter a data arrangement giving only 2 comparisons for each pair of clusters; there are $\lfloor n/4 \rfloor$ of these. An incomplete merge takes place only if a cluster of size 2 at the tail is to be merged with a cluster of size 1; that is, if the total size of the two pairs is 3 (in this case $b_1 b_0 = 1$). In the best case, this merge takes one comparison when the singleton key is smaller than the pair in the cluster of size 2. Generally, at the jth stage, there will be $\lfloor n/2^{j+1} \rfloor$ pairs of clusters of size 2^j each, and in the

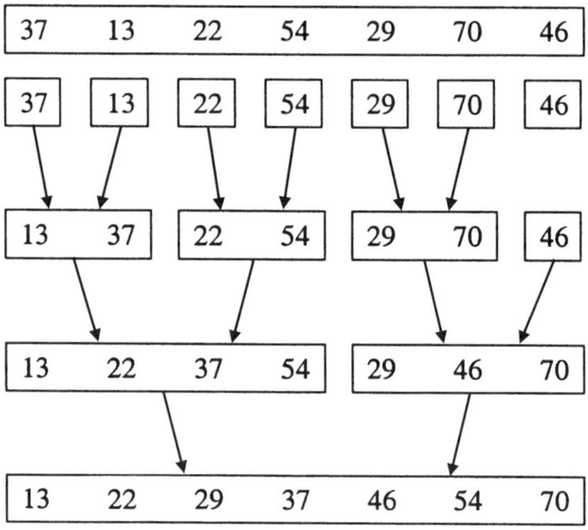

Figure 10.9. Operation of BOTTOM-UP MERGE SORT in rounds.

best case for each such pairs all the keys of one cluster in a pair are smaller than any key in the other, requiring 2^j comparisons for each pair of clusters. An additional incomplete merge will operate on a cluster of size 2^j to merge it with a smaller one. The total size of both is then at least $2^j + 1$ (b_j must be 1). In the best case, all the keys of the smaller cluster (of size $(b_{j-1} \dots b_0)_2$) will be less than any key in the other cluster, requiring only a number of comparisons equal to the size of the smaller cluster, that is, $b_j b_{j-1} \cdots b_0$ in binary representation. In other words, the incomplete merge takes $b_j(b_{j-1} \dots b_0)_2$ comparisons in the best case. The best possible number of comparisons for BOTTOM-UP MERGE SORT is therefore given by

$$\sum_{j=0}^{k} 2^j \left\lfloor \frac{n}{2^{j+1}} \right\rfloor + \sum_{j=0}^{k} (b_j b_{j-1} \dots b_1 b_0)_2.$$

This expression is a representation of the Trollope-Delange function $Z(n) = \sum_{j=0}^{n} \nu(j)$, where $\nu(j)$ is the number of 1's in the binary representation of j. For example, $\nu(14) = \nu((1110)_2) = 3$. This curious connection to digital representations also appears in the best-case performance and other analyses of the top-down version of MERGE SORT.

Exercise 1.11.3 outlines how the recurrence

$$Z(n) = Z\left(\left\lceil \frac{n}{2} \right\rceil\right) + Z\left(\left\lfloor \frac{n}{2} \right\rfloor\right) + \left\lfloor \frac{n}{2} \right\rfloor$$

is obtained for the Trollope-Delange function, and Exercise 10.5 shows how it relates to the best-case performance of the top-down version. This representation of $Z(n)$ is a prototype recurrence relation for the Flajolet-Golin paradigm and an application of Theorem 1.6 yields the following result.

Theorem 10.5 *(Panny and Prodinger, 1995). The best input instance of size n for BOTTOM-UP MERGE SORT gives a minimum number of comparisons with asymptotic value*

$$\frac{1}{2} n \lg n + n\delta_1(\lg n),$$

where δ_1 is an oscillating function of its argument.

The oscillating function in the best-case behavior of BOTTOM-UP MERGE SORT occurs at the lower-order terms. The oscillations have the shape of a self-replicating function, bounded from above by 0 and from below by -0.2. Figure 10.10 shows the fractal nature of the very small oscillations of the function $\delta_1(n)$.

By similar arguments the worst-case is obtained.

Theorem 10.6 *(Panny and Prodinger, 1995). The worst input instance of size n for BOTTOM-UP MERGE SORT gives Theorem 10.6 shows that BOTTOM-UP MERGE SORT a maximum number of comparisons with asymptotic value*

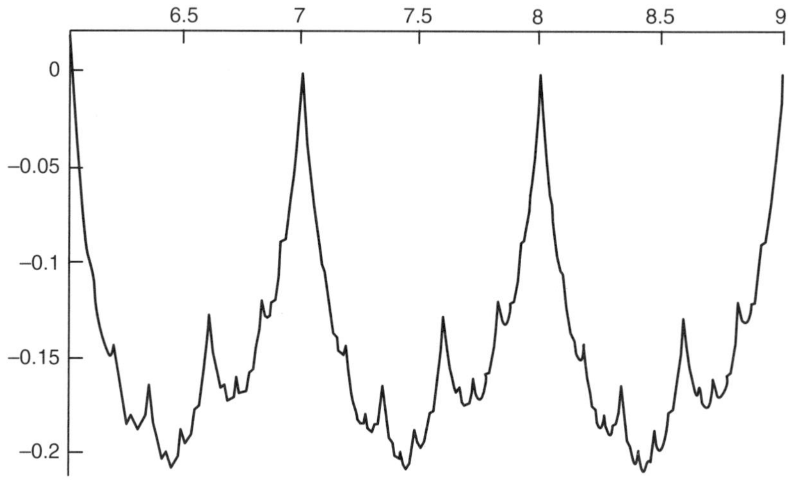

Figure 10.10. The fractal nature of the oscillations in best-case number of comparisons in BOTTOM-UP MERGE SORT.

$$n \lg n + n\delta_2(\lg n) + 1,$$

where δ_2 is an oscillating function of its argument.

Theorem 10.6 shows that BOTTOM-UP MERGE SORT remains asymptotically worst-case optimal. The next theorem shows that on average there is no noticeable difference in performance between the top-down and the bottom-up versions (compare with the result of Theorem 10.2).

Theorem 10.7 *(Panny and Prodinger, 1995). The average number of comparisons BOTTOM-UP MERGE SORT makes to sort a random input of size n has the asymptotic value*

$$n \lg n + n\delta_3(\lg n),$$

where δ_3 is an oscillating function of its argument.

EXERCISES

10.1 Construct a pair of sorted lists of sizes 3 and 4 so that when merged by *LinearMerge* (Figure 10.2) only one key will be left to be transferred in the transfer stage of the procedure. Construct a pair of sorted lists of sizes 3 and 4 so that, when merged, 4 keys will be left to be transferred. How many comparisons does *LinearMerge* make in each case? You may assume the seven keys $1, \ldots, 7$ to be your data.

10.2 Assume the integers $1, 2, \ldots, m + n$ are partitioned into two distinct sorted lists of sizes $m \le n$. How many pairs of sorted lists of these sizes are possible that leave only one key to the transfer stage of *LinearMerge* (Figure 10.2)? How many pairs will leave n keys?

10.3 Let Y_n be the number of key comparisons involved in the transfer stage of *MergeSort* algorithm (Figure 10.3) (under *LinearMerge* (Figure 10.2)) at the top level of recursion. Show that $n - Y_n$ converges in distribution to GEOMETRIC$(1/2)$, a geometric random variable with rate of success $1/2$ per trial.

10.4 Verify that $n\lceil \lg n \rceil - 2^{\lceil \lg n \rceil} + 1$ is a solution to the recurrence

$$W_n = W_{\lceil n/2 \rceil} + W_{\lfloor n/2 \rfloor} + n - 1,$$

with $W_1 = 0$. (Hint: Consider odd and even parity of n.)

10.5 Show that the best-case number of comparisons for the standard top-down version of MERGE SORT satisfies the Trollope-Delange recurrence in Exercise 1.11.3. (Refer to Exercise 1.11.4 for the asymptotic solution.)

11

Bucket Sorts

Interest in the family of bucket sort algorithms peaked in the late 1970s as they offered efficient alternatives to standard comparison-based algorithms. In Chapter 1 we extensively discussed bounds on the complexity of the class of comparison-based sorting algorithms. Theorem 1.1 states that the complexity of this class both in the worst case and on average is $\Omega(n \ln n)$ to sort n keys. DISTRIBUTIVE SORT, a bucket sort algorithm invented by Dobosiewicz in 1978, provided a basis for several sorting algorithms with only $O(n)$ average cost. The method had existed in the literature since the late 1950s in various forms and flavors. Perhaps Dobosiewicz's (1978a) description was the first to translate it into a practicable algorithm.

Bucketing is a term used for identifying numerical keys by intervals. A bucket sort algorithm sorts by "distributing" the keys into containers called buckets (hence the name DISTRIBUTIVE SORT for some flavors of the algorithm). Bucketing is usually achieved by simple arithmetic operations such as multiplication and rounding. For certain forms of input data, bucketing may be achieved at machine-level instructions such as the shifting of the contents of a register.

In this chapter we present several flavors of bucket sorting algorithms and bucket selection algorithms derived from them. The difference between various flavors of bucket sorting is in the choice of the bucket size and in what they do within a bucket targeted for subsequent sorting or selection.

11.1 THE PRINCIPLE OF BUCKET SORTING

Based on the idea of distributing keys into buckets several sorting algorithms can be designed. The idea is related to a general class of hashing algorithms, whereby one has a universe to draw data from and a hash table to store them. A typical example familiar to all of us is the personal phone directory. We draw some names of particular significance to us (friends, relatives, etc.) from the population at large to insert in our pocket phone directory, which is by necessity small. The usual alphabetical arrangement then puts all names beginning with letter A first in the A section, then all the names beginning with B in the B section, and so on. In a hashing scheme we have a hash function h, that maps sampled data into positions in the hash table. In the pocket phone directory instance, a typical hash function is one that shaves off the first letter: $h(x\sigma) = x$, for x a letter from the English alphabet and σ is a trailing string of characters. It is evident that collisions are unavoidable. In this example $h(\text{Johnson}) = h(\text{Jeffrey}) = \text{J}$ and we have to have a collision resolution algorithm. The most common collision resolution algorithm in pocket phone directories is the algorithm that considers that a page corresponds to a number of positions, and within one page we enter the data sequentially according to a first-come-first-served policy.

The assumption in this example is that we are not going to have too many friends whose names begin with a certain letter like J. Generally the success of hashing schemes depends on the assumption that there are "good" hash functions that uniformly distribute keys over hash positions.

In one flavor of bucket sorts based on hashing, keys are distributed over a *large* number of buckets by a hash function. This family of bucket sorting algorithms is called DISTRIBUTIVE SORT. In another recursive flavor, RADIX SORT, a fixed number of buckets is used. (The terms *fixed* and *large* are with respect to n, the number of keys to be sorted, as $n \to \infty$.) We shall denote the bucket size by b, which may or may not depend on n.

To unify the treatment across the various flavors of bucket sorting algorithms, we assume that n keys are drawn from the UNIFORM(0, 1] distribution. This assumption is probabilistically equivalent to the general hypothesis that underlies hashing schemes and according to which good hash functions exist to distribute keys from some domain uniformly over a hash table. The unit interval is divided into a number of equally long intervals $(0, 1/b], (1/b, 2/b], \ldots ((b-1)/b, 1]$, where b is a number to be specified later. Think of these intervals as indexed from left to right by $1, 2, \ldots, b$. In all bucketing flavors, a key K is thrown into the $\lceil bK \rceil$th bucket by an application of the hash function $h(x) = \lceil bx \rceil$, for any $x \in (0, 1]$. The application of the hash function is one type of a *bucketing operation*, but other types will be taken into consideration, too.

As an illustration, consider the 8–key input

$$X_1 = .45 \qquad X_2 = .63$$
$$X_3 = .77 \qquad X_4 = .42$$
$$X_5 = .22 \qquad X_6 = .85$$
$$X_7 = .82 \qquad X_8 = .47$$

With $b = 10$, there are 10 buckets corresponding to 10 intervals each of length 0.1. A key with most significant digit d after the decimal point falls in the $(d+1)$st bucket. For instance, $X_1 = .45$ is hashed to bucket number $\lceil 10 \times .45 \rceil = 5$. Figure 11.1 shows the distribution of these 8 keys into the 10 buckets.

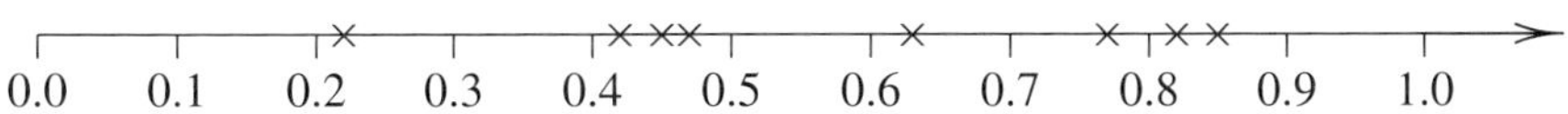

Figure 11.1. Distribution of keys into buckets.

At the outset of this presentation we write a distributional equation for all versions of bucket sorting algorithms. Suppose that within a bucket an algorithm, say A, is used to sort the contents of the bucket. Suppose such an algorithm makes Y_j operations of some kind to sort a random input of j keys. If A is some standard algorithm, the operation in question is typically a comparison of a pair of keys. Generally Y_j is a random variable having its own probability distribution. We have to take into account the possible difference between the bucketing operation, which typically involves arithmetics and ceils or floors, and the operations of A. If we take the unit cost to be that of a bucketing operation, a single operation of A may cost α. On a modern computer, a typical value of α for comparison-based sorting algorithms is somewhere between 0.01 and 0.1 (a measure of the speed of a comparison of keys relative to a bucketing operation like the hash function already considered). In recursive flavors A will be the same algorithm used at the top level, sorting the bucket by the same bucketing operations, i.e., $\alpha = 1$.

For whatever flavor of complete sorting by bucketing, the sorting process must continue in all buckets. Under our data uniformity assumption, the shares $N_1, \dots, N_b$ of the buckets $1, 2 \dots, b$ have a joint multinomial distribution on n trials and rate of success $1/b$ per trial for each bucket. In particular, the marginal distribution of any individual bucket's share is BINOMIAL$(n, 1/b)$, a binomial random variable on n trials and rate of success $1/b$ per trial.

The distributional equation for C_n, the cost of sorting n random keys, is

$$C_n \stackrel{\mathcal{D}}{=} \alpha\big(Y_{N_1}^{(1)} + Y_{N_2}^{(2)} + \cdots + Y_{N_b}^{(b)}\big) + n, \tag{11.1}$$

where for any j, $Y_j^{(k)} \stackrel{\mathcal{D}}{=} Y_j$, for $k = 1, \dots, b$, and for $k \ne \ell$, the families $\{Y_j^{(k)}\}_{j=1}^{\infty}$ and $\{Y_j^{(\ell)}\}_{j=1}^{\infty}$ are independent (but of course $(Y_{N_j}^{(j)}$ and $Y_{N_k}^{(k)}$ are dependent through the dependence of N_j and N_k). The quantity

$$W_n = Y_{N_1}^{(1)} + Y_{N_2}^{(2)} + \cdots + Y_{N_b}^{(b)}$$

is the only stochastic element in the equation. Introduce the probability generating functions $\phi_n(u) = \mathbf{E}[u^{W_n}]$ and $\psi_n(u) = \mathbf{E}[u^{Y_n}]$.

Lemma 11.1 *For $n \ge 2$,*

$$\phi_n(u) = \frac{n!}{b^n} \sum_{i_1+\cdots+i_b=n} \frac{\psi_{i_1}(u)}{i_1!} \times \cdots \times \frac{\psi_{i_b}(u)}{i_b!},$$

where the sum runs over all non-negative integer solutions of the equation $i_1 + \cdots + i_b = n$.

Proof. By conditioning on the shares of the buckets, we can write

$$\phi_n(u) = \sum_{i_1+\cdots+i_b=n} \mathbf{E}\big[u^{Y_{N_1}^{(1)} + Y_{N_2}^{(2)} + \cdots + Y_{N_b}^{(b)}} \mid N_1 = i_1, \dots, N_b = i_b\big]$$

$$\times \mathbf{Prob}\{N_1 = i_1, \cdots, N_b = i_b\}$$

$$= \sum_{i_1+\cdots+i_b=n} \mathbf{E}\big[u^{Y_{i_1}^{(1)}+\cdots+Y_{i_b}^{(b)}}\big]\binom{n}{i_1, \ldots, i_b}\frac{1}{b^n}$$

$$= \frac{1}{b^n} \sum_{i_1+\cdots+i_b=n} \mathbf{E}\big[u^{Y_{i_1}^{(1)}}\big] \times \ldots \times \mathbf{E}\big[u^{Y_{i_b}^{(b)}}\big]\binom{n}{i_1, \ldots, i_b}$$

$$= \frac{n!}{b^n} \sum_{i_1+\cdots+i_b=n} \frac{\psi_{i_1}(u)}{i_1!} \times \cdots \times \frac{\psi_{i_b}(u)}{i_b!},$$

valid for $n \ge 2$; the decomposition of $\mathbf{E}\big[u^{Y_{i_1}^{(1)}+\cdots+Y_{i_b}^{(b)}}\big]$ into the product $\mathbf{E}\big[u^{Y_{i_1}^{(1)}}\big] \times \cdots \times \mathbf{E}\big[u^{Y_{i_b}^{(b)}}\big]$ follows from the conditional independence of the action in the buckets. ■

11.1.1 Distributive Sorts

Suppose we apply bucketing to random numeric data from the interval (0, 1] to distribute them over b buckets. Several flavors along the main bucketing theme are possible. For example, the hash function

$$h(x) = \lceil bx \rceil$$

distributes the n random keys over the buckets, with n/b keys per bucket on average. Buckets can then be sorted individually by a standard sorting algorithm. One can even think of applying bucket sort recursively. However, some technical algorithmic modification will be required as discussed in Exercise 11.5.

Several bucket sorting algorithms resort to list implementation within a bucket because adding a key to an unsorted list is algorithmically trivial and because the collection at the end of sorted lists is made easy if each stretch of data belonging to a bucket has been reordered into a sorted linked list. This simplifies an initial and a final bookkeeping stage, but may complicate sorting within buckets—linked lists in bucket sorting tend to be a little bit restrictive for sorting (and more so for bucket selection algorithms derived from them) because we have to use a sorting algorithm suitable for linked lists, excluding many efficient sorting algorithms. For example, we will not be able to directly use BINARY INSERTION SORT or the standard form of QUICK SORT within buckets implemented as linked lists.

An elegant implementation called INTERPOLATION SORT manipulates only arrays. This form of bucket sorting was first given by Gonnet in 1984. Gonnet's idea is to hash every key *twice* and use a secondary array to receive the final sorted data (we can, of course, always transfer such an array back into the original array, if necessary). During the first hashing round we estimate where a key falls in the recipient array. Thus, we shall know the number of collisions at any particular slot. Suppose $A[1\,..\,n]$ is our source array of input data and $R[0\,..\,b]$ is the recipient array. The addition of a 0th position in R is only for a technical reason that has to do with the use of the sentinel version of INSERTION SORT within the buckets, as will be shortly explained. If a bucket receives k keys, we shall need k consecutive positions of R for it. The purpose of the first round is to determine the shares N_1, $N_2, \dots, N_b$, of the buckets. At the end of the first round, we know how many slots of R are needed for each bucket. But this round does not tell us which keys fall in which bucket. A second round of hashing can fill the buckets systematically. A key falling into the jth bucket belongs to a bucket that will be stored into the locations $N_1 + \cdots + N_{j-1} + 1 \mathinner{..} N_1 + \cdots + N_j$. We can therefore scan A, and systematically fill the buckets. The systematic scan can be done from the bottom of the bucket up, or vice versa. The bottom filling is convenient—we rehash each key; the first time a key falls into the jth bucket, we place it into $R[N_1 + \cdots + N_j]$, the second time a keys falls into the jth bucket, we place it into $R[N_1 + \cdots + N_j - 1]$, and so on. Figure 11.2 gives the formal code for INTERPOLATION SORT—Gonnet's implementation of bucket sorting. Assuming the input data have already been loaded into $A[1 \mathinner{..} n]$, INTERPOLATION SORT produces the final sorted output in $R[1 \mathinner{..} n]$, via the counting array $count[1 \mathinner{..} b]$. At the end of the first round $count[j]$ counts how many keys fall at the jth bucket. We then turn the count into a cumulative count, that is, the cumulative bucket sizes up to and including the jth bucket, by an accumulation of all the contents of $count$ up to its jth position, for all j. Thus, after accumulation necessarily $count[b] = n$. The contents of $count$ then guide the rehashing of the second round. The jth bucket is to fill positions $count[j - 1] + 1, \dots, count[j]$ (the number of keys in a bucket can possibly be 0). Starting at the bottom position, when a key falls in the jth bucket it is placed at position $count[j]$, and $count[j]$ is decremented by 1, thus creating a corresponding index for the next bottom-up filling

```
R[0] ← −∞;
for i ← 1 to n do
    count[j] ← 0;
for i ← 1 to n do
    begin
        j ← h(A[i]);
        count[j] ← 1 + count[j];
    end;
{Accumulate bucket sizes}
for i ← 1 to b − 1 do
    count[i + 1] ← count[i + 1] + count[i];
for i ← 1 to n do
    begin
        j ← h(A[i]);
        R[count[j]] ← A[i];
        count[j] ← count[j] − 1;
    end;
call LinearInsertSort(B);
```

Figure 11.2. INTERPOLATION SORT—an implementation of DISTRIBUTIVE SORT.

of the jth bucket. The filling proceeds in this manner until all the locations of the bucket are exhausted.

After the two rounds of hashing and rehashing, INTERPOLATION SORT may invoke any implementation of INSERTION SORT which is well suited for the purpose. Supposing bucket j starts at position $k = N_1 + \cdots + N_{j-1} + 1$, we are guaranteed that all the elements above this position (those belonging to the first $j - 1$ buckets) are smaller in value than any element in the jth bucket. Particularly, $R[k - 1]$ is at most $R[k]$. So, $R[k - 1]$ can act as a stopping sentinel in the sorting of the stretch $R[N_1 + \cdots + N_{j-1} + 1 \mathinner{..} N_1 + \cdots + N_j]$. Only the first bucket (stored in $R[1 \mathinner{..} N_1]$) is not preceded by a sentinel formed naturally by some other key in the data. That is why we introduced $R[0]$, which is to be initialized to $-\infty$ (refer to the discussion of Section 2.3 for the practical meaning of $-\infty$ in the algorithmic sense).

For instance, if INTERPOLATION SORT is applied to the 8–key input given in Section 11.1, with the hash function discussed there, the algorithm passes through the stages described in Figure 11.3.

Even though LINEAR INSERTION SORT takes average time of quadratic order $\Theta(n^2)$ to sort n *random* keys, its use within buckets still gives linear overall average time for INTERPOLATION SORT. After two hashing rounds, keys of relatively close values are clustered together; the data are no longer random—the array $R[1 \mathinner{..} n]$ does *not* follow the random permutation model, but is rather a collection of b subarrays (of total size n) each following the random permutation model. If b is large, we shall add up small quadratic forms. First the stretch $A[N_1 + \cdots + N_{b-1} + 1 \mathinner{..} n]$ is sorted. Then the algorithm moves up to sort the bucket above it (the stretch $A[N_1 + \cdots + N_{b-2} + 1 \mathinner{..} N_1 + \cdots + N_{b-1}]$, and so on. INTERPOLATION SORT makes $2n$ applications of the hash function and its cost is therefore

$$C_n = \alpha W_n + 2n \stackrel{D}{=} \alpha\big(Y_{N_1}^{(1)} + Y_{N_2}^{(2)} + \cdots + Y_{N_b}^{(1)}\big) + 2n,$$

A	After First Hashing Round *count*	After Accumulation *count*	R
	0	0	$-\infty$
.45	0	0	.22
.63	1	1	.47
.77	0	1	.42
.42	3	4	.45
.22	0	4	.63
.85	1	5	.77
.82	1	6	.82
.47	2	8	.85
	0	8	

Figure 11.3. Hashing and rehashing in INTERPOLATION SORT.

with $Y_j^{(k)} \stackrel{\mathcal{D}}{=} Y_j$, INSERTION SORT's number of comparisons when it sorts j keys, for $k = 1, \ldots, b$; the families $\{Y_j^{(k)}\}$ and $\{Y_j^{(\ell)}\}$ are independent (but of course, for $k \neq \ell$, $Y_{N_k}^{(k)}$ and $Y_{N_\ell}^{(\ell)}$ are dependent through the dependence of N_k and N_ℓ). Moreover, $Y_{N_1}^{(1)}, \ldots, Y_{N_b}^{(b)}$ are equidistributed and on average INTERPOLATION SORT costs

$$\mathbf{E}[C_n] = 2n + \alpha b \mathbf{E}[Y_{N_1}].$$

The intuition here is that, if b is large, N_1 (and subsequently $\mathbf{E}[Y_{N_1}]$) is small and the average complexity remains linear. For this form of bucket sorting we consider the equation in Lemma 11.1 when specialized to the case $b = n$. The term DISTRIBUTIVE SORT as originally used by Dobosiewicz (1987a) refers to the particular choice $b = n$. In choosing a large number b of buckets a natural value is $b = n$. The choice $b >> n$ will not help much as it gives several empty buckets, which will increase the overhead as, for instance, in an implementation like INTERPOLATION SORT at its accumulation stage. In other implementations based on lists, the check for an empty list will be done too often. The choice $b << n$ will lead to more collisions. The idea in $b = n$ is that on average a bucket contains one key.

We shall carry out the analysis for the choice $b = n$, and keep in mind that it generalizes easily to other choices of large b. Specialized to the case $b = n$, the right-hand side of the equation in Lemma 11.1 can be viewed as the nth coefficient in a generating function. We can express the specialized equation in the form

$$\phi_n(u) = \frac{n!}{n^n}[z^n]\Big(\sum_{j=0}^{\infty} \psi_j(u)\frac{z^j}{j!}\Big)^n.$$

This representation admits the following central limit result. The theorem follows from a rather general result in Flajolet, Poblete, and Viola (1998) for hashing with linear probing that broadly states that, under suitable conditions, coefficients of generating functions raised to large powers follow a Gaussian law. We work through some of the details to obtain the first two moments as well in order to completely characterize the limiting normal distribution. We shall assume that within the buckets a "reasonable" sorting algorithm is used. All standard sorting algorithms have polynomial time worst-case behavior. Thus, we assume, for example, that, uniformly over all sample space points, $Y_j \leq j^\theta$, for some fixed $\theta > 0$.

Theorem 11.1 *(Mahmoud, Flajolet, Jacquet, Régnier, 2000). Let W_n be the extra number of operations within the buckets of DISTRIBUTIVE SORT to sort n random keys. Suppose for some fixed $\theta > 0$, the algorithm applied in the buckets uses $Y_j \leq j^\theta$ operations. Then*

$$\frac{W_n - \mu n}{\sqrt{n}} \xrightarrow{\mathcal{D}} \mathcal{N}(0, \sigma^2),$$

where

$$\mu = e^{-1} \sum_{j=0}^{\infty} \frac{\mathbf{E}[Y_j]}{j!};$$

$$\sigma^2 = e^{-1} \sum_{j=0}^{\infty} \frac{\mathbf{E}[Y_j^2]}{j!} - \mu^2.$$

Proof. All derivatives in this proof are with respect to z. Let $\psi_n(e^{it})$ be the characteristic function of Y_n (we consider characteristic functions to avoid existence problems of moment generating functions). Let us denote the bivariate generating function $\sum_{j=0}^{\infty} \psi_j(u) z^j / j!$ by $\Psi(u, z)$. Then by Cauchy's formula,

$$\phi_n(u) = \frac{n!}{2\pi i n^n} \oint_{\Gamma} \frac{\Psi^n(u, z)}{z^{n+1}}\, dz,$$

where Γ is any closed contour enclosing the origin. We choose Γ to be a particular contour consisting of the line segment connecting $c - iM$ to $c + iM$ (for some fixed $c > 0$ and some large M) and a closing (left) arc of the circle centered at the origin and passing through these two points. As $M \to \infty$, one can check that the integral on the arc approaches 0 (see the proof of Lemma 1.4 for this type of argument).

To evaluate the remaining line integral asymptotically by the saddle point method, write the last equation in the form:

$$\phi_n(u) = \frac{n!}{2\pi i n^n} \int_{c-i\infty}^{c+i\infty} e^{ng(u,z)} \frac{dz}{z}, \tag{11.2}$$

where by definition $g(u, z) = \ln\{\Psi(u, z)/z\}$. We shall eventually let $u \to 1$. The saddle point is the special value of z that solves the saddle point equation $g'(u, z) = 0$, or

$$z\Psi'(u, z) = \Psi(u, z). \tag{11.3}$$

One can verify that near $u = 1$, $z = 1 + O(1 - u)$ is a saddle point of the integrand in (11.2). Therefore, we deform the line of integration to become one connecting $c - i\infty$ to $c + i\infty$ and going through $z = 1$ through an angle picking up the steepest descent of the function g.

Replacing $n!$ by its asymptotic Stirling's approximation, and using the saddle point method (as $u \to 1$):

$$\phi_n(u) \sim \frac{\Psi^n(u, 1)}{e^n \sqrt{g''(u, 1)}}.$$

Let us now set $u = e^{it}$, and expand the characteristic functions $\psi_j(e^{it})$ around $t = 0$ (i.e., around $u = 1$) in powers of t with coefficients that are moments of the cost.

Denoting the first two moments of Y_j respectively by μ_j and s_j, near $t = 0$ we obtain

$$\Psi^n(e^{it}, z) = \exp\Big\{n \ln\Big(\sum_{j=0}^{\infty} \psi_j(e^{it})\frac{z^j}{j!}\Big)\Big\}$$

$$= \exp\Big\{n \ln\Big[\sum_{j=0}^{\infty}\Big(\frac{z^j}{j!} + \frac{\mu_j z^j}{j!}it - \frac{s_j z^j}{2!\, j!}t^2\Big) + O(t^3)\Big]\Big\}.$$

Letting $u \to 1$ implies $z \to 1$, and we have

$$\Psi^n(e^{it}, 1) = \exp\Big\{n \ln\Big[e\Big(1 + \mu it - s\frac{t^2}{2}\Big) + O(t^3)\Big]\Big\},$$

where

$$\mu \stackrel{\text{def}}{=} e^{-1}\sum_{j=0}^{\infty}\frac{\mu_j}{j!},$$

$$s \stackrel{\text{def}}{=} e^{-1}\sum_{j=0}^{\infty}\frac{s_j}{j!}.$$

Expanding the logarithm with the usual calculus equality $\ln(1 + x) = x - x^2/2 + O(x^3)$, we get

$$\phi_n(e^{it}) = \mathbf{E}\big[e^{W_n it}\big] \sim \frac{e^{\mu nit - \sigma^2 nt^2/2 + O(nt^3)}}{\sqrt{g''(e^{it}, 1)}},$$

where $\sigma^2 = s - \mu^2$. From the continuity of g, we can easily verify that $g''(e^{it}, 1) \to g''(1, 1) = 1$, as $t \to 0$. The assumption that the sorting algorithm within buckets is reasonable, with worst-case polynomial time, guarantees $\mu_j = O(j^{\theta})$ and $s_j = O(j^{2\theta})$. Thus, the series in μ and σ^2 converge.

Finally, set $t = v/\sqrt{n}$ for a fixed v and let $n \to \infty$ (so indeed $u \to 1$). So,

$$\mathbf{E}\Big[\exp\Big\{\frac{W_n - \mu n}{\sqrt{n}}iv\Big\}\Big] \to e^{-\sigma^2 v^2/2};$$

the right hand side is the characteristic function of $\mathcal{N}(0, \sigma^2)$ and of course convergence of characteristic functions implies weak convergence. ■

Theorem 11.1 covers a wide class of bucket sorting algorithms. We instantiate it here with INTERPOLATION SORT and leave a number of other instances for the exercises. INTERPOLATION SORT (see Figure 11.2) uses LINEAR INSERTION SORT (with sentinel) in the buckets. For LINEAR INSERTION SORT Y_j, the number of comparisons it performs to sort a file of j random keys, is a random variable whose exact and limit distributions are completely characterized (cf. Section 2.3). We recall here results concerning the first two moments:

$$\mathbf{E}[Y_j] = \frac{j(j+3)}{4},$$

$$\mathbf{Var}[Y_j] = \frac{j(j-1)(2j+5)}{72}.$$

So, $\mu = \frac{5}{4}$, and $\sigma^2 = \frac{91}{36}$, and the particular central limit theorem for INTERPOLATION SORT's cost is

$$\frac{C_n - (2 + 5\alpha/4)n}{\sqrt{n}} \xrightarrow{\mathcal{D}} \mathcal{N}\Big(0, \frac{91}{36}\alpha^2\Big).$$

11.1.2 Radix Sorting

The bucketing operation in RADIX SORT is based on the digital composition of data giving rise to a recursive flavor of bucket sorting with a *fixed* number b of buckets. For example, if our data are distinct keys that are strings of 0's and 1's, the algorithm separates those data into two groups (buckets): Those strings beginning with 0 go into one bucket; those beginning with 1 go into the other bucket. RADIX SORT continues its action recursively in each bucket, but at the jth level of recursion it uses the jth character of the strings to further separate the data. The process continues until each key falls in a bucket by itself.

In practical terms, the "separation" of data may be done by moving data around within a host data structure, say like an array. For instance, if our data are distinct keys that are strings of 0's and 1's, the first layer of bucketing according to 0's and 1's may be achieved by a partitioning algorithm that emulates the essence of the algorithm PARTITION that we used with QUICK SORT. The difference here is that partitioning is not based on a pivot, but rather on the first character of the keys. Figure 11.4 illustrates the principle of digital bucketing: After one application of the partitioning algorithm (based on the most significant bit) a splitting position s identifies the boundaries—the s keys beginning with 0 are now in the subarray $A[1 \mathinner{..} s]$; the subarray $A[s+1 \mathinner{..} n]$ contains keys beginning with 1.

A partitioning algorithm similar to Sedgewick's algorithm (see Exercise 7.2) is easy to implement. In the language of numerics, if we wish to treat our data as numeric keys, say from the interval $(0, 1]$, data grouping according to 0's and 1's at the first layer amounts to creating the two intervals $(0, 1/2]$ and $(1/2, 1]$, then placing keys beginning with 0 (whose numerical value is $\le 1/2$) in the left bucket and placing keys beginning with 1 (whose numerical value is at least $1/2$) in the right bucket.

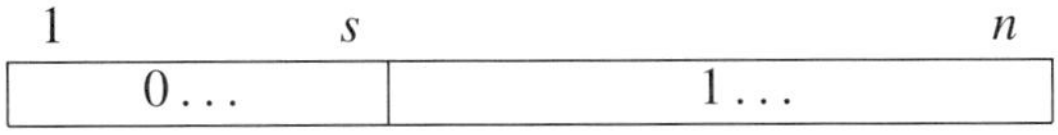

Figure 11.4. Digital bucketing.

This bucketing strategy of RADIX SORT continues recursively on nonempty buckets after rescaling, that is, working on diminishing bucket sizes $1/2$, $1/4$, etc. until enough information is gathered for complete sorting. One sees that RADIX SORT is a digital counterpart of QUICK SORT.

At the lowest level of machine computing, data appear in binary form (a representation of numbers and other data in the *base* 2 number system). The base of the number system is also called its *radix*. Other common forms, at a slightly higher level of interaction with machines, are the octal (base 8), decimal (base 10), and hexadecimal (base 16). Other forms of data suitable for direct application of RADIX SORT are text and DNA strands.

In these forms, data are strings over a finite alphabet. In particular, numbers from the interval $(0, 1]$ may be thought of as infinite expansions. For rational numbers there are two representations; we choose the infinite. For example, $1/2$ in binary is either 0.1 or $0.011111\ldots$ and we choose the latter. Generally, irrational numbers have infinite expansion and some rational numbers have two possible representations, one finite and one infinite. For rational numbers with two binary representations we shall choose the infinite expansion so that all our numeric data have the same uniform infinite representation. In practice, real numbers are approximated by cutting off the expansion at some finite precision. If a key is available as a binary expansion of bits (binary digits):

$$K = 0.d_1d_2d_3\ldots,$$

with every $d_j \in \{0, 1\}$, one can immediately access d_j by direct indexing. For example, the ith key may be the ith member of an array $A[1\,..\,n]$, and may be a finite string of type **string**$[L]$, where L is a predesignated maximum key length. (In many programming languages the type **string**$[L]$ is used as a shorthand notation for **array** $[1\,..\,L]$ of characters.)

We shall discuss the algorithm and its analysis in the practicable case of binary data, the case $b = 2$ (generalization of the algorithm and its analysis to higher b does not pose any additional difficulty). The bucketing operations of RADIX SORT are bit extractions performed at the top level of recursion on the most significant bit, then again recursively in the buckets on less significant bits. That is to say, the comparative factor α that measures the units of operations in the buckets relative to the top level operations is 1.

We can let the partitioning algorithm access position bit, where bit progressively assumes the values $1, 2, \ldots, L$. An implementation, $RadixSort$, may proceed as follows. The assignment of the initial value 1 to bit is passed on to the algorithm by the external call

call $RadixSort(1, n, 1)$;

then the QUICK-SORT-like algorithm (Figure 11.5) first partitions (splits at position s, say) then recursively invokes itself twice: one call taking care of the left bucket, and one call taking care of the right bucket. Each call passes on a higher bit position to further recursive levels.

```
procedure RadixSort(ℓ, u, bit: integer);
    local s: integer;
    begin
        if ℓ < u then
            begin
                call Partition(ℓ, u, s, bit);
                call RadixSort(ℓ, s, bit + 1);
                call RadixSort(s + 1, u, bit + 1);
            end;
    end;
```

Figure 11.5. The RADIX SORT algorithm.

The bit accessing operation can be done arithmetically by integer division by 2. For example, at the jth level of recursion in RADIX SORT, the jth bit is used for further data separation and this value is passed via the variable bit to the partitioning algorithm. The partition algorithm can take this value and extract the jth bit from any key K by the operation $\lfloor 2^j K \rfloor$ **mod** 2. Arithmetic bucketing operations like the one just discussed may be simplified in practice to hardwired machine arithmetic such as register shift instructions. The algorithm is also directly applicable to nonnumeric data, as in the lexicographic sorting of text or DNA strands.

As the bucketing operations of RADIX SORT are bit extractions, we shall measure the cost C_n of RADIX SORT by the total number of bits it extracts from the n keys of the file presented to the algorithm to operate on. Equation (11.1) then becomes a distributional recurrence (with $\alpha = 1$):

$$C_n \stackrel{D}{=} C_{N_1}^{(1)} + C_{N_2}^{(2)} + n,$$

where for each j, $C_j^{(1)} \stackrel{D}{=} C_j^{(2)} \stackrel{D}{=} C_j$, and the two families $\{C_j^{(1)}\}_{j=1}^{\infty}$ and $\{C_j^{(2)}\}_{j=1}^{\infty}$ are independent, but again $C_{N_1}^{(1)}$ and $C_{N_2}^{(2)}$ are dependent. Let $\xi_n(u) = \mathbf{E}[u^{C_n}]$ be the probability generating function of the cost, as measured by the number of bit inspection operations performed. Note that we are considering the *total* cost and not the extra cost in the buckets; to the extra cost in the buckets we add n (the number of bit inspections at the top level of recursion) to get the total number of bit inspections. This is reflected as an additional u^n factor in the formula of Lemma 11.1, which now uses recursively the probability generating function in the buckets, too. For $n \geq 2$, this gives us the functional equation

$$\frac{\xi_n(u)}{n!} = \frac{u^n}{2^n} \sum_{i_1+i_2=n} \frac{\xi_{i_1}(u)}{i_1!} \times \frac{\xi_{i_2}(u)}{i_2!},$$

where the sum runs over all non-negative integer solutions of the equation $i_1+i_2 = n$.

Introduce the super generating function $\Xi(u, z) = \sum_{j=0}^{\infty} \xi_j(u) z^j / j!$. To get a functional equation on $\Xi(u, z)$, multiply the last recurrence by z^n and sum over

$n \ge 2$, the range of validity of the recurrence:

$$\sum_{n=2}^{\infty} \frac{\xi_n(u)}{n!} z^n = \sum_{n=2}^{\infty} \frac{u^n z^n}{2^n} \sum_{j=0}^{n} \frac{\xi_j(u)}{j!} \times \frac{\xi_{n-j}(u)}{(n-j)!}.$$

Extend the sums on n from 0 to ∞ (to complete the super moment generating function):

$$\Xi(u,z) - \frac{\xi_0(u)}{0!} - \frac{\xi_1(u)}{1!} z = \sum_{n=0}^{\infty} \sum_{j=0}^{n} \frac{\xi_j(u)}{j!} \Big(\frac{uz}{2}\Big)^j \times \frac{\xi_{n-j}(u)}{(n-j)!} \Big(\frac{uz}{2}\Big)^{n-j}$$
$$- \sum_{n=0}^{1} \frac{u^n z^n}{2^n} \sum_{j=0}^{n} \frac{\xi_j(u)}{j!} \times \frac{\xi_{n-j}(u)}{(n-j)!}.$$

The boundary conditions $C_0 = C_1 \equiv 0$ (that is, $\xi_0(u) = \xi_1(u) \equiv 1$), come in handy to simplify the functional equation to

$$\Xi(u,z) - 1 - z = \Xi^2\Big(u, \frac{uz}{2}\Big) - 1 - 2 \times \frac{uz}{2},$$

which is

$$\Xi(u,z) = \Xi^2\Big(u, \frac{uz}{2}\Big) + z(1-u). \tag{11.4}$$

Derivatives of this functional equation give us functional equations on the moments. For instance, the kth derivative with respect to u, at $u = 1$, gives

$$\frac{\partial^k}{\partial u^k} \Xi(1,z) = \sum_{j=0}^{\infty} \frac{d^k \xi_j(u)}{du^k} \times \frac{z^j}{j!}\Big|_{u=1}$$
$$= \sum_{j=0}^{\infty} \mathbf{E}\big[C_j(C_j-1)\ldots(C_j-k+1)\big]\frac{z^j}{j!},$$

an exponential generating function for the kth factorial moment of the cost. In particular, taking the first and second derivatives of the functional equation gives functional equations on exponential generating functions of the first and second moments as will be utilized in the following theorems to derive the mean and variance of the cost of RADIX SORT.

For the sequence of means we define the exponential generating function

$$M(z) = \frac{\partial}{\partial u} \Xi(1,z) = \sum_{j=0}^{\infty} \mathbf{E}[C_j] \frac{z^j}{j!}.$$

If we take the derivative of (11.4) with respect to u once, at $u = 1$, we obtain

$$M(z) = 2\Xi\Big(1, \frac{z}{2}\Big) \frac{\partial}{\partial u} \Xi\Big(u, \frac{uz}{2}\Big)\Big|_{u=1} - z;$$

the boundary values in the functional equation are given by

$$\Xi\big(1, z\big) = \sum_{j=0}^{\infty} \xi_j(1) \frac{z^j}{j!} = e^z,$$

and

$$\begin{aligned}\frac{\partial}{\partial u} \Xi(u, uz)\big|_{u=1} &= \frac{\partial}{\partial u} \sum_{j=0}^{\infty} \xi_j(u) \frac{u^j z^j}{j!}\Big|_{u=1} \\ &= \sum_{j=0}^{\infty} \xi_j'(1) \frac{z^j}{j!} + \sum_{j=0}^{\infty} \xi_j(1) \frac{j z^j}{j!} \\ &= \sum_{j=0}^{\infty} \mathbf{E}[C_j] \frac{z^j}{j!} + z \sum_{j=1}^{\infty} \frac{z^{j-1}}{(j-1)!} \\ &= M(z) + z e^z.\end{aligned}$$

In terms of the mean generating function:

$$M(z) = 2e^{z/2}\Big[M\Big(\frac{z}{2}\Big) + \frac{z}{2} e^{z/2}\Big] - z.$$

In fact, $e^{-z}M(z)$ has a Poissonization interpretation—had we started with $N(z) =$ POISSON(z) random number of keys, instead of a fixed n, the average cost of sorting would have been

$$\begin{aligned}\mathbf{E}\big[C_{N(z)}\big] &= \sum_{j=0}^{\infty} \mathbf{E}\big[C_{N(z)} \mid N(z) = j\big] \mathbf{Prob}\{N(z) = j\} \\ &= \sum_{j=0}^{\infty} \mathbf{E}[C_j] \frac{z^j e^{-z}}{j!} \\ &= e^{-z} M(z).\end{aligned}$$

Thus, if we let $A(z) = \mathbf{E}\big[C_{N(z)}\big] = e^{-z}M(z)$, the Poissonized average, we have the functional equation

$$A(z) = 2A\Big(\frac{z}{2}\Big) + z - ze^{-z}. \tag{11.5}$$

The situation here has led us in a natural way to a functional equation for the Poissonized version of the problem. It is not as direct to get a functional equation for the fixed population model. So, we can attempt to first solve the Poissonized problem, then revert back to the fixed population model, the model of our prime interest. This amounts to working with a Poisson transform, then inverting it. We can of course do the same for higher moments, but expect the functional equations to require tedious handling. Indeed, that is how we shall get the variance, later in this section.

Let us start with the average cost of RADIX SORT, which will shed light on the general approach. Once the reader appreciates the mechanics of the Poisson transform and inversion process, the more tedious variance will appear to follow the same mechanical path, only requiring more computation.

The average cost of RADIX SORT is presented next. Knuth (1973) attributes the theorem to joint discussions with De Bruijn. Their proof uses the so-called Gamma function method (an alternative nomenclature for the Mellin transform) and employs approximating functions by their expansions. These basic ideas received much attention and inspired a lot of research that resulted in a refined toolkit of analytical methods (discussed in Section 1.11). The proof we present is based on that analytical toolkit.

Theorem 11.2 *(Knuth, 1973). Let C_n be the cost of RADIX SORT (number of bit extractions) to sort n random keys. Then*

$$\mathbf{E}[C_n] = n \lg n + \Big(\frac{\gamma}{\ln 2} + \frac{1}{2} - Q(\lg n)\Big)n + O(n^{\varepsilon}),$$

where $\gamma = 0.5772156\ldots$ is Euler's constant, $0 < \varepsilon < 1$, and the function Q is periodic in its argument and is given by the Fourier expansion

$$Q(x) = \frac{1}{\ln 2} \sum_{\substack{k=-\infty \\ k\neq 0}}^{\infty} \Gamma\Big(\frac{2\pi i k}{\ln 2}\Big) e^{2\pi i k x}.$$

Proof. Let $A(z) = \mathbf{E}\big[C_{N(z)}\big] = e^{-z}M(z)$. For this Poissonized cost we have derived the functional equation (11.5). Iterating we get

$$\begin{aligned} A(z) &= z(1 - e^{-z}) + 2A\Big(\frac{z}{2}\Big) \\ &= z(1 - e^{-z}) + z(1 - e^{-z/2}) + 4A\Big(\frac{z}{4}\Big) \\ &\ \ \vdots \\ &= z\sum_{k=0}^{K}(1 - e^{-z/2^k}) + 2^{K+1}A\Big(\frac{z}{2^{K+1}}\Big), \end{aligned}$$

which is valid for any positive integer K. Because $\mathbf{E}[C_0] = \mathbf{E}[C_1] \equiv 0$, the series

$$A(z) = e^{-z}M(z) = \sum_{j=2}^{\infty} \mathbf{E}\big[C_j\big] \frac{z^j e^{-z}}{j!} = O(z^2),$$

as $z \to 0$. Thus, for any fixed z, the remainder term is $2^{K+1}O(2^{-2K}) = O(2^{-K})$, as $K \to \infty$. The remainder approaches 0, as $K \to \infty$. Subsequently, by letting $K \to \infty$, we can then write

$$A(z) = z\sum_{k=0}^{\infty}\left(1 - e^{-z/2^k}\right). \tag{11.6}$$

The sum in this formula is one of our examples on the Mellin transform and its inversion (see Example 1.1 of Subsection 1.11.2). The Poissonized average is therefore

$$\mathbf{E}\left[C_{N(z)}\right] = A(z) = z\lg z + \left[\frac{\gamma}{\ln 2} + \frac{1}{2} - Q(\lg z)\right]z + O(z^{-d}),$$

for any arbitrary $d > 0$; the function $Q(\lg z)$ is periodic and of a (small) bounded absolute value.

What is left is to de-Poissonize the result back to the original fixed population model. All the conditions for the de-Poissonization lemma are valid—the Poissonized average $A(z) \sim z\lg z$ is $O(z^{1+\varepsilon})$ for any $\varepsilon > 0$, and apart from $O(n^{1/2+\varepsilon}\ln n)$ error, the fixed population model has the same average as the Poisson model:

$$\mathbf{E}[C_n] \sim \mathbf{E}\left[C_{N(n)}\right] + O\left(n^{\varepsilon'}\right) = n\lg n + \left[\frac{\gamma}{\ln 2} + \frac{1}{2} - Q(\lg n)\right]n + O\left(\frac{1}{n^d}\right) + O\left(n^{\varepsilon'}\right);$$

the statement of the theorem follows because $d > 0$, and we can take $\varepsilon' < 1$. ■

Two small error terms appear in the last display in the proof of Theorem 11.2: one coming from the Mellin transform inversion when the line of integration is shifted to the right, the other from de-Poissonization. Combined, the two errors are bounded by $o(n)$.

We can now turn to the variance $\mathbf{Var}[C_n]$ of the number of bit inspections in RADIX SORT while it sorts n random keys. The machinery of the computation has been demonstrated in some slow-paced detail for the mean. We shall only sketch this lengthy variance computation. First introduce the Poisson generating function of the second factorial moment $S(z)$:

$$S(z) = e^{-z}\frac{\partial^2}{\partial u^2}\Xi(1, z).$$

After somewhat lengthy algebra on (11.4) one finds

$$S(z) = 2S\left(\frac{z}{2}\right) + 4zA\left(\frac{z}{2}\right) + 2zA'\left(\frac{z}{2}\right) + 2A^2\left(\frac{z}{2}\right) + z^2$$

We introduce the function $V(z)$ defined by $V(z) = S(z) + A(z) - A^2(z)$ which, in other words, is the variance of the Poissonized cost.

Plugging in the recursive representation (11.5) for $A(z)$ gives the functional equation:

$$V(z) = 2V\Big(\frac{z}{2}\Big) + 2zA'\Big(\frac{z}{2}\Big) + 2ze^{-z}A(z) + z - ze^{-z} + z^2e^{-2z}.$$

Next, introduce the Mellin transform $V^*(s)$ of $V(z)$. To go further one needs the Mellin transform of $A'(z)$. One can get it by differentiating (11.6) to have a series representation of $A'(z)$, whose Mellin transform is easily obtained:

$$\mathcal{M}\{A'(z); s\} = \frac{(s-1)\Gamma(s)}{1-2^s} \qquad (11.7)$$

in the vertical strip $-1 < \Re s < 0$. We leave it as an exercise to derive the Mellin transform:

$$V^*(s) = \frac{1}{1-2^{s+1}}\Big(\frac{2^{s+2}s\Gamma(s+1)}{1-2^{s+1}} - \Gamma(s+1) + D(s)\Big),$$

with

$$D(s) = \Gamma(s+2)\left[\frac{1}{2^{2+s}} + 2\sum_{k=0}^{\infty}\Big(1 - \frac{1}{(1+2^{-k})^{s+2}}\Big)\right].$$

From the singularity analysis of the above we may obtain the asymptotic expansion of $V(z)$. The asymptotic expansion of $\mathbf{Var}[C_n]$ is then obtained by de-Poissonization:

$$\mathbf{Var}[C_n] = V(n) - n\big(A'(n)\big)^2 + O(n\ln^2 n).$$

The asymptotic equivalent of $A'(n)$ is obtained by inverting the Mellin transform of $A'(z)$; cf. (11.7). The inversion gives an asymptotic approximation for $A'(z)$, which is de-Poissonized into $A'(n)$. This lengthy process culminates into the result for the fixed population variance.

Theorem 11.3 *(Mahmoud, Flajolet, Jacquet, and Régnier, 2000). Let C_n be the cost of RADIX SORT (number of bit extractions) to sort n random keys. Then*

$$\mathbf{Var}[C_n] \sim \big(4.3500884\ldots + \delta(\lg n)\big)n,$$

where δ is a periodic function of its argument, and with a small amplitude.

We note in passing that the periodic fluctuations in the variance occur in its leading term, unlike those in the mean, which only occur in the lower-order term. The order of the variance is remarkably smaller than the order of the mean, indicating good concentration around average value—as n increases the mean of the distribution shifts up, and the length of Chebyshev's 3-standard-deviation interval that contains at least 89% of the distribution grows at a slower rate. This shows that the rate of convergence to a limit is fast.

We are now in a position to turn to limit distributions. We have the mean and standard deviation of the number of bit inspections. We use the leading terms in these factors for centering and scaling and we expect the resulting normed random variable

$$C_n^* = \frac{C_n - n \lg n}{\sqrt{\big((4.3500884\ldots) + \delta(\lg n)\big)n}}$$

to have a limit distribution. The proof technique is similar to what we did with QUICK SORT; we guess a functional equation on the limit then use it to characterize that limit. The success of this technique hinges on finding suitable representations. Recall the fundamental recurrence

$$C_n \stackrel{\mathcal{D}}{=} C_{N_1}^{(1)} + C_{N_2}^{(2)} + n,$$

where $C_j^{(i)} \stackrel{\mathcal{D}}{=} C_j$, for $i = 1, 2$, and the two families $\{C_j^{(1)}\}_{j=1}^{\infty}$ and $\{C_j^{(2)}\}_{j=1}^{\infty}$ are independent. The share of the left bucket (keys beginning with 0) is $N_1 = \text{BINOMIAL}(n, 1/2)$, and that of the right (keys beginning with 1) is $N_2 = n - N_1$. Set $v(n) = \mathbf{Var}[C_n]$, and think of it as a function of one argument, so $v(N_1)$ and $v(N_2)$ are random variables. We have

$$\frac{C_n - n \lg n}{\sqrt{v(n)}} = \frac{C_{N_1}^{(1)} - N_1 \lg N_1}{\sqrt{v(N_1)}} \times \sqrt{\frac{v(N_1)}{v(n)}} + \frac{C_{N_2}^{(2)} - N_2 \lg N_2}{\sqrt{v(N_2)}} \times \sqrt{\frac{v(N_2)}{v(n)}} + \frac{g_n(N_1)}{\sqrt{v(n)}},$$

where

$$g_n(u) \stackrel{\text{def}}{=} n - n \lg n + u \lg u + (n - u) \lg(n - u).$$

In terms of normed random variables

$$C_n^* \stackrel{\mathcal{D}}{=} C_{N_1}^* \sqrt{\frac{v(N_1)}{v(n)}} + \tilde{C}_{N_2}^* \sqrt{\frac{v(N_2)}{v(n)}} + \frac{g_n(N_1)}{\sqrt{v(n)}},$$

where, for all j, $\tilde{C}_j^* \stackrel{\mathcal{D}}{=} C_j^*$, and the families $\{C_j^*\}_{j=1}^{\infty}$ and $\{\tilde{C}_j^*\}_{j=1}^{\infty}$ are independent families.

Though technically involved, what is left is conceptually easy and similar to work we already encountered in the analysis of QUICK SORT. After all, RADIX SORT is the digital twin of QUICK SORT, but based on digital partitioning. So, RADIX SORT has similar recurrences, but with different probabilities for partitioning. We shall only outline the proof. If as $n \to \infty$, a limit random variable C exists for C_n^*, both $C_{N_1}^*$ and $\tilde{C}_{N_2}^*$ will approach copies of the limit C, as both N_1 and N_2 approach infinity almost surely. The fact that N_1 is a BINOMIAL$(n, 1/2)$ plays a crucial role here. As a consequence, $N_1/n \xrightarrow{P} 1/2$ and $N_2/n \xrightarrow{P} 1/2$. Asymptotically, both $v(N_1)$ and $v(N_2)$ are concentrated around $v(n/2)$. The periodic part of v replicates itself when its argument is doubled. That is,

$$\frac{v(N_1)}{v(n)} \sim \frac{(4.3500884\ldots)n/2 + \delta(n/2)n/2}{(4.3500884\ldots)n + \delta(n)n} \xrightarrow{P} \frac{1}{2}.$$

Similarly $v(N_2)/v(n) \xrightarrow{P} 1/2$. Furthermore, $g_n(N_1)/\sqrt{v(n)} \xrightarrow{P} 0$. These sketchy calculations are treated more rigorously in the exercises.

Though $C^*_{N_1}$ and $\tilde{C}^*_{N_2}$ are dependent through N_1 and N_2, their dependence gets weaker as n goes up (the asymptotic tendency of N_1/n and N_2/n to $1/2$ kicks in) and the dependence between $C^*_{N_1}$ and $\tilde{C}^*_{N_2}$ gets weaker, a phenomenon we are already familiar with from the analysis of QUICK SORT. One therefore expects that, if a limit C exists, it would satisfy a functional equation of the form

$$C \overset{\mathcal{D}}{=} \frac{C + \tilde{C}}{\sqrt{2}}, \tag{11.8}$$

where $\tilde{C} \overset{\mathcal{D}}{=} C$, and is independent of it. To actually prove that a limit C exists requires some additional technical work; one can show that such a limit exists from recurrence equations on the moment generating function.

What remains to be shown is that a random variable satisfying (11.8) must necessarily be a standard normal variate. This is intuitively true because the convolution of $\mathcal{N}(0, 1/\sqrt{2}\,)$ with an independent copy of itself satisfies

$$\frac{1}{\sqrt{2}}\mathcal{N}(0, 1) \oplus \frac{1}{\sqrt{2}}\mathcal{N}(0, 1) \overset{\mathcal{D}}{=} \mathcal{N}\Big(0, \frac{1}{2}\Big) \oplus \mathcal{N}\Big(0, \frac{1}{2}\Big) \overset{\mathcal{D}}{=} \mathcal{N}(0, 1).$$

This argument only shows that $\mathcal{N}(0, 1)$ satisfies the equation. One needs an additional argument to show that it is the only distribution that does. This follows from a standard argument presented next. (This proof is different from the one in the original study.)

Theorem 11.4 *(Mahmoud, Flajolet, Jacquet, and Régnier, 2000). The cost C_n of RADIX SORT (number of bit extractions) to sort n random keys satisfies a Gaussian limit law:*

$$C^*_n = \frac{C_n - n\lg n}{\sqrt{\big((4.3500884\ldots) + \delta(\lg n)\big)n}} \xrightarrow{\mathcal{D}} \mathcal{N}(0, 1).$$

Proof. We have shown that the limit random variable C satisfies the distributional equation:

$$C \overset{\mathcal{D}}{=} \frac{C + \tilde{C}}{\sqrt{2}}.$$

Let $\phi_X(t)$ be the characteristic function of a random variable X. The convolution in the distributional equation implies that the characteristic function satisfies a form that can be iterated. For any fixed real t,

$$\begin{aligned}
\phi_C(t) &= \phi_{C/\sqrt{2}}(t)\phi_{\tilde{C}/\sqrt{2}}(t) \\
&= \phi_C^2\Big(\frac{t}{\sqrt{2}}\Big) \\
&= \phi_C^4\Big(\frac{t}{2}\Big) \\
&\ \vdots \\
&= \phi_C^{2^n}\Big(\frac{t}{2^{n/2}}\Big) \\
&= \Big[1 + \mathbf{E}[C]\frac{t}{2^{n/2}} - \mathbf{Var}[C]\frac{t^2}{2^n\,2!} + o\Big(\frac{1}{2^n}\Big)\Big]^{2^n}.
\end{aligned}$$

The limit C is centered and normalized, with 0 mean and variance 1. That is, as $n \to \infty$

$$\phi_C(t) = \Big[1 - \frac{t^2}{2^{n+1}} + o\Big(\frac{1}{2^n}\Big)\Big]^{2^n} \to e^{-t^2/2};$$

the characteristic function of C converges to that of the standard normal variate. ■

11.2 BUCKET SELECTION

Several bucket sorting algorithms lend themselves to an intuitive adaptation to become single or multiple selection algorithms. As in most selection problems, we are given a rank $m \in \{1, \ldots, n\}$ and we wish to identify the numeric key among n data with rank m, i.e., the mth order statistic. Suppose N_j is the share of the jth bucket, $j = 1, \ldots, b$. Under the data uniformity assumption, the shares $N_1, \ldots, N_b$ have a joint multinomial distribution on n trials and rate of success $1/b$ per trial for each bucket. In particular, the marginal distribution of any individual bucket's share is BINOMIAL$(n, 1/b)$.

A bucket selection algorithm then continues its search for the desired order statistic by considering the keys in the bucket indexed i, for an index i satisfying

$$N_1 + \cdots + N_{i-1} < m \le N_1 + \cdots + N_i; \tag{11.9}$$

interpret a sum as 0 when it is empty.

If the correct bucket containing the desired order statistic is found to contain only one key, the process then terminates: The only key in the bucket must be the mth order statistic, otherwise a selection algorithm is invoked to find the key ranked $m - \sum_{j=1}^{i-1} N_j$ within the N_j keys that landed in the jth bucket. Probabilistically, very few keys fall in the correct bucket and either a recursive action or switching to some other selection algorithm should lead to rather fast termination. The bucketing operation is performed n times on the initial list of keys, then followed by a stochastically small number of operations. A bucket selection distributing the n keys over a large number of buckets at the first step, such as the case $b = n$, will be called a DISTRIBUTIVE SELECTION algorithm.

In all variations considered we shall regard m itself is chosen randomly according to a uniform distribution on its range. Thus, m becomes a random variable M_n = UNIFORM[1 .. n], independent of the shares of the buckets. We have already used this averaging technique in analyzing QUICK SELECT. This averaging works as a smoothing operator over all the fixed cases. It gives tractable recurrences and at the same time brings out the character of the distributions involved.

Let C_n be the number of bucketing operations involved in the selection of a rank M_n = UNIFORM[1 .. n]. After the first layer of bucketing, the n keys have been deposited in the buckets. Once the correct bucket containing the required order statistic is identified, some algorithm takes over to continue the search. (This algorithm is possibly the same top-level bucket selection algorithm used recursively.) Suppose, as we already did in bucket sorting, that the algorithm used in the buckets selects via operations (possibly comparisons) that take α units of time relative to one bucketing operation at the top level. Again, we assume that the selection algorithm in the correct bucket makes Y_j operations to select a randomly selected order statistic (of rank UNIFORM[1 .. j]) from among j random keys.

Let I_j be the indicator of the event (11.9). The recurrence for C_n is just like (11.1), with each term carrying an indicator—the indicators pick a term corresponding to the number of operations in the correct bucket to resume searching in and truncate all others:

$$C_n \stackrel{D}{=} \alpha\big(I_1 Y_{N_1}^{(1)} + I_2 Y_{N_2}^{(2)} + \cdots + I_b Y_{N_b}^{(b)}\big) + n,$$

where, for each $j \ge 1$, $Y_j^{(k)} \stackrel{D}{=} Y_j$, for $k = 1, \ldots, b$, and for $k \ne \ell$, the families $\{Y_j^{(k)}\}_{j=1}^{\infty}$ and $\{Y_j^{(\ell)}\}_{j=1}^{\infty}$ are independent (but of course $Y_{N_j}^{(j)}$ and $Y_{N_k}^{(k)}$ are dependent through the dependence of N_j and N_k). The quantity

$$Z_n \stackrel{\text{def}}{=} I_1 Y_{N_1}^{(1)} + I_2 Y_{N_2}^{(2)} + \cdots + I_b Y_{N_b}^{(b)} \tag{11.10}$$

is the only stochastic component of the cost and captures the essence of the extra cost after the first layer of bucketing. Introduce the probability generating functions $\phi_n(u) = \mathbf{E}[u^{Z_n}]$ and $\psi_n(u) = \mathbf{E}[u^{Y_n}]$.

Lemma 11.2 *For $n \ge 2$,*

$$\phi_n(u) = \frac{1}{nb^{n-1}} \sum_{j=1}^{n} j\psi_j(u)(b-1)^{n-j}\binom{n}{j}.$$

Proof. Let $\mathcal{P}_n$ be the probability $\mathbf{Prob}\{N_1 = i_1, \dots, N_b = i_b, M_n = m\}$. We have assumed M_n and the shares of the buckets to be distributed like UNIFORM[1 .. n] and an independent multinomial random vector on n trials with $1/b$ chance for each bucket. According to this independence and the respective distributions,

$$\mathcal{P}_n = \mathbf{Prob}\{N_1 = i_1, \dots, N_b = i_b\} \times \mathbf{Prob}\{M_n = m\} = \frac{1}{b^n}\binom{n}{i_1, \dots, i_b} \times \frac{1}{n}.$$

By conditioning the representation (11.10) on the shares of the buckets, we can write

$$\begin{aligned}\phi_n(u) &= \sum_{\substack{i_1+\cdots+i_b=n\\ 1\le m\le n}} \mathcal{P}_n \mathbf{E}\big[u^{I_1 Y_{N_1}^{(1)} + I_2 Y_{N_2}^{(2)} + \cdots + I_b Y_{N_b}^{(b)}} \mid N_1 = i_1, \dots, N_b = i_b, M_n = m\big]\\ &= \sum_{\substack{i_1+\cdots+i_b=n\\ 1\le m\le i_1}} \mathcal{P}_n \mathbf{E}\big[u^{Y_{i_1}^{(1)}} \mid M_n = m\big] + \sum_{\substack{i_1+\cdots+i_b=n\\ i_1< m\le i_1+i_2}} \mathcal{P}_n \mathbf{E}\big[u^{Y_{i_2}^{(2)}} \mid M_n = m\big]\\ &\qquad + \cdots + \sum_{\substack{i_1+\cdots+i_b=n\\ i_1+i_2+\cdots+i_{b-1}<m\le i_1+i_2+\cdots+i_b}} \mathcal{P}_n \mathbf{E}\big[u^{Y_{i_b}^{(b)}} \mid M_n = m\big],\end{aligned}$$

valid for $n \ge 2$; the sums above run over all nonnegative integer solutions of the equation $i_1 + \cdots + i_b = n$. By symmetry, the b sums are identical; we can use b copies of the first:

$$\begin{aligned}\phi_n(u) &= b \sum_{i_1+\cdots+i_b=n} i_1 \mathbf{E}[u^{Y_{i_1}}]\binom{n}{i_1, \dots, i_b}\frac{1}{nb^n}\\ &= \frac{1}{nb^{n-1}} \sum_{j=1}^{n} j\psi_j(u)\binom{n}{j} \sum_{i_2+\cdots+i_b=n-j} \frac{(n-j)!}{i_2! \dots i_b!}.\end{aligned}$$

When expanded by the multinomial theorem, the expression

$$\underbrace{(1+1+\cdots+1)}_{(b-1)\text{ times}}{}^{n-j} = (b-1)^{n-j}$$

coincides with the multifolded sum. ∎

11.2.1 Distributive Selection

A distributive sorting algorithm can be turned into a distributive selection algorithm by focusing on only the correct bucket containing the desired order statistic and ignoring all the other buckets.

For example, one can derive INTERPOLATION SELECT from INTERPOLATION SORT as follows. The algorithm sets up the same data structures: $A[1\,..\,n]$ for input data, $R[1\,..\,n]$ a recipient (or $R[0\,..\,n]$ if the selection algorithm in the bucket plans on using a sentinel), and $count[1\,..\,b]$ ($count[j]$ corresponds to the jth bucket,

$j = 1, \ldots, b$), as in INTERPOLATION SORT (the algorithm of Figure 11.2). INTERPOLATION SELECT goes through two rounds of hashing. The first hashing round followed by the accumulation stage are the same as in INTERPOLATION SORT. The difference between sorting and selection begins at this point. After accumulation, $count[j]$ contains exactly the number of keys that hash to a position less than or equal to j. By examining the array $count$ after accumulation, we know how many elements lie in the bucket containing the desired order statistic—if our order statistic is m, we shall look for the largest index i in $count$ such that

$$count[i] \leq m;$$

in spirit we are implementing the event (11.9) on the accumulated counts. Determining the correct position i can be done in linear time in b by a scan of $count$ as may be done by a **while** loop that continues to advance an index i until the condition is satisfied (i is initialized to 1). Of course, only one index value satisfies the condition (the algorithm is specified in Figure 11.6).

In the second hashing round we only transfer the elements of the correct bucket say to the top positions of the recipient array R. The elements that hash to position i are "hits" and all the other are "misses." The index of the correct bucket is stored in hit. Input data whose hash position matches hit are moved sequentially to R. The first of these is moved to $R[1]$, the second to $R[2]$ and so on. There are $count[i] - count[i-1]$ keys that hash to position j. The algorithm therefore must sequentially fill $R[1 \mathinner{..} size]$ (the size of the correct bucket containing the desired order statistic) with those keys, that is, $size = count[i] - count[i-1]$. We can reuse i to step through the input array A rehashing its keys. We designate the variable k to step ascendingly through the filling process; every time a key is a hit, the key is copied into the recipient array R, and k is advanced by 1, so that the next hit goes into a new slot of R. The rehashing and data transfer can be stopped right after moving $size$ keys from A to R. So, we can iterate a **while** loop, diminishing $size$ from the size of the correct bucket by 1 each time we transfer a key and stop the **while** loop when $size$ becomes 0.

After the entire correct bucket has been moved to $R[1 \mathinner{..} size]$, the selection process can proceed with this portion of data. Any reasonable selection algorithm may be used in the bucket, but now the key whose rank is m among all input keys will have relative rank among the members of the bucket moved to R, shifted down by the cumulative number of keys that fall in lower buckets, that is, a shift down by $\sum_{j=1}^{hit-1} N_j$. This cumulative share can be found in $count[hit-1]$, and the size can be extracted from this information by assigning $count[hit] - count[hit-1]$ to $size$. Any selection algorithm can then be applied to find the key ranked $m - count[hit-1]$ among $R[1 \mathinner{..} size]$. This selection algorithm does not have to be sophisticated for the same reason discussed in bucket sorting: The bucket containing the order statistic we are after has few keys, and any reasonable selection algorithm will terminate quickly (even a complete sorting algorithm may be used).

To illustrate the general idea, let us assume that a selection algorithm based on SELECTION SORT is commissioned with the task; let us call this algorithm CHOOSE;

```
R[0] ← −∞;
for i ← 1 to b do
      count[i] ← 0;
for i ← 1 to n do
      begin
            j ← h(A[i]);
            count[j] ← 1 + count[j];
      end;
{Accumulate bucket sizes}
for i ← 1 to b − 1 do
      count[i + 1] ← count[i + 1] + count[i];
i ← 1;
while count[i] < m do
      i ← i + 1;
hit ← i;
size ← count[i] − count[i − 1];
i ← 1;
k ← 1;
while size > 0 do
      begin
            j ← h(A[i]);
            if j = hit then
                  begin
                        R[k] ← A[i];
                        size ← size − 1;
                        k ← k + 1;
                  end;
            i ← i + 1;
      end;
call Choose(R, m − count[hit − 1], size);
```

Figure 11.6. INTERPOLATION SELECT—an implementation of DISTRIBUTIVE SELECT.

CHOOSE is capable of finding the mth order statistic among n keys, for any $1 \leq m \leq n$ (CHOOSE is the algorithm that chooses the first m minima, then stops). We now compose the algorithm from the first two stages (first round of hashing and accumulation) taken as is from INTERPOLATION SORT (Figure 11.2), then followed by a stage of identification of the correct bucket and continuation of the selection process with CHOOSE within that bucket.

For instance, if INTERPOLATION SELECT is applied to the 8–key input data given in Section 11.1 to identify their 7th order statistic, with the hash function discussed there which uses 10 buckets, the rehashing stage will discover that only two keys fall in the bucket containing the 7th order statistic and will place them in the recipient array. CHOOSE will then operate on the recipient array to determine which among the two keys is the 7th order statistic; see Figure 11.7.

	After First Hashing Round	After Accumulation	
A	*count*	*count*	R
	0	0	
.45	0	0	.85
.63	1	1	.82
.77	0	1	
.42	3	4	
.22	0	4	
.85	1	5	
.82	1	6	
.47	2	8	
	0	8	

Figure 11.7. Hashing and rehashing in INTERPOLATION SELECT.

The analysis of INTERPOLATION SELECT begins with Lemma 11.2. Extract the coefficient of u^k from the probability generating function in Lemma 11.2 when specialized to the case $b = n$:

$$\begin{aligned}\textbf{Prob}\{Z_n = k\} &= \frac{1}{n^n}\sum_{j=1}^{\infty} j[u^k]\psi_j(u)(n-1)^{n-j}\binom{n}{j} \\ &= \frac{1}{n^n}\sum_{j=1}^{\infty} j\textbf{Prob}\{Y_j = k\}(n-1)^{n-j}\binom{n}{j}. \qquad (11.11)\end{aligned}$$

Let $B_n = \text{BINOMIAL}(n, 1/n)$. The term $n^{-n}(n-1)^{n-j}\binom{n}{j} = n^{-j}(1-1/n)^{n-j}\binom{n}{j}$ is the probability that $B_n = j$, which for any given j converges to $e^{-1}/j!$ by the standard approximation of the binomial distribution of B_n to POISSON(1). At any fixed k, passing to the limit (as $n \to \infty$) gives us

$$\lim_{n\to\infty} \textbf{Prob}\{Z_n = k\} = \sum_{j=1}^{\infty} \textbf{Prob}\{Y_j = k\}\frac{e^{-1}}{(j-1)!}.$$

The Poisson probabilities appearing on the right hand side indicate that the number of additional operations after the first level of bucketing is like the behavior of the selection algorithm within the correct bucket on POISSON(1) + 1 random number of keys. As the last limiting calculation is true for any fixed k, we can express the behavior of Z_n simply as convergence in distribution.

Theorem 11.5 *(Mahmoud, Flajolet, Jacquet, and Régnier, 2000). Suppose INTERPOLATION SELECT uses a selection algorithm that makes Y_j extra operations to select a uniformly chosen random order statistic from among j random keys. Let Z_n be the number of these extra operations when INTERPOLATION SELECT is applied* *to n keys. Then Z_n satisfies the compound Poisson law:*

$$Z_n \xrightarrow{\mathcal{D}} Y_{\mathrm{POISSON}(1)+1}.$$

For instance, suppose CHOOSE is the algorithm used within the bucket targeted for further search. To find the kth order statistic among j keys, CHOOSE deterministically makes

$$(j-1)+(j-2)+\cdots+(j-k) = jk - \frac{1}{2}k(k+1)$$

additional comparisons after the first layer of bucketing. To find the average of random selection in the bucket (all order statistics are equally likely), given that the bucket size is j, we average the last expression by conditioning on the rank $R_j = \mathrm{UNIFORM}[1\,..\,j]$ of the randomly selected order statistic

$$\begin{aligned}\mathbf{E}[Y_j] &= \sum_{k=1}^{j} \mathbf{E}[Y_j \mid R_j = k]\,\mathbf{Prob}\{R_j = k\}\\ &= \frac{1}{j}\sum_{k=1}^{j}\Big[jk - \frac{1}{2}k(k+1)\Big]\\ &= \frac{j^2-1}{3}.\end{aligned}$$

From the exact distribution in (11.11) and approximation of binomial probabilities by their Poisson limits (as we did for the limit distribution), we now have:

$$\begin{aligned}\mathbf{E}[Z_n] &= \sum_{k=0}^{\infty} k\mathbf{Prob}\{Z_n = k\}\\ &= \sum_{k=0}^{\infty}\frac{k}{n^n}\sum_{j=1}^{\infty} j\mathbf{Prob}\{Y_j = k\}(n-1)^{n-j}\binom{n}{j}\\ &\to \sum_{j=1}^{\infty}\sum_{k=0}^{\infty} k\mathbf{Prob}\{Y_j = k\}\frac{e^{-1}}{(j-1)!}\\ &= \sum_{j=1}^{\infty}\mathbf{E}[Y_j]\frac{e^{-1}}{(j-1)!}\\ &= \frac{1}{3}\sum_{j=1}^{\infty}\frac{e^{-1}(j^2-1)}{(j-1)!}\\ &= \frac{4}{3};\end{aligned}$$

only slightly more than one additional comparisons will be needed in the bucket containing the order statistic and the average cost of the whole process is then

$$\mathbf{E}[C_n] = 2n + \frac{4\alpha}{3} + o(1).$$

Similarly, by a slightly lengthier calculation of second moments, one finds

$$\mathbf{Var}[C_n] \to \frac{589}{180}\alpha^2.$$

Note that the variance is $O(1)$ and does not grow with n, a highly desirable feature in algorithmic design.

11.2.2 Radix Selection

As RADIX SORT is a digital analog of QUICK SORT, so is RADIX SELECT in relation to QUICK SELECT (Hoare's FIND algorithm). The digital analog follows the same principle of truncating the side of the array that does not contain the sought order statistic. Owing to the lack of a pivot in digital partitioning, the boundary stopping conditions are only slightly different from QUICK SELECT.

Assume n digital keys are stored in the array $A[1 \mathinner{..} n]$ and the algorithm is to select the key whose rank is m. Like RADIX SORT, RADIX SELECT splits the data (at position s) into two groups based on the first bit: $A[1 \mathinner{..} s]$ contains all keys beginning with 0 most significant bit, and $A[s+1 \mathinner{..} n]$ contains all keys beginning with 1 (see Figure 11.4). At a higher level of recursion RADIX SELECT uses less significant bits, and the bit position employed for the separation is incremented and passed on in the variable bit to the next higher level of recursion. The algorithm is recursive and takes in the two indexes ℓ and u to select from a subarray $A[\ell \mathinner{..} u]$ with delimiting lower index ℓ and upper index u. Unlike QUICK SELECT, RADIX SELECT has no pivot that may happen to match the required order statistic. This actually simplifies the stopping condition slightly—the recursion must continue all the way down to a subarray $A[\ell \mathinner{..} u]$ of size one (whose upper and lower limits are the same, i.e., $\ell = u$). The only element of that subarray is the required order statistic. Figure 11.8 shows $RadixSelect$, a radix selection algorithm that uses $Partition$, an implementation of the partitioning algorithm; the latter can be identically the same one we used for the two-sided digital sorting algorithm RADIX SORT. The algorithm may be initiated from outside by a call to $RadixSelect(1, n)$.

To analyze this version of bucket selection, we want to develop asymptotics for the probability generating function of the total cost. Consider the *total* cost (not the extra cost in the buckets); to the extra cost in the buckets we add n (the number of bit inspections at the top level of recursion). This is reflected as an additional u^n factor in the formula of Lemma 11.2, when $b = 2$. Note that RADIX SELECT is recursive and bit inspections are its operation in the buckets, the same operation used for data separation at the top level of recursion. That is, the probability generating function $\psi_j(u)$ for the total number of bit inspections in the buckets is itself $\phi_j(u)$, the prob-

```
function RadixSelect (ℓ, u: integer) : integer;
    local s: integer;
    begin
        if ℓ = u then return(A[ℓ])
        else begin
                Partition(ℓ, u, s, bit);
                if s > m then return(RadixSelect(ℓ, s, bit + 1))
                else return(RadixSelect(s + 1, u, bit + 1));
            end;
    end;
```

Figure 11.8. A digital algorithm for finding the mth order statistic.

ability generating function of the data separation at the top level. It is usually the case that one develops ordinary or exponential generating functions for the sequence $\phi_n(u)$; however, in our case, the recurrence suggests developing a generating function for $n\phi_n(u)$, as it is the combination that appears on both sides of a rearranged version of the probability generating function of Lemma 11.2. So, we introduce the bivariate exponential generating function

$$\Phi(u, z) = \sum_{n=0}^{\infty} n\phi_n(u)\frac{z^n}{n!}.$$

It then follows from Lemma 11.2, by multiplying its two sides by z^n and summing over $n \ge 2$ (the range of validity of the recurrence), that

$$\sum_{n=2}^{\infty} n\phi_n(u)\frac{z^n}{n!} = 2\sum_{n=2}^{\infty}\frac{u^n z^n}{2^n}\sum_{j=1}^{n}\frac{j\phi_j(u)}{j!} \times \frac{1}{(n-j)!},$$

or

$$\Phi(u, z) - \frac{\phi_1(u)}{1!}z = 2\sum_{n=0}^{\infty}\sum_{j=0}^{n} j\phi_j(u)\frac{(uz/2)^j}{j!} \times \frac{(uz/2)^{n-j}}{(n-j)!}$$
$$-2\sum_{n=0}^{1}\Big(\frac{uz}{2}\Big)^n\sum_{j=0}^{n}\frac{j\phi_j(u)}{j!} \times \frac{1}{(n-j)!}.$$

Observing the boundary values $C_0 = C_1 = 0$, we have the adjusting boundary probability generating functions $\phi_0(u) = \phi_1(u) \equiv 1$, and it follows that

$$\Phi(u, z) - z = 2\Big(\sum_{j=0}^{\infty} j\phi_j(u)\frac{(uz/2)^j}{j!}\Big)\Big(\sum_{k=0}^{\infty}\frac{(uz/2)^k}{k!}\Big) - uz,$$

which we can reorganize as

$$\Phi(u, z) = 2\Phi\Big(u, \frac{uz}{2}\Big)e^{uz/2} + z(1 - u).$$

The exponential factor on the right-hand side suggests that we can "linearize" this functional equation by introducing a suitable multiplier:

$$e^{uz/2}e^{u^2z/4}e^{u^3z/8}\ldots = \exp\Big(\frac{uz}{2-u}\Big),$$

so that when we divide both sides by that multiplier, the exponential term on the right-hand side will "clip" the first term in the infinite product. Subsequently, the recursive term on the right-hand side will have the same nature as that on the left, only with its second argument scaled:

$$\frac{\Phi(u, z)}{e^{uz/2}e^{u^2z/4}e^{u^3z/8}\ldots} = 2\frac{\Phi(u, uz/2)}{e^{u^2z/4}e^{u^3z/8}e^{u^4z/16}\ldots} + \frac{z(1-u)}{e^{uz/2}e^{u^2z/4}e^{u^3z/8}\ldots}.$$

So, if we let

$$h(u, z) = \frac{\Phi(u, z)}{e^{uz/2}e^{u^2z/4}e^{u^3z/8}\ldots},$$

we have the equivalent simpler form

$$h(u, z) = 2h\Big(u, \frac{uz}{2}\Big) + z(1-u)\exp\Big(-\frac{uz}{2-u}\Big).$$

Think of $h(u, z)$ as an exponential generating function in its own right, producing a sequence of coefficients $h_k(u)$. These coefficients can be extracted from the last recurrence upon expanding its terms into series:

$$\sum_{k=0}^{\infty} h_k(u)\frac{z^k}{k!} = 2\sum_{k=0}^{\infty} h_k(u)\frac{u^k z^k}{(k!)2^k} + z(1-u)\sum_{k=0}^{\infty}(-1)^k\Big(\frac{u}{2-u}\Big)^k\frac{z^k}{k!}.$$

One has

$$h_k(u) = (-1)^{k-1}\frac{k(1-u)}{1 - 2^{1-k}u^k}\left(\frac{u}{2-u}\right)^{k-1},$$

and

$$\begin{aligned}\frac{n\phi_n(u)}{n!} &= [z^n]h(u, z)\exp\Big(\frac{uz}{2-u}\Big)\\ &= [z^n]\Big(\sum_{k=0}^{\infty} h_k(u)\frac{z^k}{k!}\Big)\Big(\sum_{j=0}^{\infty}\frac{1}{j!}\Big(\frac{uz}{2-u}\Big)^j\Big).\end{aligned}$$

Thus,

$$\phi_n(u) = \frac{1}{n} \sum_{k=0}^{n} \binom{n}{k} \left(\frac{u}{2-u} \right)^{n-k} h_k(u).$$

Inserting the expression for $h_k(u)$, we get

$$\phi_n(u) = \frac{1}{n} \left(\frac{u}{2-u} \right)^{n-1} \sum_{k=1}^{n} \binom{n}{k} (-1)^{k-1} k \frac{1-u}{1-2^{1-k}u^k}. \tag{11.12}$$

For instance, $\phi_4(u)$ is given by the rational function

$$\phi_4(u) = \frac{u^4(8 + 8u^2 + 4u^3 + u^5)}{(2-u^2)(4-u^3)(8-u^4)}.$$

The moment generating function of C_n is $\phi_n(e^t)$. One does not expect $\phi_n(e^t)$ to converge, but rather the moment generating function $\phi_n^*(e^t)$ of a suitably normed random variable (obtained from C_n by the usual centering by subtracting the mean and by scaling by the standard deviation). It is sufficient for convergence of moment generating functions to establish the existence of a limit in a neighborhood of the origin. Thus, we shall look into the behavior of $\phi_n^*(e^t)$ near $t = 0$ (or equivalently that of $\phi_n^*(u)$ near $u = 1$). Note that the first term in the sum in (11.12) is simply n and does not depend on u, whereas the rest of the finite series is multiplied by $1 - u$. Therefore, near $u = 1$, the dominant factor of the probability generating function is given by a simple expression:

$$\phi_n(u) \sim \left(\frac{u}{2-u} \right)^{n-1}.$$

Easy calculus, involving only finding the first three derivatives of the right-hand side of the last expression then yields:

$$\phi_n(e^t) \sim 1 + 2nt + 2n(2n+1)\frac{t^2}{2!} + 2n(4n^2 + 6n + 3)\frac{t^3}{3!} + O(t^4).$$

The coefficient-transfer principle of Flajolet and Odlyzko (1990) broadly states that under suitable (and rather mild conditions often met by generating functions underlying algorithms), asymptotic equivalents of functions translate into asymptotic equivalent of their coefficients. This general principle suggests that in our case the coefficient of t should asymptotically be equivalent to the mean, and the coefficient of $t^2/2!$ should asymptotically be equivalent to the second moment. This heuristically gives us the mean value

$$\mathbf{E}[C_n] \sim 2n,$$

and the variance

$$\mathbf{Var}[C_n] \sim 2n.$$

This heuristic guides the intuition to the correct normed random variable $(C_n - 2n)/\sqrt{2n}$, which we can tackle rigorously to derive a limit distribution.

Theorem 11.6 *(Mahmoud, Flajolet, Jacquet, and Régnier, 2000). Let C_n be the number of bucket operations (digit extractions) performed by RADIX SELECT using two buckets to find a randomly chosen order statistic among n keys. As $n \to \infty$,*

$$\frac{C_n - 2n}{\sqrt{n}} \xrightarrow{\mathcal{D}} \mathcal{N}(0, 2).$$

Proof. We have developed the asymptotic relation

$$\mathbf{E}[e^{C_n t}] = \phi_n(e^t) \sim \Big(\frac{1}{2e^{-t} - 1}\Big)^{n-1},$$

and guessed that the mean and variance are both asymptotically equivalent to $2n$. This suggests centering C_n by $2n$ and scaling by $\sqrt{2n}$. Let us consider

$$\mathbf{E}\Big[\exp\Big(\frac{C_n - 2n}{\sqrt{n}}t\Big)\Big] \sim e^{-2t\sqrt{n}} \frac{1}{\big(2e^{-t/\sqrt{n}} - 1\big)^{n-1}}.$$

We can now develop local expansions:

$$\begin{aligned}\mathbf{E}\Big[\exp\Big(\frac{C_n - 2n}{\sqrt{n}}t\Big)\Big] &\sim e^{-2t\sqrt{n}} \exp\Big\{-(n-1) \\ &\qquad \times \ln\Big[2\Big(1 - \frac{t}{\sqrt{n}} + \frac{t^2}{2n} + O\Big(\frac{1}{n^{3/2}}\Big)\Big) - 1\Big]\Big\} \\ &\sim e^{-2t\sqrt{n}} \exp\Big\{-(n-1)\ln\Big[1 - \frac{2t}{\sqrt{n}} + \frac{t^2}{n} + O\Big(\frac{1}{n^{3/2}}\Big)\Big]\Big\}.\end{aligned}$$

Using the calculus equality $\ln(1-x) = -x + \frac{1}{2}x^2 + O(x^3)$, we further have

$$\begin{aligned}\mathbf{E}\Big[\exp\Big(\frac{C_n - 2n}{\sqrt{n}}t\Big)\Big] &\sim e^{-2t\sqrt{n}} \exp\Big\{-(n-1)\Big[-\Big(\frac{2t}{\sqrt{n}} - \frac{t^2}{n} - O\Big(\frac{1}{n^{3/2}}\Big)\Big) \\ &\qquad + \frac{1}{2}\Big(\frac{2t}{\sqrt{n}} - \frac{t^2}{n} - O\Big(\frac{1}{n^{3/2}}\Big)\Big)^2 + O\Big(\frac{1}{n^{3/2}}\Big)\Big]\Big\} \\ &\sim e^{t^2 + O(1/\sqrt{n})} \\ &\to e^{t^2}, \qquad \text{as } n \to \infty.\end{aligned}$$

The right-hand side of the last relation is the moment generating function of $\mathcal{N}(0, 2)$. ■

We conclude this section with a few words about possible generalization of both RADIX SELECT and its analysis. We have considered the case of a fixed number of buckets $b = 2$, which suits binary data, the most common type of internal computer data. To handle digital data for a higher fixed radix $b > 2$, one can come up with a "b-sided algorithm," which partitions the data according to the b digits of the b–ary number system: All data beginning with 0 go together into one segment, all data beginning with 1 go into the next, and so on until the last segment which contains all data beginning with the $(b-1)$st digit. The algorithm then invokes itself recursively to select within the bucket in which the required order statistic has fallen, but higher levels of recursion will use less significant digits of the keys. The algorithm for $b = 2$ is the most practicable. The partitioning process becomes more complicated if $b > 2$.

For the mathematical analysis, the binary case has already illustrated simply all the principles involved in the analysis for higher b. Extending the result to higher (fixed) b follows the same route and poses no particular difficulty. The proof of Theorem 11.6 generalizes immediately to the case of a general bucket size b fixed.

EXERCISES

11.1 What is the probability that the hash function $h(x) = \lceil bx \rceil$ maps n given independent keys from the UNIFORM(0, 1] distribution into a particular bucket among b buckets?

11.2 Consider a RECURSIVE BUCKET SORT, when the sorting within a bucket is done by bucket sorting itself. Give an example where RECURSIVE BUCKET SORT may take at least 2^n levels of recursion to sort n uniform random keys.

11.3 The saddle point argument becomes slightly more delicate with RECURSIVE BUCKET SORT. Justify the following technical points in proving a central limit theorem for RECURSIVE BUCKET SORT:

(a) The function $\Phi(u, z) \stackrel{\text{def}}{=} \sum_{j=0}^{\infty} \phi_j(u) z^j / j!$ involves the *unknown* functions $\phi_j(u)$. Why does the argument that the saddle point is at $z = 1$ still hold?

(b) As Exercise 11.2 shows, the condition that $Y_j < j^{\theta}$, for some $\theta > 0$, uniformly for all sample space points is no longer valid. Argue that the proof of central tendency still holds for RECURSIVE BUCKET SORT.

11.4 Derive a central limit theorem for RECURSIVE BUCKET SORT.

11.5 (Dobosiewicz, 1987a) In view of the practical difficulty of possible nontermination in the direct application of recursion (Exercise 11.2), describe at a high level a RECURSIVE BUCKET SORT algorithm with guaranteed successful termination. Your algorithm must avoid the degenerate case when all the keys fall into one bucket, then recursively again they fall together in one bucket, and so on. (Hints: Your algorithm can make sure that at least two keys fall in two different buckets by adjusting the boundaries of the interval (0,1] to $(min, max]$, where min and max are the minimal and maximal keys. Your algorithm also needs to rescale the hash function to work on diminishing interval lengths.)

11.6 Let $B_n = \text{BINOMIAL}(n, 1/n)$. Show that

$$\mathbf{Prob}\{B_n = j\} = \frac{e^{-1}}{j!} - \frac{e^{-1}(j^2 - 3j + 1)}{2n(j!)} + O\Big(\frac{1}{n^2}\Big).$$

11.7 Consider the distributive selection algorithm INTERPOLATION SELECT, that uses n buckets and a selection algorithm based on SELECTION SORT within a bucket, under the standard probability models (UNIFORM(0, 1) data and discrete UNIFORM$[1\,..\,n]$ order statistic). In the text we developed the cost's mean and variance down to $o(1)$. Based on the finer approximation of binomial random variables by Poisson random variables, as in Exercise 11.6:

(a) Develop an asymptotic expansion for $\mathbf{E}[C_n]$, down to $O(n^{-2})$ in the form $\mathbf{E}[C_n] = n + a_0 + a_1/n + O(n^{-2})$. Your answer should of course match the text's asymptotic result, which is developed only down to $o(1)$ terms.

(b) Develop a series expansion for the variance down to $O(n^{-2})$.

11.8 Consider RECURSIVE BUCKET SELECT, when distributive selection itself is used recursively in the bucket identified as the one containing the desired order statistic and is targeted for subsequent search. Give a good approximation by a series expansion in inverse powers of n for the mean of the cost of RECURSIVE BUCKET SELECT, when it operates to find a uniformly chosen order statistic. Show that the variance is $O(1)$.

11.9 Let C_n be the number of comparisons of RADIX SORT when it sorts n elements. Let $v(n)$ be the function that agrees with $\mathbf{Var}[C_n]$ at each n. It is proved in Theorem 11.3 that the leading term of the variance is a periodic function of its argument with an oscillatory behavior that repeats as n is doubled. So, $v(n)/v(2n) \to 1$, as $n \to \infty$. Show that this is true in the probabilistic sense, too, that is, if Y_n is a random variable such that Y_n/n converges in probability to $1/2$, then $v(Y_n)/v(n) \xrightarrow{P} 1/2$, too.

11.10 In the text, the average number of bits inspected by the digital algorithm RADIX SELECT to choose a randomly determined order statistic in a file of size n is shown to be asymptotically $2n$. Using the analytic toolkit find the more accurate asymptotic expansion

$$\mathbf{E}[C_n] = 2n + \lg n - \frac{3}{2} + \frac{\gamma}{\ln 2} - \frac{1}{\ln 2} \sum_{\substack{k=-\infty \\ k \neq 0}}^{\infty} \Gamma\Big(-\frac{2\pi i k}{\ln 2}\Big) e^{2\pi i k \lg n} + o(1).$$

12

Sorting Nonrandom Data

Throughout the previous chapters the main focus has been the analysis of sorting algorithms under certain models of randomness. For comparison-based algorithms, the random permutation probability model was taken as our gold standard as it covers a wide variety of real-life situations, for example, data sampled from any continuous distribution. For the family of bucket sorting algorithms, uniform data models were considered. Again, sorting by bucketing is based on hashing, for which several practical hash functions are known to produce the required uniformity.

What if the data are not random? In this chapter we address questions on the robustness of our sorting algorithms when the data deviate from the assumed probability model. What constitutes nonrandom data? There are many varieties of data classes that are nonrandom. The most important families of data groups include almost deterministic files, nearly sorted files, and almost random files. Data that are nonrandom according to one standard model, may still be random under another. The models considered in previous chapters did not allow positive probability for key repetition. Data replication appears in samples from a finite discrete distribution.

The first group of almost deterministic files includes files that are almost known in advance apart from some perturbation. The second group of almost sorted files includes files that are very nearly sorted apart from some perturbation. For example, a database manager may decide to maintain files in sorted order. In a dynamic system, insertions and deletions occur. The manager may decide to keep the file as it is and add new entries at the bottom. The manager may perform periodic sorting updates on such a file whenever the collection of new entries adds up to a certain proportion of the whole file. Among the many sorting algorithms we have discussed, which is favorable to this situation?

"Nearly random data" is a term that refers to data that are presumed to follow some probability model, but for some reason or another they deviate slightly from that model. For instance, the keys in digital models may be assumed to come from an ergodic source that emits independent equiprobable bits of data. Such a hardware device may age over time and lose some of its specifications. For instance, owing to the aging of components, the bit probabilities may change to

$$\mathbf{Prob}\{bit = 1\} = p, \qquad \mathbf{Prob}\{bit = 0\} = 1 - p,$$

for $p \neq 1/2$. Does the quality of a digital algorithm like RADIX SORT or RADIX SELECT change substantially by a small perturbation in the probability model? In other words, are radix computation algorithms robust?

Questions regarding algorithm robustness or suitability for the prominent classes of nonrandom data are highlighted in this chapter. This is a huge, newly evolving domain by itself. We shall only try to bring the issues to light through tidbits of information.

12.1 MEASURES OF PRESORTEDNESS

The amount of "disorder" in data files can be quantified in many possible ways and according to many measures. Since these are measures of inherent disorder prior to the application of any sorting algorithm, they are often called *measures of presortedness*. No single algorithm is universally superior across all measures of presortedness. Take, for example, the number of inversions in a permutation as a measure of presortedness. The number of comparisons made by LINEAR INSERTION SORT (the sentinel version) to sort a permutation Π_n of $\{1, 2, \ldots, n\}$ is

$$C(\Pi_n) = n + I(\Pi_n),$$

where $I(\Pi_n)$ is the number of inversions in the permutation Π_n. Accordingly, LINEAR INSERTION SORT performs well on permutations with a low number of inversions. An already sorted, or nearly sorted file (one with say $o(n)$ inversions) will make only an asymptotically linear number of comparisons. Inverted files (files sorted backwards) and nearly inverted files (files with $\Omega(n^2)$ inversions) will require the onerous $\Theta(n^2)$ number of comparisons.

LINEAR INSERTION SORT is good with a low number of inversions. But, will it be as good according to a different measure of presortedness? If we consider the number of runs instead, a permutation like

$$\Pi_n = (n + 1, n + 2, n + 3, \ldots, 2n, 1, 2, 3, \ldots, n)$$

has only two runs. Yet, it has n^2 inversions and will still require a quadratic number of comparisons under LINEAR INSERTION SORT and under QUICK SORT. By contrast, Π_n is one of MERGE SORT's best-case inputs.

Other natural measures of the presortedness of a permutation are the minimal number of data that if removed will leave behind a sorted list of numbers, and the smallest number of data exchanges needed to sort the given permutation. Both measures are related to fundamental properties of permutations.

12.2 DATA RANDOMIZATION

Starting with any permutation, $\Pi_n = (\pi_1, \ldots, \pi_n)$ of $\{1, \ldots, n\}$, one can perform a series of selections that will effectively obliterate any nonrandomness in the choice of Π_n, even if it is deterministic. The selection process needs good random number generation (or a simulator of such an operation), which is not too much a restriction with modern pseudo-number generation technology. A random permutation $\Xi_n = (\xi_1, \dots, \xi_n)$ can be generated in linear time and used for randomization as follows. Starting with the ascending arrangement $\Xi_n = (\xi_1, \dots, \xi_n) = (1, 2, \dots, n)$, the random number generator is invoked to produce U, a UNIFORM$[1 \,..\, n]$ random variable. Positions U and n are swapped in Ξ_n. The initial segment $(\xi_1, \dots, \xi_{U-1}, \xi_n, \xi_{U+1}, \dots, \xi_{n-1})$ now contains integers whose relative ranks are a permutation of $\{1, \dots, n-1\}$. A penultimate index is then generated as U, a UNIFORM$[1 \,..\, n-1]$ random variable, and the corresponding entry at position U is swapped with ξ_{n-1}, and so on. By the arguments of Proposition 1.6 it is not hard to see that $(\pi_{\xi_1}, \dots, \pi_{\xi_n})$ is a random permutation.

If an algorithm is known to perform well on random data, we can randomize the data first. "Working well" should typically mean achieving the average lower bound $\Theta(n \ln n)$. The randomization stage adds only $O(n)$ preprocessing cost and will not affect the asymptotic standing of the algorithm.

The randomization process can be integrated in some algorithms like the family of QUICK SORT and all its sampling derivatives. QUICK SORT works well when its pivot evenly splits the file under sorting at most levels of recursion. Slow quadratic behavior occurs when bad pivots persist as in the cases of sorted and inverted files, or files that are nearly so. By choosing a random pivot at every step of the recursion, QUICK SORT will exhibit a behavior on sorted and nearly sorted files close to the ideal $\Theta(n \ln n)$ average speed of QUICK SORT on random data. Of course, this will be achieved only "most of the time" because the random process of pivot selection may still produce bad pivots by chance (but with rather low probability). Only a minor modification of QUICK SORT is required to achieve a randomized version of QUICK SORT. The algorithm is still based on PARTITION (Figure 7.2), but when it sorts $A[\ell \,..\, u]$ it first generates a random position uniformly from $\{\ell, \dots, u\}$. The randomized QUICK SORT is shown in Figure 12.1. In the figure, $Uniform(\ell, u)$ is an implementation of the random number generator that generates an integer from the set $\{\ell, \dots, u\}$, with all numbers being equally likely. All variations of QUICK SORT, like one-sided selection versions, multiple quick selection versions, and sampling

```
procedure RandomizedQuickSort(ℓ,u: integer);
    local p: integer;
    begin
        if ℓ < u then
            begin
                p ← Uniform(ℓ, u);
                call Partition(ℓ, u, p);
                call RandomizedQuickSort(ℓ, p − 1);
                call RandomizedQuickSort(p + 1, u);
            end;
    end;
```

Figure 12.1. Randomized QUICK SORT algorithm.

versions, can be modified in a similar way to randomize the pivot and eliminate nonrandomness in any given data file.

12.3 GUARANTEED PERFORMANCE

The case may arise when users demand guaranteed high-speed performance. A user may insist on having $O(n \ln n)$ sorting algorithm for all inputs, regardless of the model of randomness and regardless of the particular n-key input at hand. We have already seen algorithms that accomplish that. HEAP SORT and MERGE SORT are two paramount examples of practical fast algorithms with asymptotic $n \lg n$ performance. The question naturally arises, for any given n, what particular algorithm sorts all inputs of that size with the fewest number of comparisons. A parallel question arises for selection of one rank and other varieties of partial selection problems.

12.3.1 The FORD-JOHNSON Algorithm

It had been believed for quite some time that the FORD-JOHNSON algorithm (presented below) sorts with the fewest number of comparisons. This hypothesis checked out for $n = 1, 2, \ldots, 11$; the FORD-JOHNSON algorithm sorts any input of size $n \le 11$ with at most $\lceil \lg n! \rceil$ comparisons, the theoretical lower bound that cannot be improved. No other comparison-based algorithm will do better for $n \le 11$. For $n = 12$ the FORD-JOHNSON algorithm sorts with 30 comparisons. The carefully constructed experiments of Wells (1965) showed that for $n = 12$ no comparison-based algorithm sorts all 12–key inputs with less than $30 > 29 = \lceil \lg 12! \rceil$ comparisons, raising the prospect that, after all, the FORD-JOHNSON algorithm may be worst-case optimal in the exact sense. The exact sense of worst-case optimality would mean here that, for all inputs, the FORD-JOHNSON algorithm sorts with the fewest possible number of comparisons; in other words, the number of comparisons the FORD-JOHNSON makes to sort its worst-case input of size n defines the true lower bound on comparison-based algorithms, which is a little over $\lceil \lg n! \rceil$.

This remained a tantalizing open question for about 20 years until Manacher in 1979 published his findings that algorithms exist that can in the worst case sort with fewer comparisons than the FORD-JOHNSON algorithm does. By the mid 1980s, other algorithms were discovered that can slightly improve on that.

In any case, all the designs that followed the FORD-JOHNSON algorithm were based upon that algorithm. They were variations that looked out for boundary conditions and similar things. The FORD-JOHNSON logarithm and these offshoots are not very practical algorithms. They are only of theoretical interest because they are the best known algorithms (counting pair comparisons) that are closest to the theoretical $\lceil \lg n! \rceil$ bound. The FORD-JOHNSON algorithm therefore remains the core for algorithmic studies of asymptotic lower bounds and we present it next.

Because the FORD-JOHNSON algorithm is only of theoretical value, we shall not present an implementation; we shall give only an outline. Assuming there are n keys to be sorted, the steps of the algorithm are as follows:

- Step 1: Divide the keys into $\lfloor n/2 \rfloor$ pairs, leaving one key out if n is odd. Call the larger key of the ith pair a_i, and the smaller b_i.
- Step 2: Sort the $\lfloor n/2 \rfloor$ a_i's recursively by the FORD-JOHNSON algorithm. Keep the pair correspondence, if a_i is renamed a_j, reindex b_i as b_j.
- Step 3: We now have a known partial order captured essentially by the inequalities

$$b_1 \le a_1 \le a_2 \le a_3 \ldots \le a_{\lfloor n/2 \rfloor}; \tag{12.1}$$
$$b_i \le a_i, \qquad i = 1, 2, \ldots, \lfloor n/2 \rfloor.$$

We call (12.1) the main chain and we shall continue to grow it until it becomes a complete order. Using the standard BINARY SEARCH algorithm, insert the b's into the main chain in groups (b_3, b_2), then (b_5, b_4), $(b_{11}, b_{10}, \ldots, b_6), \ldots, (b_{t_k}, b_{t_k-1}, \ldots, b_{t_{k-1}+1}), \ldots$, where t_k is $\frac{1}{3}(2^{k+1} + (-1)^k)$ for increasing values of k. (Only the last group may not be complete.)

Figure 12.2 sketches the main chain after Step 2, and the chosen grouping within the b's. The algorithm would work for any choice of t_k forming an increasing sequence. The choice $\frac{1}{3}(2^{k+1} + (-1)^k)$ is for worst-case efficiency. Let us develop the insight for this choice. It is critical for fast insertion that we insert b_3 before b_2—in this order, b_3 is inserted within the chain $b_1 \le a_1 \le a_2$ (no need to consider a_3 as we have the prior information $b_3 \le a_3$). Inserting an element in a sorted list of size 3 takes at most 2 comparisons in the worst case for BINARY SEARCH (probe the median of the three elements, then one of the two extrema). When we insert b_2 next, we need only consider the elements preceding a_2 in the partial order as we have the prior information $b_2 \le a_2$. The two elements b_1 and a_1 precede a_2. We have no prior information relating b_2 and b_3. In the worst case, b_3 will precede a_2 and the insertion of b_2 among the three elements $\{b_1, a_1, b_3\}$ will again require 2 comparisons. In total, this order of insertion requires 4 comparisons in the worst case.

Had we followed the natural sequence, inserting $b_2, b_3, \ldots$, the worst-case input arrangement would have required 5 comparisons for b_2 and b_3 instead of the 4 comparisons we get in the FORD-JOHNSON algorithm—the element b_2 is inserted in two elements, requiring two comparisons in BINARY SEARCH's worst case. The portion of the main chain relevant to the insertion of b_3 has been lengthened by 1, and

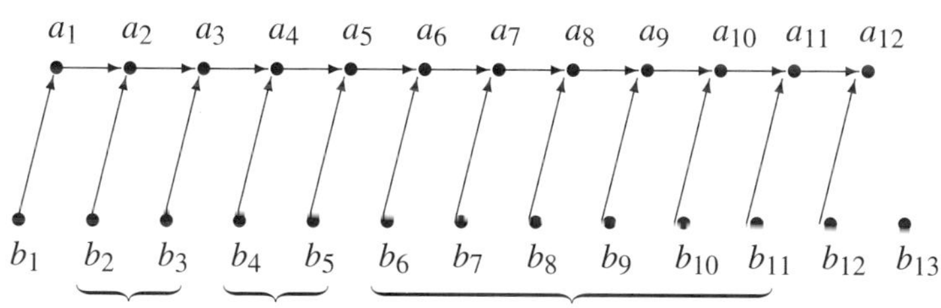

Figure 12.2. The poset discovered by the first two stages of the FORD-JOHNSON algorithm.

we have no prior information on the relative order of b_2 and b_3. So, b_2 goes anywhere on the chain $b_1 \le a_1 \le a_2$. But then b_3 must be inserted among $\{b_1, a_1, a_2, b_2\}$, requiring 3 comparisons in the worst case (for example, when b_3 is second ranked in the set $\{b_1, a_1, a_2, b_2, b_3\}$).

By the same token, for any group $(b_{t_k}, b_{t_k-1}, \ldots, b_{t_{k-1}+1})$, we insert keys in the group starting with the highest index and coming down to the least index in the group. Inserting a key in a list of $2^k - 1$ elements by BINARY SEARCH requires k comparisons; the decision tree for this search is the perfect tree. If we proceed with the keys in their natural indexing order $1, 2, 3, \ldots$, the list is lengthened after each step. The corresponding decision tree grows by one node and is not perfect any more; some leaves in the tree require k comparisons, others require $k+1$ comparison. An adversary insisting on challenging the algorithm will choose the worst case for each tree. The adversary will always choose the $k + 1$ case. The FORD-JOHNSON algorithm takes advantage of the partial order obtained in its first two stages and makes an insertion of all the keys indexed $t_k, t_{k-1}, \ldots, t_{k-1} + 1$ in a list of size $2^k - 1$, each key requiring k comparisons in the worst case. This gives us a clue for finding t_k. Half of the nodes in the insertion tree are b's, the other half are a's. Up to (and including) a_{t_k}, the pairs (a_i, b_i), $i = 1, \ldots, t_{k-1}$ have been inserted ($2t_{k-1}$ elements) as well as $a_{t_{k-1}+1}, a_{t_{k-1}+2}, \ldots, a_{t_k}$ (these are $t_k - t_{k-1}$ elements). So, up to (and including) a_{t_k} there are $2t_{k-1} + (t_k - t_{k-1}) = t_k + t_{k-1}$ elements. Each of $b_{t_k}, b_{t_k-1}, \ldots, b_{t_{k-1}}$ is to be inserted with k comparisons. The first of these, b_{t_k} (which is known to be $\le a_{t_k}$), need not be compared with a_{t_k}. It need only be inserted among $t_k + t_{k-1} - 1$ elements. Without lengthening the chain, in the worst case each inserted b will fall bellow the next inserted b in the chain, but the next b excludes the a corresponding to it, maintaining the number of elements to insert among constant (or the shapes of the insertion trees constant; all are the perfect tree of height k). We want the constant number involved on the chain in the binary insertion to be $2^k - 1$, the number of nodes in a perfect tree of height k. This maximizes the number of nodes inserted with k comparisons. Setting

$$t_k + t_{k-1} - 1 = 2^k - 1$$

gives a recurrence (with $t_0 = 1$). This recurrence is easy to unwind:

$$\begin{aligned} t_k &= 2^k - (2^{k-1} - t_{k-2}) \\ &= 2^k - 2^{k-1} + (2^{k-2} - t_{k-3}) \\ &\vdots \\ &= 2^k - 2^{k-1} + 2^{k-2} + \cdots + (-1)^k. \end{aligned}$$

This alternating geometric series sums up to $t_k = \frac{1}{3}(2^{k+1} + (-1)^k)$.

The recursive structure of the FORD-JOHNSON algorithm gives a recurrence. Suppose for a given n the index k is found from the inequality $t_{k-1} < \lceil \frac{n}{2} \rceil \le t_k$, and let $F(n)$ be the number of comparisons the FORD-JOHNSON algorithm makes to sort n items in the worst case. The first step requires $\lfloor n/2 \rfloor$ comparisons. The algorithm then invokes itself recursively on $\lfloor n/2 \rfloor$ elements, requiring $F(\lfloor n/2 \rfloor)$ comparisons in the worst case. By construction, the worst-case number of comparisons for each of $b_{t_j}, b_{t_j-1}, \ldots, b_{t_{j-1}+1}$ keys is j, for $j = 1, 2, \ldots, k-1$. Further, the algorithm makes k comparisons to insert each of $b_{\lceil n/2 \rceil}, \ldots, b_{t_{k-1}+1}$. In the worst case, Step 3 requires

$$\sum_{j=1}^{k-1} j(t_j - t_{j-1}) + k\Big(\Big\lceil \frac{n}{2} \Big\rceil - t_{k-1}\Big) = k\Big\lceil \frac{n}{2} \Big\rceil - (t_0 + t_1 + \cdots + t_{k-1}).$$

The sum $t_0 + \cdots + t_{k-1}$ is given by

$$\sum_{j=0}^{k-1} t_j = \sum_{j=0}^{k-1} \frac{2^{j+1} + (-1)^j}{3} = \begin{cases} \dfrac{2^{k+1} - 2}{3}, & \text{if } k \text{ is even;} \\[2ex] \dfrac{2^{k+1} - 1}{3}, & \text{if } k \text{ is odd.} \end{cases}$$

This two-line formula can be written succinctly as

$$\sum_{j=0}^{k-1} t_j = \Big\lfloor \frac{2^{k+1}}{3} \Big\rfloor.$$

So, for $t_{k-1} < \lceil n/2 \rceil \le t_k$, we have the recurrence

$$F(n) = F\Big(\Big\lfloor \frac{n}{2} \Big\rfloor\Big) + \Big\lfloor \frac{n}{2} \Big\rfloor + k\Big\lceil \frac{n}{2} \Big\rceil - \Big\lfloor \frac{2^{k+1}}{3} \Big\rfloor. \tag{12.2}$$

The recurrence is a bit unwieldy to be handled directly. However, we shall be able to solve it from another consideration underlying the algorithm. That consideration will allow us to find and solve a simpler recurrence.

When $t_{k-1} < \lceil \frac{1}{2} n \rceil \le t_k$, the goal of the FORD-JOHNSON algorithm is for successive insertions of $b_{t_k}, \ldots, b_{t_{k-1}+1}$ to add only k comparisons each. That is, the goal of the algorithm is to have

$$F(n) - F(n-1) = k.$$

Lemma 12.1 *(Knuth, 1973).*

$$F(n) - F(n-1) = k \qquad \textit{iff} \qquad \Big\lfloor \frac{2^{k+1}}{3} \Big\rfloor < n \le \Big\lfloor \frac{2^{k+2}}{3} \Big\rfloor.$$

Proof. We prove the lemma by induction on n. The proof is a little different according to parity. The sequence $t_k = \frac{1}{3}(2^{k+1} + (-1)^k)$ is a sequence of odd numbers.

When n is even, it cannot coincide with any t_k. Thus, $t_{k-1} < \lceil n/2 \rceil = n/2 = \lceil (n-1)/2 \rceil < t_k$, for some k. The parity of k determines the form of t_k. If k is even,

$$\left\lfloor \frac{2^{k+1}}{3} \right\rfloor = \frac{2^{k+1}-2}{3} = 2\frac{2^k-1}{3} = 2t_{k-1} < n;$$

if k is odd,

$$\left\lfloor \frac{2^{k+1}}{3} \right\rfloor = \frac{2^{k+1}-1}{3} < \frac{2^{k+1}+2}{3} = 2\frac{2^k+1}{3} = 2t_{k-1} < n.$$

For all parities of k,

$$\left\lfloor \frac{2^{k+1}}{3} \right\rfloor < n.$$

Similarly,

$$n \le \left\lfloor \frac{2^{k+2}}{3} \right\rfloor.$$

Hence, for n even,

$$\left\lfloor \frac{2^{k+1}}{3} \right\rfloor < n \le \left\lfloor \frac{2^{k+2}}{3} \right\rfloor.$$

Since, for any j,

$$\left\lfloor \frac{2^j}{3} \right\rfloor = \left\lfloor \frac{1}{2} \left\lfloor \frac{2^{j+1}}{3} \right\rfloor \right\rfloor$$

(if j is even, both sides are $\frac{1}{3}(2^j - 1)$; if j is odd, both sides are $\frac{1}{3}(2^j - 2)$), we also have

$$\left\lfloor \frac{2^k}{3} \right\rfloor \le \frac{1}{2} \left\lfloor \frac{2^{k+1}}{3} \right\rfloor < \frac{n}{2} \le \frac{1}{2} \left\lfloor \frac{2^{k+2}}{3} \right\rfloor \le \frac{2^{k+1}}{3},$$

so that

$$\left\lfloor \frac{2^k}{3} \right\rfloor < \frac{n}{2} \le \left\lfloor \frac{2^{k+1}}{3} \right\rfloor.$$

Further, from the recurrence (12.2),

$$\begin{aligned} F(n) - F(n-1) &= \left(F\left(\left\lfloor \frac{n}{2} \right\rfloor \right) + \left\lfloor \frac{n}{2} \right\rfloor + k \left\lceil \frac{n}{2} \right\rceil - \left\lfloor \frac{2^{k+1}}{3} \right\rfloor \right) \\ &\quad - \left(F\left(\left\lfloor \frac{n-1}{2} \right\rfloor \right) + \left\lfloor \frac{n-1}{2} \right\rfloor + k \left\lceil \frac{n-1}{2} \right\rceil - \left\lfloor \frac{2^{k+1}}{3} \right\rfloor \right) \\ &= F\left(\frac{n}{2} \right) + \frac{n}{2} + \frac{kn}{2} - F\left(\frac{n}{2} - 1 \right) - \left(\frac{n}{2} - 1 \right) - \frac{kn}{2} \end{aligned}$$

$$= F\left(\frac{n}{2}\right) - F\left(\frac{n}{2} - 1\right) + 1.$$

Then by induction, the difference $F(n/2) - F(n/2 - 1)$ is $k - 1$, and the lemma follows.

For n odd, a similar argument applies. We leave it to the reader to verify this case.

The opposite direction of the theorem is proved by tracing the above proof path backward. ■

Lemma 12.1 provides an easy expression to unfold $F(n)$ recursively.

Proposition 12.1 *(Hadian, 1969). The worst-case number of comparisons $F(n)$ for the FORD-JOHNSON algorithm to sort n keys is*

$$F(n) = \sum_{j=1}^{n} \left\lceil \lg \frac{3j}{4} \right\rceil.$$

Proof. The proof uses a form equivalent to Lemma 12.1. The floors may be removed as n is an integer and 2^j is not a multiple of 3 for any positive integer j, that is, $F(n) - F(n-1) = k$ iff

$$\frac{2^{k+1}}{3} < n \le \frac{2^{k+2}}{3}.$$

Taking base-2 logarithms

$$k + 1 < \lg(3n) \le k + 2,$$

or

$$k = \left\lceil \lg \frac{3n}{4} \right\rceil,$$

is an equivalent condition. We have a new simple recurrence

$$F(n) = F(n-1) + \left\lceil \lg \frac{3n}{4} \right\rceil,$$

which subsequently unwinds:

$$\begin{aligned} F(n) &= F(n-2) + \left\lceil \lg \frac{3(n-1)}{4} \right\rceil + \left\lceil \lg \frac{3n}{4} \right\rceil \\ &\vdots \\ &= \sum_{j=1}^{n} \left\lceil \lg \frac{3j}{4} \right\rceil. \end{aligned}$$

■

Theorem 12.1 *(Knuth, 1973). In the worst case, the FORD-JOHNSON algorithm requires*

$$n \lg n + (\lg 3 - \alpha_n - 1 - 2^{1-\alpha_n})n + O(\ln n)$$

comparisons to sort an input of size n, where $\alpha_n = \lg(3n) - \lfloor \lg(3n) \rfloor$.

Proof. Starting with the exact expression of Proposition 12.1, we can divide up the sum over ranges of j for which the ceiled logarithm assumes the same value; over the range $2^{k-1} < 3j < 2^k$, we have $\lceil \lg(3j) \rceil = k$. Let L_n denote $\lfloor \lg(3n) \rfloor$. Note that $3j$ is never of the form 2^k; $\lceil \lg(3n) \rceil = L_n + 1$. Write

$$\begin{aligned} F(n) &= \sum_{j=1}^{n} \bigl(\lceil \lg(3j) \rceil - 2\bigr) \\ &= \sum_{k=1}^{L_n} \sum_{2^{k-1}<3j<2^k} \lceil \lg(3j) \rceil + \sum_{j=\frac{1}{3}(2^{L_n}+b_n)}^{n} \lceil \lg(3j) \rceil - 2n \\ &\stackrel{\text{def}}{=} Q(n) + R(n) - 2n, \end{aligned}$$

where $b_n = 1$, if L_n is odd, and $b_n = 2$, if L_n is even. All the terms in the remainder summation $R(n)$ are the same, all equal to $\lceil \lg(3n) \rceil$. So,

$$\begin{aligned} R(n) &= \lceil \lg(3n) \rceil \Bigl(n - \frac{2^{L_n}}{3} + 1 - \frac{b_n}{3}\Bigr) \\ &= (L_n + 1)\Bigl(n - \frac{2^{L_n}}{3}\Bigr) + O(\ln n). \end{aligned}$$

The main sum $Q(n)$ can be partitioned into k odd and even:

$$\begin{aligned} Q(n) &= \sum_{k=1}^{L_n} \sum_{2^{k-1}<3j<2^k} k \\ &= \sum_{k=1}^{L_n} k\Bigl(\Bigl\lfloor \frac{2^k}{3} \Bigr\rfloor - \Bigl\lceil \frac{2^{k-1}}{3} \Bigr\rceil + 1\Bigr) \\ &= \sum_{\substack{k=1 \\ k \text{ even}}}^{L_n} k\Bigl(\frac{2^k - 1}{3} - \frac{2^{k-1} + 1}{3} + 1\Bigr) + \sum_{\substack{k=1 \\ k \text{ odd}}}^{L_n} k\Bigl(\frac{2^k - 2}{3} - \frac{2^{k-1} + 2}{3} + 1\Bigr) \\ &= \frac{1}{3} \sum_{k=1}^{L_n} 2^{k-1} k + \frac{1}{3} \sum_{\substack{k=1 \\ k \text{ even}}}^{L_n} k - \frac{1}{3} \sum_{\substack{k=1 \\ k \text{ odd}}}^{L_n} k. \end{aligned}$$

Using the identity $\sum_{k=1}^{m} 2^{k-1}k = 2^m(m-1)+1$, we can simplify:

$$Q(n) = \left(\frac{2^{L_n}}{3}(L_n - 1) + \frac{1}{3}\right) + \frac{(-1)^{L_n}}{3}\left\lceil \frac{1}{2}L_n \right\rceil$$
$$= \frac{2^{L_n}}{3}(L_n - 1) + O(\ln n).$$

Putting $Q(n)$ and $R(n)$ together,

$$F(n) = n(L_n + 1) - \frac{2^{L_n+1}}{3} + O(\ln n)$$
$$= n\big(\lg(3n) - \alpha_n - 1\big) - \frac{2^{\lg(3n)-\alpha_n+1}}{3} + O(\ln n),$$

which further simplifies to the asymptotic expression in the statement of the theorem. ■

The factor $\lg 3 - \alpha_n - 1 - 2^{1-\alpha_n} = \lg 3 - 3 + \delta(n) = -1.415\ldots + \delta(n)$, where $\delta(n)$ is a positive oscillating function of small magnitude (not exceeding 0.087). For example, $\delta(4) = \delta(8) = 0.081704166\ldots$. The coefficient of n in $F(n)$ oscillates between -1.329 and -1.415. Note how close $F(n)$ is to the theoretical lower bound

$$\lceil \lg n! \rceil = n \lg n - \frac{n}{\ln 2} + O(\ln n),$$

with $1/\ln 2 \approx 1.44$. Probably that is why the FORD-JOHNSON algorithm remained unchallenged for about 20 years. Manacher (1979b) discovered an improving algorithm. The improvements are only in small bands, and within a band where improvement is possible it is mostly by only a few comparisons. For example, the smallest band for which Manacher's algorithm offers an improvement is 189–191. Sorting 191 keys by the FORD-JOHNSON algorithm requires 1192 comparisons on its worst input; sorting 191 keys by Manacher's algorithm requires 1191 comparisons.

The oscillations are the key point that led Manacher to discover the improvement, to which we hint in passing. The FORD-JOHNSON algorithm is based on BINARY MERGE. A more flexible algorithm like the HWANG-LIN merging algorithm can provide fewer comparisons for the right sizes of lists. The minima of $\delta(x)$, the continuation of $\delta(n)$ to the positive real line, occur very close to $u_k = \lfloor \frac{2^{k+2}}{3} \rfloor$, $k = 1, 2, \ldots$. It is then conceivable that sorting $u_j + u_k$ by first sorting u_j keys by the FORD-JOHNSON algorithm, then sorting u_k keys by the FORD-JOHNSON algorithm, then merging by a good algorithm may constitute an algorithm that betters the FORD-JOHNSON algorithm. The idea is that in Manacher's algorithm, each invocation of the FORD-JOHNSON algorithm operates on a file of size that allows the FORD-JOHNSON algorithm to use the minimal value of its oscillating function. Admittedly, the minimal value is not much less than other typical values of the function. However, these values appear in the linear coefficient of the worst-case function F. This may lead to some rather marginal savings, if the merging is done by a sensitive algorithm, one that senses the right way for each possible pair of sorted lists. Versions of the HWANG-LIN algorithm can be designed to address just that.

12.3.2 Linear-Time Selection

The discussion in this section is on selecting the ith smallest in a given set. Prior to the discovery of linear-time selection algorithms, it had been believed that the complexity of the comparison-based selection of the ith item in a list of size n is $\Omega(n \ln n)$. This is a problem of partial sorting. However, the only known solution to it was to do complete sorting followed by identifying the ith element. The first worst-case linear-time selection algorithm was invented by Blum, Floyd, Pratt, Rivest, and Tarjan in 1973. This algorithm selects the ith item in $O(n)$, for any fixed $i = 1, \ldots, n$. For large n, this algorithm finds the key ranked i with at most cn comparisons, for a constant $c > 0$. Variations on this algorithm are then possible for lowering the constant c. We are interested in presenting the idea in its simplest form and shall leave the improving variations as exercises.

The idea in a linear-time selection algorithm is somewhat similar to Hoare's FIND algorithm, a one-sided QUICK SORT. In particular, it resembles a sampling version of QUICK SELECT. The key point in linear-time selection is the choice of a "good" pivot. Linear-time selection makes certain that a nonextremal pivot is used as the basis for splitting. We saw that a SAMPLE SORT algorithm based on pivoting on median-of-$(2k + 1)$ can improve QUICK SORT. The idea used here is to divide the input into a large number of small groups and find the median of each by a simple algorithm. The median of medians must offer good pivoting. To illustrate the idea in its simplest form, we shall use median of medians of groups of size 5 for splitting. It should be immediately obvious that median of medians of groups of size 7, 9, etc. will lead to successive improving variations.

The linear-time selection algorithm will be called SELECT, and will find the ith ranked key in the host structure $A[1\,..\,n]$. Its implementation, *Select*, will receive as input i, the rank required, and $A[\ell\,..\,u]$, the stretch of A on which the algorithm will act. The lower end of that subarray is demarcated by the index ℓ, the upper end by u. The outside call is

$$\textbf{call } Select(A[1\,..\,n], i);$$

the implementation *Select* is shown in Figure 12.3. Note that the algorithm has to pass the array it is selecting from as a parameter, because various stages of the algorithm apply the algorithm recursively to different arrays. Technically, in a programming language, the call will be more like **call** $Select(A, 1, n, i)$, but we shall stick here to the more expressive notation chosen in the last display. This creates large overhead in space, as the various calls will tend to build up a large stack of recursion. Other time-related overhead is also inherent. The number of keys n will have to be astronomically large before this algorithm outperforms other practical selection algorithms. As mentioned above the algorithm is not particularly practical, but is, of course, of interest from a theoretical point of view.

```
function Select(A[ℓ .. u], i) : integer;
    local j, p, s, m: integer;
          Med : array[1 .. s] of integer;
    begin
          s ← u − ℓ + 1;
          if s < 5 then
                return (Median(A[ℓ .. u]));
          else
                begin
                      for j ← 0 to ⌊n/5⌋ − 1 do
                            Med[j + 1] ← Median(A[ℓ + 5j .. ℓ + 5j + 4]);
                      if s mod 5 ≠ 0 then
                            Med[s/5 + 1] ← Median(A[u + 1 − (s mod 5) .. u]);
                      m ← Select(Med[1 .. ⌈s/5⌉], ⌈1/2⌈s/5⌉⌉);
                      call Partition(A[ℓ .. u], m, p);
                      if p = i then return(A[p])
                      else if p > i then return(Select(A[ℓ .. p − 1], i));
                            else return(Select(A[p + 1 .. u], i));
                end;
    end;
```

Figure 12.3. A guaranteed linear-time selection algorithm.

The n-key input is divided into $\lceil n/5 \rceil$ groups of size five each, except possibly the last. (If n is not a multiple of 5, the last group is not a complete quintuple; it may then have up to 4 keys.) The median of each quintuple is found by a worst-case optimal median-finding algorithm for 5 elements with no more than 6 comparisons (see Exercise 1.9.5). For the incomplete group, if it exists, at most 5 comparisons are sufficient for the complete sorting of up to 4 keys. (In Figure 12.3 the implementation *Median* is assumed to be a worst-case optimal function for finding the median of up to 5 keys.) Finding all $\lceil n/5 \rceil$ medians of the groups therefore makes no more than $6\lceil n/5 \rceil$ comparisons. These medians are kept in temporary storage, called *Med*. In a call to SELECT on $A[\ell \mathinner{..} u]$, $Med[1 \mathinner{..} \lceil (u - \ell + 1)/5 \rceil]$ is constructed. The median of medians is then found by a call to SELECT on this array to find its median.

The median of group medians, m, can provide our good pivot. After finding m, the information gathered so far is a partial order. Figure 12.4 illustrates the partial order discovered after finding the median of medians of 29 keys (self-loops are not shown). In the figure the group medians are drawn as squares and the black box indicates the median of medians. There are about $n/5$ group medians; about half of them are below m. We can see right away that when used for splitting, this median of medians pushes at least about $n/10$ keys below the pivot. In the worst case, when no other keys will adjoin to the subfile of size $n/10$ to balance the situation even more, the two final subfiles will be of sizes $n/10$ and $9n/10$; both sizes grow linearly with n. The algorithm will not face the bad cases that QUICK SORT occasionally suffers from, where one side is very large, and the other very small because the pivot tended to be an extreme value.

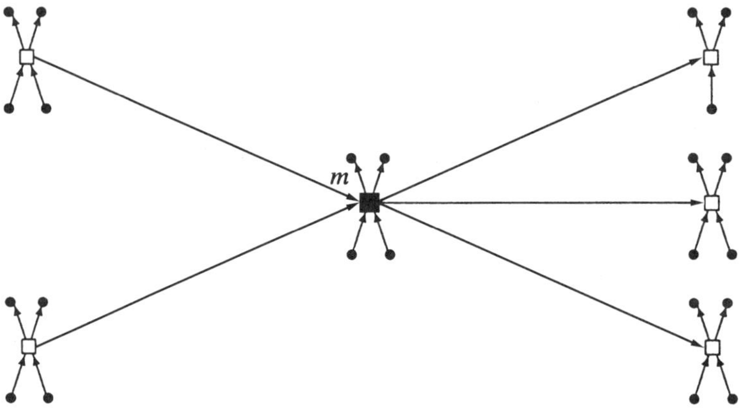

Figure 12.4. The poset found by SELECT after it has determined the group medians (square nodes).

Lemma 1.1 ensures that when we identify the median of a quintuple, we also know the pair of elements below it and the pair above it. Every quintuple with a median below m forces two other keys to lie below m; a total of three keys in each such group are below m.

The size of the remaining portion under selection is denoted by s (thus, $s = u - \ell + 1$). The median of medians is found by a recursive call to SELECT itself; the call

$$\textbf{call } Select\left(Med\left[1 \mathinner{..} \left\lceil \frac{s}{5} \right\rceil\right], \left\lceil \frac{1}{2}\left\lceil \frac{s}{5} \right\rceil\right\rceil\right)$$

finds the median element within the array of medians for this stage. Upon returning with a median of medians, the algorithm goes through a splitting process similar to QUICK SORT's PARTITION algorithm (see, for example, Figure 7.2), but slightly adapted to pivot on m. That algorithm returns p, the correct position of m. The splitting median of medians will go to its correct position (call it p) in the list; all the elements less than the splitter will be transferred to the left of it, and all elements larger than the splitter will be transferred to positions to the right of it.

If the splitter's position p is i, we have found the ith order statistic in our input; it is the splitter (now element $A[p]$ in the array). If the splitter's position is greater than i, our required order statistic is to the left of the splitter. A recursive call on the left data group $A[\ell \mathinner{..} p - 1]$ will complete the selection. Conversely, if the splitter's position is less than i, a recursive call on $A[p + 1 \mathinner{..} u]$ will complete the selection.

Let $W(n)$ be the uniform worst-case number of comparisons of SELECT over all i. The various components of SELECT contribute comparisons as follows. Finding the $\lceil n/5 \rceil$ group medians requires at most $6\lceil n/5 \rceil$ comparisons. The recursive call for finding the median of medians in the worst case requires $W(\lceil n/5 \rceil)$ comparisons. The partitioning process requires $n - 1$ comparisons. If i is below m, the algorithm operates recursively on all the keys below m. How many keys are below m in the worst case? There are $\lceil \frac{1}{2} \lceil n/5 \rceil \rceil - 1$ medians below m. Two keys in each group are known to be at most the median of the group; three keys in each group are known to be less than m. The partial order discovered so far does not tell us anything about the other two keys. In the worst case they will be added to the keys below m. That is, in the worst case, all 5 keys from the $\lceil \frac{1}{2} \lceil n/5 \rceil \rceil - 1$ groups will be below m. Similarly, if n is not a multiple of 5, there are $\lceil n/5 \rceil - \lceil \frac{1}{2} \lceil n/5 \rceil \rceil - 1$ complete quintuples whose medians are above m. Each such group has two keys known to be above m (keys above the group median); the other two keys (keys below the group median) are not comparable to m. In the worst case each such group adds two keys to the group below m. The remaining group adds

$$g(n) = \left\lceil \frac{n \textbf{ mod } 5}{2} \right\rceil - 1, \qquad \text{if } (n \textbf{ mod } 5) > 0.$$

Only a small adjustment is needed for (n **mod** $5 = 0$). The middle group itself provides 2 extra keys. For example, if the pivot is at position $m > i$, the recursion on the left data group will be on at most

$$L_n = \begin{cases} 5\left(\left\lceil \frac{1}{2} \times \frac{n}{5} \right\rceil - 1\right) + 2\left(\frac{n}{5} - \left\lceil \frac{1}{2} \times \frac{n}{5} \right\rceil\right) + 2; & \text{if } (n \textbf{ mod } 5) = 0; \\ 5\left(\left\lceil \frac{1}{2} \left\lceil \frac{n}{5} \right\rceil \right\rceil - 1\right) + 2\left(\left\lceil \frac{n}{5} \right\rceil - \left\lceil \frac{1}{2} \left\lceil \frac{n}{5} \right\rceil \right\rceil - 1\right) & \\ \qquad + g(n) + 2; & \text{otherwise.} \end{cases}$$

This function is upper bounded by the simpler expression $\lceil \frac{1}{10}(7n - 5) \rceil$.

If $i > m$, a worst-case recursive call on the right side $A[p + 1 \,..\, n]$ will operate on R_n keys, a function that differs slightly from L_n because of the effect of ceils and floors. By the same counting arguments we find that the right call operates on at most $R_n \leq \lceil \frac{1}{10}(7n - 5) \rceil$ keys.

Whether the recursion is invoked on the left side or the right side, we have an upper bound on $W(n)$. Uniformly in i,

$$W(n) \leq 6\left\lceil \frac{n}{5} \right\rceil + W\left(\left\lceil \frac{n}{5} \right\rceil\right) + n - 1 + W\left(\left\lceil \frac{7n - 5}{10} \right\rceil\right). \tag{12.3}$$

Theorem 12.2 *(Blum, Floyd, Pratt, Rivest, and Tarjan, 1973). In a list of n keys, any order statistic $i \in \{1, \ldots, n\}$ can be found by making at most $23n$ comparisons.*

Proof. We need to show that the inequality (12.3) can be satisfied if $W(n)$ is linearly bounded in n: That is, for some positive constant c, $W(n) \leq cn$, and identify c.

Assume by induction that $W(k) \leq ck$, for $k = 1, \ldots, n - 1$. We can then replace the two occurrences of W on the right-hand side of (12.3) by c times their argument:

$$F(n) \leq 6\left\lceil \frac{n}{5} \right\rceil + c\left\lceil \frac{n}{5} \right\rceil + n - 1 + c\left\lceil \frac{7n - 5}{10} \right\rceil$$

$$\leq 6\Big(\frac{n}{5}+1\Big) + c\Big(\frac{n}{5}+1\Big) + n - 1 + c\Big(\frac{7n-5}{10}+1\Big)$$
$$\leq \Big(\frac{9c+22}{10}\Big)n + \frac{3}{2}c + 5.$$

The induction would be completed if the right-hand side of the inequality is upper bounded by cn, for some induction basis. The question then is the following. Is there a value of c such that

$$\Big(\frac{9c+22}{10}\Big)n + \frac{3}{2}c + 5 \leq cn,$$

for all n greater than or equal to some n_0? We are asking what c satisfies

$$\Big(\frac{c}{10} - \frac{22}{10}\Big)n \geq \frac{3}{2}c + 5?$$

Clearly for any $\varepsilon > 0$, $c = 22 + \varepsilon$ can satisfy the inequality for large enough n. For simplicity, we can take $c = 23$, and the inequality is valid for all $n \geq 395$. Hence we can take $n_0 = 395$, as the basis of induction.

For $n \leq 395$, we can use, for example, MERGE SORT to sort the file, then identify the ith key with at most $n\lceil \lg n \rceil - 2^{\lceil \lg n \rceil} + 1 \leq 23n$ comparisons (refer to Theorem 10.1 for MERGE SORT's worst-case number of comparisons). ■

The bound of $23n$ in the proof of Theorem 12.2 is rather crude. In the proof, the number 23 is chosen for simplicity; we could have chosen, for instance, 22.5. More importantly, the algorithm based on medians of quintuples was given for its simplicity to prove worst-case linearity of selection. Slightly more complex algorithms based on medians of groups of size 7, 9, etc. can be employed to obtain better bounds. The best known uniform bound is $3n$.

12.4 PRESORTING

Presorting is a term that refers to preprocessing data to improve their profile for later sorting by a standard algorithm. The idea has been recently introduced by Hwang, Yang, and Yeh (1999). Naturally, presorting must be simpler than sorting and simple enough not to introduce a higher order of magnitude, or even add too much overhead to the existing orders of magnitude. For example, if we have at hand a sorting algorithm that is $O(n \ln n)$ for all inputs of size n, we may consider presorting algorithms of up to $O(n)$ operations and we would require those to be based on simple operations. One expects the effects of presorting to ripple through lower orders of magnitude, for if it affects higher orders, that presorting algorithm together with the sorting that follows would amount to a new major family of sorting algorithms.

Some rather simple operations are capable of reducing one measure of presortedness or another. For example, a single exchange of a pair that is out of its natural sequence may reduce the number of inversions in the input. We have an input permutation $\Pi_n = (\pi_1, \dots, \pi_n)$ on which we define an operation $Sw_{ij}(\Pi_n)$, for $1 \le i < j \le n$, that transforms the given input into a new permutation $\Pi'_n = (\pi'_1, \dots, \pi'_n)$, with $\pi'_k = \pi_k$, for all $k \notin \{i, j\}$, and the pair $(\pi'_i, \pi'_j) = (\pi_i, \pi_j)$, if $\pi_i < \pi_j$, and $(\pi'_i, \pi'_j) = (\pi_j, \pi_i)$, if $\pi_i > \pi_j$. In words, the swapping operation Sw_{ij} switches π_i and π_j if they are out of natural sequence. Clearly, the resulting permutation has fewer inversions. How many inversions are eliminated on average by Sw_{ij}? The next lemma answers this question. The answer will give us a clue on which pair will be the optimal choice.

Lemma 12.2 *(Hwang, Yang, and Yeh, 1999). Operating on a random permutation Π_n of $\{1, \dots, n\}$, the single operation $Sw_{ij}(\Pi_n)$ reduces the number of inversions by $\frac{1}{3}(j - i) + \frac{1}{6}$ on average.*

Proof. Let the given permutation be $\Pi_n = (\pi_1, \dots, \pi_n)$. By basic properties of random permutations, the triple π_i, π_j, and π_k have relative ranks that are random permutations of $\{1, 2, 3\}$. Let Z_{ikj} be the number of inversions reduced in the triple (π_i, π_k, π_j), $i < k < j$, by the swap. The three subarrangements (π_i, π_k, π_j) with relative ranks

1 2 3

1 3 2

2 1 3

are not changed by an application of Sw_{ij} (for these $\pi_i < \pi_j$); with probability $1/2$, Sw_{ij} offers no reduction in inversions; $Z_{ikj} = 0$.

For subarrangements with relative ranks

2 3 1

3 1 2

Sw_{ij} swaps π_i and π_j reducing inversions by 1 (event with probability $1/3$); $Z_{ikj} = 1$. For subarrangements with relative ranks

3 2 1

the swap reduces inversions by 3 (event with probability $1/6$); $Z_{ikj} = 3$.

Conditioned on the event that Sw_{ij} induces a swap ($\pi_i > \pi_j$),

$$Z_{ikj} \mid \pi_i > \pi_j = 1 + \begin{cases} 2, & \text{if } \pi_i > \pi_k > \pi_j; \\ 0, & \text{otherwise.} \end{cases}$$

If there is a swap at all, positions k with $i < k < j$ share the same reduction of inversions. Hence $Z_{ij} = \sum_{i<k<j} Z_{ikj}$, the total number of inversions eliminated by Sw_{ij} is

$$\begin{aligned}\mathbf{E}[Z_{ij}] &= \frac{1}{2}\mathbf{E}[Z_{ij} \mid \pi_i < \pi_j] + \frac{1}{2}\mathbf{E}[Z_{ij} \mid \pi_i > \pi_j] \\ &= 0 + \frac{1}{2}\Big[1 + \sum_{i<k<j} \Big(0 \times \frac{2}{3} + 2 \times \frac{1}{3}\Big)\Big]. \\ &= \frac{1}{2} + \frac{1}{3}(j - i - 1).\end{aligned}$$

■

Lemma 12.2 unequivocably recommends that a single swap will be most effective if the swapped elements are as far apart as possible (maximizing the distance $j - i$). The most effective operation is therefore $Sw_{1,n}$, and the largest average reduction in the number of inversions possible by a single swap is $\frac{1}{3}n - \frac{1}{6}$. Since the average number of inversions is quadratic on average, a linear amount of reduction will only affect the lower-order terms in the average value. The operation Sw_{1n} slightly speeds up an algorithm like INSERTION SORT, which depends in a direct way on the number of inversions.

Intuitively, the same kinds of Gaussian limits for plain inversions will prevail for random permutations preprocessed by Sw_{1n}. Only the rate of convergence to such a Gaussian law will change. We can develop this result technically by a simple manipulation.

Theorem 12.3 *(Hwang, Yang, and Yeh, 1999). Let $\tilde{I}_n$ be the number of inversions in a random permutation of $\{1, \dots, n\}$ preprocessed by the switching operation Sw_{1n}. The probability generating function of $\tilde{I}_n$ is*

$$\frac{2}{n!}\Big[\frac{1 + 2z + 3z^2 + \cdots + (n-1)z^{n-2}}{(1-z)^{n-2}} \prod_{j=1}^{n-2}(1 - z^j)\Big].$$

Proof. Let Sw_{1n} operate on $\Pi_n = (\pi_1, \dots, \pi_n)$, a random permutation of $\{1, \dots, n\}$, to produce $\tilde{\Pi}_n = (\pi'_1, \pi_2, \pi_3, \dots, \pi_{n-1}, \pi'_n)$. Suppose $\tilde{\Pi}_n$ has $\tilde{I}_n$ inversions. The number of inversions among members of the subarrangement $(\pi_2, \dots, \pi_{n-1})$ is distributed like I_{n-2}, the number of inversions of a random permutation of $\{1, \dots, n-2\}$, with probability generating function (see (1.13))

$$\frac{1}{(n-2)!} \prod_{j=1}^{n-2} \frac{1 - z^j}{1 - z}.$$

Independently π'_1 and π'_n will contribute additional inversions. Let us denote this random number of additional inversions by T_n. (To visualize the independence, it is helpful to take a view similar to Proposition 1.6, whereby we can think of $(\pi_2, \ldots, \pi_{n-1})$ to have arrived first, then $\pi_1 \stackrel{\mathcal{D}}{=}$ UNIFORM$[1\,..\,n-1]$, and $\pi_n \stackrel{\mathcal{D}}{=}$ UNIFORM$[1\,..\,n]$.)

The total number of inversions after the swap is the convolution

$$\tilde{I}_n \stackrel{\mathcal{D}}{=} I_{n-2} + T_n. \tag{12.4}$$

Each of the numbers $1, 2, \ldots, \pi_1' - 1$ causes an inversion with π_1' in $\tilde{\Pi}_n$, and each of the numbers $\pi_n' + 1, \pi_n' + 2, \ldots, n$ causes an inversion with π_n' in $\tilde{\Pi}_n$. So, $T_n = (\pi_1' - 1) + (n - \pi_n')$. For a fixed k, there are $k + 1$ possibilities that achieve $T_n = (\pi_1' - 1) + (n - \pi_n') = k$; these possibilities are the feasible solutions to the equation $\pi_n' - \pi_1' = n - k - 1$. Namely, these solutions are $(\pi_1', \pi_n') = (1, n - k), (2, n - k + 1), \ldots, (k + 1, n)$. A permutation $\tilde{\Pi}_n = (\pi_1', \pi_2, \pi_3, \ldots, \pi_{n-1}, \pi_n')$ arises from either the permutation $(\pi_1', \pi_2, \pi_3, \ldots, \pi_{n-1}, \pi_n')$, in which Sw_{1n} effects no change, or from $(\pi_n', \pi_2, \pi_3, \ldots, \pi_{n-1}, \pi_1')$, on which Sw_{1n} swaps the first and last elements. Hence

$$\mathbf{Prob}\{\tilde{\Pi}_n = (\pi_1', \pi_2, \pi_3, \ldots, \pi_{n-1}, \pi_n')\} = 2\frac{(n-2)!}{n!},$$

which induces the probability generating function

$$\mathbf{E}\big[z^{T_n}\big] = 2\frac{1 + 2z + 3z^2 + \cdots + (n-1)z^{n-2}}{n(n-1)}. \tag{12.5}$$

The probability generating function of the convolution (12.4) is the product of the two generating functions of its independent ingredients. ■

Asymptotic normality of the number of inversions in preprocessed random permutations follows as an easy corollary. The convolution in (12.4) provides a direct bridge to limit distributions. If the (asymptotically) centered random variable $T_n - \frac{2}{3}n$ is normalized by n, it converges to a limit T (Exercise 12.6 establishes the limit distribution). However, to get a limit for I_{n-2}, we must normalize $I_{n-2} - \mathbf{E}[I_{n-2}]$ by $n^{3/2}$, to get the normal limit $\mathcal{N}(0, \frac{1}{36})$. Norming $T_n - \frac{2}{3}n$ by $n^{3/2}$ results in a random variable that behaves like $T/\sqrt{n}$, which of course converges almost surely to 0. That is,

$$\frac{\tilde{I}_n - \frac{1}{4}n^2}{n^{3/2}} \xrightarrow{\mathcal{D}} \mathcal{N}\Big(0, \frac{1}{36}\Big),$$

the same law as that of the number of inversions in a raw permutation.

In the preceding paragraphs we discussed the effects of a single swapping operation. The ideas and results extend easily to a sequence of such operations. A natural sequence of operations to consider is $Sw_{i,n-i+1}$, for $i = 1, 2, \ldots, \lfloor n/2 \rfloor$. Such a sequence will produce a permutation with a (normalized) number of inversions having the same limit distribution as a raw random permutation. The $\lfloor n/2 \rfloor$ operations require $O(n)$ running time, and will not interfere with high-order asymptotics of a sorting algorithm that follows. For example, if the presorting is followed by the quadratic-time INSERTION SORT, the net result will be a quadratic-time algorithm, still having a Gaussian law for its number of comparisons, but differing in its rate of convergence.

EXERCISES

12.1 A university's Admissions Officer maintains a sorted data file $A[1\,..\,n]$ of size n. New records are placed in the file by LINEAR INSERTION SORT. Assume the rank of a new entry to be uniformly distributed over the range of possible gaps between existing keys. What is the distribution of the number of comparisons needed to insert the k additional entries, for k fixed as $n \to \infty$?

12.2 (Mannila, 1985) Let $M(\Pi_n)$ be the minimal number of data that if removed from a permutation Π_n of $\{1, 2, \ldots, n\}$ will leave behind a sorted list of numbers. Show that $M(\Pi_n) = n -$ the length of the longest ascending subsequence of Π_n.

12.3 (Mannila, 1985) Let $X(\Pi_n)$ be the minimal number of pairwise exchanges in the permutation Π_n of $\{1, 2, \ldots, n\}$ that result in a sorted list. Show that $X_n = X(\Pi_n) = n - C(\Pi_n)$, where $C(\Pi_n)$ is the number of cycles in Π_n. Hence, derive a central limit theorem for X_n.

12.4 A data source was designed to emit sequences of independent equiprobable bits. Over time, the device calibration has slightly deteriorated. It still produces independent bits, but now the 1's and 0's have probabilities p and $1 - p$, respectively. What is the average number of bits inspected by RADIX SORT (the algorithm of Figure 11.5) during the course of sorting n independent digital keys taken from this source? Refer to the discussion of Subsection 11.1.2 for the model of randomness of digital keys. Comparing the result with Theorem 11.2, would you consider RADIX SORT a robust algorithm?

12.5 (Hwang, Yang, and Yeh, 1999) Let $I_n^{(k)}$ be the number of inversions in a random permutation of $\{1, \ldots, n\}$ preprocessed by the sequence of switching operations $Sw_{i,n-i+1}$, for $i = 1, \ldots, k \le \lfloor n/2 \rfloor$ (see Section 12.4 for a definition of these operations). Show that the probability generating function of $I_n^{(k)}$ is

$$\frac{2^k}{n!} \prod_{j=1}^{k} \left(1 + 2z + 3z^2 + \cdots + (n - 2j + 1)z^{n-2j}\right) \prod_{j=1}^{n-2k} \frac{1 - z^j}{1 - z}.$$

Hence, or otherwise, derive the mean and variance of $I_n^{(k)}$. (Hint: Argue inductively, building on the case of $k = 1$ (Theorem 12.3) for positions 1 and n, then move on toward the "inner cycles.")

12.6 Let $\tilde{I}_n$ be the number of inversions in a random permutation $(\pi_1, \ldots, \pi_n)$ of $\{1, \ldots, n\}$ after it has been preprocessed by the single switching operations $Sw_{1,n}$ (see Section 12.4 for a definition of this operation). As in the text, $\tilde{I}_n$ is distributed like a convolution $I_{n-2} + T_n$, where I_{n-2} stands for the inversions already present among $(\pi_2, \ldots, \pi_{n-1})$, and T_n is the number of additional inversions caused by the keys at positions 1 and n.

(a) Derive the exact mean and variance of T_n.

(b) Show that

$$\frac{T_n - \frac{2}{3}n}{n} \xrightarrow{\mathcal{D}} T,$$

where the limiting random variable T has the density $2t + \frac{4}{3}$, for $-\frac{2}{3} < t < \frac{1}{3}$.

13

Epilogue

After taking the grand tour of standard sorting algorithms, a natural question to ask is, Which method is recommended? The analysis shows that there is no simple answer to this question. No one single sorting algorithm appears to be universally superior to all other sorting algorithms across all situations. An algorithm excelling in its average running time, like QUICK SORT, may have bad input cases on which its running time is much higher than the average; an algorithm with guaranteed worst-case performance, like MERGE SORT, does not operate in situ; MERGE SORT demands about double the space that QUICK SORT needs. Arguing further, one finds that every algorithm may have some shortcoming for a particular situation.

Table 13.1 summarizes the analysis of the standard sorting methods discussed in the book. For every sorting algorithm discussed earlier, the table lists the leading asymptotic equivalent of some dominant characteristic of the algorithm. For instance, for comparison-based sorting algorithms, the number of comparisons is the selected characteristic. Of course, in practice several other factors may alter the constants of the actual running time. For example, while QUICK SORT's average number of comparisons is about $1.44n \lg n$ and HEAP SORT's average number of comparisons is only $n \lg n$, in practice the average *running time* of practicable forms of HEAP SORT is nearly twice QUICK SORT's average running time.

The choice is situation dependent. The analysis, however, gives some general guidelines:

- For very large random inputs following the random permutation probability model, parsimonious divide-and-conquer methods are asymptotically fast. Both MERGE SORT and HEAP SORT have guaranteed $n \lg n$ performance (comparisons that dominate the running time). QUICK SORT is fast on average requiring only $2n \lg n$ comparisons asymptotically. While MERGE SORT requires twice the space of the input, QUICK SORT and HEAP SORT run in situ. The downside is that QUICK SORT has a few atypical inputs on which it runs in $\Theta(n^2)$ time, and HEAP SORT has overhead that manifests itself if n is not large enough.

- For medium and small files following the random permutation probability model, some naive methods prove to be serious competitors to parsimonious sorting algorithms. SHELL SORT is not a bad algorithm for this application.

TABLE 13.1. Probabilistic Analysis of Standard Sorting Algorithms

Sorting Algorithm	Mean	Variance	Limit Distribution
LINEAR INSERTION	$\frac{1}{4}n^2$	$\frac{1}{36}n^3$	normal
BINARY INSERTION	$n \lg n$	$n \times oscillation$	normal
(2,1)-SHELL SORT	$\frac{1}{8}n^2$	$const \times n^3$	normal; lower-order terms form a Brownian bridge
Pratt's SHELL	$\Theta(n \ln^2 n)$	unknown	unknown
BUBBLE	$\frac{1}{2}n^2$	—	passes: Rayleigh comparisons: about $\frac{n^2}{2}$ (high probability)
SELECTION	$\frac{1}{2}n^2$	0	degenerate
COUNT	$\frac{1}{2}n^2$	0	degenerate
QUICK	$2n \ln n$	$const \times n^2$	fixed-point solution of a functional equation
MEDIAN-3-QUICK SORT	$\frac{12}{7}n \ln n$	$const \times n^2$	fixed-point solution of a functional equation
HEAP	$n \lg n$	unknown	unknown
MERGE	$n \lg n$	$n \times oscillation$	normal
INTERPOLATION	$const \times n$	$const \times n$	normal
RADIX	$n \lg n$	$n \times oscillation$	normal

- For nearly sorted inputs an algorithm like SHELL SORT with very few increments, or even INSERTION SORT (the simplest SHELL SORT using only one increment), is probably among the best methods for the task.
- BUCKET SORT and its derivatives have an inviting $O(n)$ average time. This class of algorithms is based on smashing the input keys into their bare atomic digital units. The average $O(n)$ time is attained under the assumption of uniform data, or a hashing scheme that achieves that. The class of algorithms is not in situ and their running time may tend to be a bit too high in the worst case. The number crunching required by BUCKET SORT algorithms suggests that one should select them with caution. A BUCKET SORT algorithm may be good on mainframe computers that have fast registers for performing arithmetic operations; the same algorithm on the same input may require a lot longer time on a personal computer.
- The situation is less clear for inputs not following the random permutation probability models. Not too much analysis has been done on these models. The general rule of thumb may be: When in doubt, use a member of the parsimonious class of sorting algorithms with guaranteed performance.

Answers to Exercises

CHAPTER 1.1

1.1.1 (i) 1.3 2.6 2.9 3.4 4.4 7.1.

(ii) 17 29 33 47 56 60 71 84.

(iii) The sorting here is lexicographic:

Alice Beatrice Elizabeth Jim Johnny Philip.

1.1.2 (i) 2 1 6 5 4 3.

(ii) 3 6 2 5 8 1 7 4.

(iii) 4 2 6 1 5 3.

1.1.3 No special interpretation is needed because the ordering of the constraint (1.2) includes the"less than" option. However, the permutation satisfying (1.2) may not be unique. For instance, for

$$(X_1, X_2, X_3, X_4, X_5, X_6, X_7) = (6, 2, 6, 6, 8, 5, 8)$$

we have

$$X_2 \le X_6 \le X_1 \le X_3 \le X_4 \le X_5 \le X_7$$

and

$$X_2 \le X_6 \le X_3 \le X_4 \le X_1 \le X_7 \le X_5$$

both satisfy (1.2). That is, the two permutations (2, 6, 1, 3, 4, 5, 7) and (2, 6, 3, 4, 1, 7, 5) both qualify as a map.

1.2.1

```
max ← A[1];
for i ← 2 to n do
    if A[i] > max then
        max ← A[i];
```

Clearly, this algorithm makes $n - 1$ data comparisons.

1.2.2 The algorithm compares successive pairs and tosses winners into a bag W and losers into another bag L. For n even, this requires $n/2$ comparisons. For n odd, $\lfloor n/2 \rfloor$ comparisons pair all but the last element. We can then match the last element against a winner. If it loses, it gets thrown into L. If it wins, it replaces the person it defeats, who then goes into L. For n odd this requires $\lfloor n/2 \rfloor + 1$ comparisons. For all parities, this initial classification requires $\lceil n/2 \rceil$ comparisons. The maximum must be in W and the minimum in L. We can then apply an algorithm similar to that of Exercise 1.2.1 to find the maximum in W and a mirror image algorithm to find the minimum in L. The algorithm of Exercise 1.2.1 can be easily turned into a procedure, say *max*, that is called to find the maximum in a set $B[1 \mathinner{..} s]$, of size s. The mirror image for minimum finding can be implemented likewise and these two procedures will be called. One of the two bags has size $\lceil n/2 \rceil$, the other has size $\lfloor n/2 \rfloor$. The identification of the two extremal elements in the two bags combined requires $(\lceil n/2 \rceil - 1) + (\lfloor n/2 \rfloor - 1) = n - 2$. The total number of comparisons to find the maximum and the minimum by this algorithm requires $\lceil n/2 \rceil + n - 2 < 3n/2$. Here is an implementation that assumes the initial data in the array $A[1 \mathinner{..} n]$:

```
for i ← 1 to ⌊n/2⌋ do
    if A[2i − 1] > A[2i] then
        begin
            W[i] ← A[2i − 1];
            L[i] ← A[2i];
        end
    else
        begin
            W[i] ← A[2i];
            L[i] ← A[2i − 1];
        end;
if (n mod 2) = 1 then
    if A[n] > W[1] then
        begin
            L[⌈n/2⌉] ← W[1];
            W[1] ← A[n];
        end
    else L[⌈n/2⌉] ← A[n];
print(min(L, ⌈n/2⌉), max(W, ⌊n/2⌋));
```

1.3.1 The algorithm is comparison based and stable. The issue of "asymptotic" extra space does not arise as the algorithm is specifically designed for three keys, and not for sets of arbitrary size. Apart from compiled code and the variables, no extra space is needed and we may consider the algorithm to be in situ.

1.4.1

```
if X1 ≤ X2 then
     if X2 ≤ X3 then
               print(X1, X2, X3)
     else if X1 ≤ X3 then
               print(X1, X3, X2)
          else print(X3, X1, X2)
else
     if X2 > X3 then
          print(X3, X2, X1)
     else if X1 ≤ X3 then
               print(X2, X1, X3)
          else print(X2, X3, X1);
```

1.4.2 If we want to switch the contents of two glasses, we need a third empty glass. The same principle applies to the swapping of two computer variables. The procedure is quite simple. It is written here for real numbers (the temporary container must also be real). Similar algorithms will work for any type by changing the declaration of the type of parameters and matching the change in the temporary variable. In modern object-oriented programming systems, one algorithm can be written for all feasible data types.

```
procedure swap(var x, y : real);
     local t: real;
     begin
          t ← x;
          x ← y;
          y ← t;
     end;
```

1.5.1 The minima are the prime numbers. This infinite partial order has no maxima. The chains are all the multiples of a prime—for every prime p_1, any sequence of the form

$$p_1, p_1p_2, p_1p_2p_3, \dots,$$

where all p_i are primes (not necessarily distinct), is a chain associated with p_1. The set $\mathcal{O}$ of odd numbers is not compatible with the partial order; for instance, 5 divides 20, yet $20 \in A - \mathcal{O}$.

1.6.1 Revisit the proof of Proposition 1.1. The height of the complete binary tree of order n is $h_n^* = \lceil \lg(n+1) \rceil$. The leaves therefore live on the two levels $h_n^* - 1$ and h_n^*. The lower of the two levels has an equivalent but more succinct representation

$$h_n^* - 1 = \lceil \lg(n+1) \rceil - 1 = \lfloor \lg n \rfloor.$$

The latter expression is used elsewhere in the book, whenever convenient.

1.6.2 Let the height of the complete binary tree of order n be h_n^*, and let Z_n be the number of leaves at level h_n^*. As argued in the proof of Proposition 1.1, levels $0, 1, \ldots, h_n^* - 2$ are saturated, and level $h_n^* - 1$ may or may not be saturated. The number of internal nodes on level $h_n^* - 1$ is $n - (1 + 2 + \cdots + 2^{h_n^*-2}) = n - 2^{h_n^*-1} + 1$. So, there are $2^{h_n^*-1} - n + 2^{h_n^*-1} - 1$ leaves on level $h_n^* - 1$. The total number of leaves is $n+1$. Hence $Z_n = n+1-(2^{h_n^*-1} - n + 2^{h_n^*-1} - 1) = 2(n + 1 - 2^{h_n^*-1})$.

1.6.3 Suppose T_n is an extended binary tree with external path length X_n. If its height h_n exceeds h_n^*, the height of a complete binary tree on n nodes, there must be a leaf on some level $\ell < h_n - 1$ (otherwise, all levels preceding $h_n - 1$ are saturated with internal nodes and T_n is complete, in which case h_n cannot exceed h_n^*). There must also be two sibling leaves on level h_n (otherwise, every leaf on level h_n has an internal node sibling; those siblings will force leaves on levels higher than h_n, a contradiction). Transfer the parent of the two sibling leaves from level h_n to replace a leaf at level ℓ, producing T_n', a new tree whose external path length is $X_n + 2(\ell + 1) - \ell - 2h_n + (h_n - 1) = X_n + \ell - h_n + 1 < X_n$. If all levels except possibly the last are saturated in T_n', then T_n' is complete; we are done. Otherwise, repeat the whole operation on T_n' to produce T_n'', whose external path length is less than that of T_n'. Continue the process, always producing trees with smaller path length, until it is no long possible (the final tree is complete).

1.6.4 The result can be proved by induction on n. The basis of this induction is trivial. Assume the assertion to be true for all trees of size less than n. Suppose the given binary search tree is T_n, of size n, and let L_i and R_j be its left and right subtrees respectively (of sizes $i, j < n$). The key K_{n+1} of rank r_{n+1} (among all $n + 1$ keys) is to be inserted in T_n. If $r_{n+1} \le i + 1$, insertion falls in L_i. Let $Keys(T)$ denote the set of keys in the nodes of a binary search tree T. Ranked among the keys of $Keys(L_i) \cup K_{n+1}$, the key K_{n+1} is the r_{n+1}st largest. By induction, the insertion of K_{n+1} hits the r_{n+1}st leaf in L_i (in a left-to-right labeling of leaves of L_i), which is the r_{n+1}st largest in T_n.

If $r_{n+1} > i + 1$, insertion falls in R_j. Ranked among the keys of $Keys(R_j) \cup K_{n+1}$, the key K_{n+1} is the $(r_{n+1} - i - 1)$st largest and by induction its insertion hits the $(r_{n+1} - i - 1)$st leaf in R_j (in a left-to-right labeling of leaves of R_j), which is the r_{n+1}st largest in T_n.

1.6.5 Proposition 1.4 gives the exact distribution $\mathbf{Prob}\{S_n^{(j)} = k\}$ in five ranges of k. For $j \le (n+1)/2$ the minimum in the first range $k \le \min\{j-2, n-j-1\}$ is $j - 2$ and in this range

$$\mathbf{Prob}\{S_j^{(n)} = k\} = \frac{2}{(k+2)(k+3)}.$$

In the second range $j - 1 \le k \le n - j - 1$ the probability is

$$\mathbf{Prob}\{S_j^{(n)} = k\} = \frac{1}{(k+1)(k+2)} + \frac{2(j-1)}{(k+1)(k+2)(k+3)}.$$

The third range $n - j \le k \le j - 2$ does not exist. In the fourth range $\max\{j-1, n-j\} \le k \le n-2$, the maximum is $n - j$ and the probability is

$$\mathbf{Prob}\{S_j^{(n)} = k\} = \frac{2}{(k+1)(k+2)} + \frac{2(n-k-2)}{(k+1)(k+2)(k+3)}.$$

The fifth range is the singleton point $k = n - 1$, where the probability is

$$\mathbf{Prob}\{S_j^{(n)} = n-1\} = \frac{1}{n}.$$

Putting it together, one finds the rth moment:

$$\begin{aligned}\mathbf{E}\big[(S_j^{(n)})^r\big] &= \sum_{k=0}^{n} k^r \mathbf{Prob}\{S_j^{(n)} = k\} \\ &= 2\sum_{k=0}^{j-2} \frac{k^r}{(k+2)(k+3)} + \sum_{k=j-1}^{n-j-1} \frac{k^r(k+2j+1)}{(k+1)(k+2)(k+3)} \\ &\qquad +2(n+1)\sum_{k=n-j}^{n-2} \frac{k^r}{(k+1)(k+2)(k+3)} + (n-1)^r \frac{1}{n}.\end{aligned}$$

Simplify (computer algebra may be quite helpful here) for $r = 1$ and $r = 2$ to get the first two moments. The case $j \ge (n+1)/2$ can be argued by the symmetry of $S_n^{(j)}$ and $S_n^{(n-j)}$, or directly as we did for $j \le (n+1)/2$ (here one finds the second range empty).

1.6.6 The depth of a node is also its number of ancestors, and the required result will establish an average symmetry principle between ancestors and descendants. let Π_n be a permutation of $\{1, \dots, n\}$ that gives a binary search tree where k is an ancestor of j. Swap j and k in Π_n, to obtain the new permutation Ξ_n. The tree constructed from Ξ_n has k as a descendant of j. We thus find a one-to-one correspondence between the set of permutations giving binary search trees with k as ancestor of j, and the set of permutations giving binary search trees with k as descendant of j.

Let $A_j^{(n)}$ and $S_j^{(n)}$ be respectively the number of ancestors and descendants of j in a binary search tree of order n. Further, it follows that

$$\begin{aligned}\mathbf{E}\big[S_j^{(n)}\big] &= \sum_{\substack{1\le k\le n\\ k\ne j}} \mathbf{Prob}\{k \text{ is a descendant of } j\} \\ &= \sum_{\substack{1\le k\le n\\ k\ne j}} \mathbf{Prob}\{k \text{ is an ancestor of } j\}\end{aligned}$$

$$= \mathbf{E}\big[A_j^{(n)}\big]$$
$$= H_{n-j+1} + H_j - 1,$$

as given in Proposition 1.4.

Even though the number of ancestors of j has the same average as that of the number of descendants, these random variables have different distributions. For instance,

$$\mathbf{Prob}\{A_1^{(3)} = 0\} = \frac{1}{3},$$

whereas

$$\mathbf{Prob}\{S_1^{(3)} = 0\} = \frac{1}{2}.$$

1.6.7 Let T_Ξ be the binary search tree constructed from a permutation Ξ, and let $\mathcal{P}_{T_\Xi}(j)$ be the path from the root to j in that tree.

We first show that (a) implies (b). The descendants of j must follow j in Π_n. We show that if $d_1 - 1$ is feasible (in $\{1, \ldots, n\}$), it must be on $\mathcal{P}_{T_{\Pi_n}}(j)$. Toward a contradiction, suppose $d_1 - 1$ is feasible but not on $\mathcal{P}_{T_{\Pi_n}}(j)$. If d_1 precedes $d_1 - 1$ in Π_n, the insertion of $d_1 - 1$ must compare it with d_1, and $d_1 - 1$ is a descendant of d_1, whence also a descendant of j, a contradiction. Since $d_1 - 1$ cannot be a descendant of j and does not lie on $\mathcal{P}_{T_{\Pi_n}}(j)$, the insertion of d_1 must fall in the subtree rooted at $d_1 - 1$, and d_1 is not a descendant of j, another contradiction. The argument for $d_{k+1} + 1$ when it is feasible is similar. We have shown that (a) implies (b).

Assuming condition (b), let Π_j be a permutation obtained by truncating Π_n right before j. This truncated permutation comprises the integers preceding j in Π_n. The tree T_{Π_j} constructed from Π_j contains whichever of $d_1 - 1$ and $d_{k+1} + 1$ when feasible. Also, if $d_1 - 1$ is feasible, it is the largest number in the tree T_{Π_j} that is less than j. When j is inserted into T_{Π_n}, j must be compared with $d_1 - 1$, and $d_1 - 1$ lies on $\mathcal{P}_{T_{\Pi_n}}(j)$. A symmetric argument shows that if $d_{k+1} + 1$ is feasible, it must lie on $\mathcal{P}_{T_{\Pi_n}}(j)$.

When members of the set $\mathcal{D} - \{j\}$ are inserted, they will be sandwiched between either:

(i) $d_1 - 1$ and some member of the subtree rooted at j;

(ii) $d_{k+1} + 1$ and some member of the subtree rooted at j;

(iii) two members of the subtree rooted at j.

In any of the three cases, the insertion must fall in the subtree rooted at j. So, that subtree includes $\mathcal{D}$. Since both feasible $d_1 - 1$ and feasible $d_{k+1} + 1$ are on $\mathcal{P}_{T_{\Pi_n}}(j)$, that subtree includes only $\mathcal{D}$.

1.6.8 By an argument similar to that of Exercise 1.6.3, we can demonstrate that the complete binary tree has the least external path length among all binary trees of the same order.

Let Z_n be the number of leaves on level $\lfloor \lg n \rfloor + 1$ in a complete tree of order n, and let X_n be its external path length; Z_n is given in Exercise 1.6.2. The complete tree of order n has $n + 1$ leaves. There are $n + 1 - Z_n$ leaves at level $\lfloor \lg n \rfloor$. Hence,

$$\begin{aligned} X_n &= \big(\lfloor \lg n \rfloor + 1\big) Z_n + \lfloor \lg n \rfloor (n + 1 - Z_n) \\ &= Z_n + (n+1)\lfloor \lg n \rfloor \\ &= n\lfloor \lg n \rfloor + 2n + \lfloor \lg n \rfloor + 2 - 2^{\lfloor \lg n \rfloor + 1}. \end{aligned}$$

Let $f_n \overset{\text{def}}{=} \lg n - \lfloor \lg n \rfloor$. Write

$$\begin{aligned} X_n &= n(\lg n - f_n) + 2n + (\lg n - f_n) + 2 - 2^{\lg n - f_n + 1} \\ &= n \lg n + \lg n + 2n + 2 - (n+1) f_n - \frac{2n}{2^{f_n}}. \end{aligned}$$

This expression exceeds $n \lg n$, iff

$$\lg n + 2n + 2 - (n+1) f_n - \frac{2n}{2^{f_n}} > 0,$$

which is true because, for any $n \ge 3$ and any $x \in [0, 1]$, the function

$$g_n(x) = \lg n + 2n + 2 - (n+1)x - \frac{2n}{2^x}$$

is concave, with negative second derivative (with respect to x). We have $g_n(0) = \lg n + 2 > 0$ and $g_n(1) = \lg n + 1$, with $g_n'(0)$ positive, and $g_n'(1)$ negative—$g_n(x)$ reaches a peak between 0 and 1, and if it has roots they fall outside the interval $[0, 1]$. Whatever the fractional part of $\lg n$ is, $g_n(f_n) > 0$.

1.6.9 Extend the m-ary tree by adding the appropriate number of leaves as sons of nodes to make all internal nodes uniformly have m children. The propositions now read:

Proposition 1.1′ *The height h_n of an m-ary tree on n nodes satisfies:*

$$\lceil \log_m (n+1) \rceil \le h_n \le n.$$

Proposition 1.2′ *An extended binary tree on n internal vertices has $(m - 1)n + 1$ leaves.*

Proposition 1.3′ *Let I_n and X_n be the internal and external path lengths of an m-ary tree of size n. Then*

$$X_n = (m-1) I_n + mn.$$

The proofs are essentially the same as in the binary case. For **Proposition 1.3′**, one can guess the formula by generalizing the binary case ($X_n = I_n + 2n$) to a case $X_n = k_m I_n + a_m n$, and let the proof guide us to the right choices of the coefficients k_m and a_m.

1.7.1 Let the four numbers be a, b, c, d. Write a function $min2$ that returns the minimum of two numbers:

```
function min2(u, v : integer) : integer;
    begin
        if u ≤ v then return(u)
        else return(v);
    end;
```

Now invoke this function to pick the smaller of a and b; then invoke it again to pick the smaller of c and d, then again to find the smaller of the two minima:

```
print(min2(min2(a, b), min2(c, d)));
```

Assuming our programming language evaluates the argument of a function of two variables from left to right, the decision tree is:

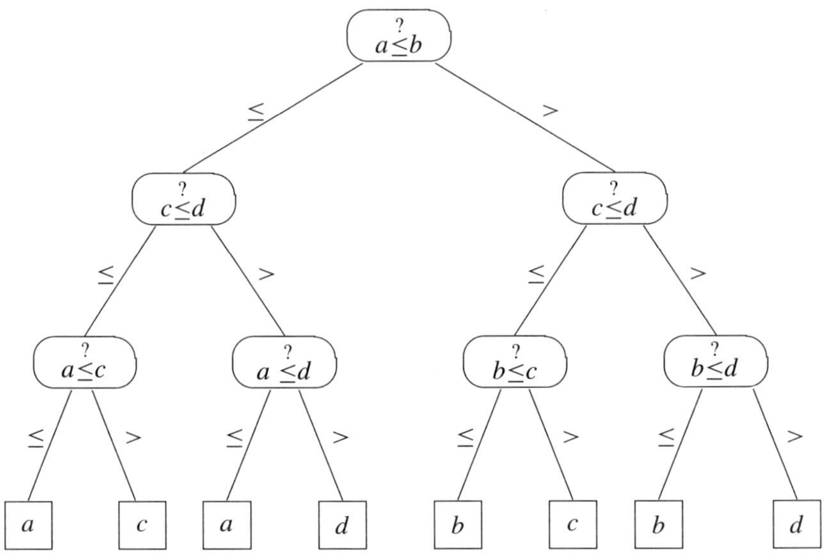

1.7.2 A total order.

1.7.3 There are $n!$ permutations (leaves) in the decision tree. The minimum height is $\lceil \lg(n!) \rceil$ by Proposition 1.1.

1.7.4 Consider the leaf w in Figure 1.6 and the path from the root to it. The root is associated with no prior information. The only order known there is the a priori reflexive partial order

$$\mathcal{R} = \{X_1 \le X_1, X_2 \le X_2, X_3 \le X_3\}.$$

The left child of the root (let us call it node u; see Figure 1.6) induces the partial order

$$P_u = \mathcal{R} \cup \{X_1 \le X_2\}.$$

The right child of u, that is, node v in Figure 1.6, induces the partial order

$$P_v = \mathcal{R} \cup \{X_1 \le X_2, X_3 \le X_2\}.$$

The chosen leaf induces the total order

$$P_w = \{X_1 \le X_1, X_2 \le X_2, X_3 \le X_3, X_1 \le X_2, X_3 \le X_1, X_3 \le X_2\}.$$

1.8.1 Yes, the algorithm of Figure 1.6 is a worst-case optimal algorithm for sorting three numbers. This can be argued exhaustively. Any one query involving a pair of numbers will leave us with no sufficient information on the third (of course a single query involving the same key twice is redundant and will leave us with more uncertainty). Any two nonredundant questions must involve a key twice and each of the other two keys once. In a worst-case scenario the repeat key will outrank the other two and a third question must address the relation between the two non-repeat keys. The given algorithm asks only three questions in its worst case.

1.8.2 Indeed, the algorithm makes $n - 1$ inputs on a particular input. However, Theorem 1.1 does not fail; *some other* input will force the algorithm to switch to a standard comparison-based sorting algorithm. If that other input is the worst-case input for the standard sorting algorithm, it will force $\Omega(n \ln n)$ comparisons. The proposed algorithm then makes $\Omega(n \ln n)$ comparisons in addition to the $n - 1$ comparisons of the preprocessing stage.

1.8.3 Pair the first two keys and rank them $a \le b$, then pair and rank the third and fourth keys as $c \le d$. Compare b and d so that $b \le d$; relabel if necessary to maintain the correspondence $a \le b$ and $c \le d$. The fifth key, e, is incomparable so far with any other. We have an induced poset (transitive information is suppressed for clarity and self-reflexive loops are not shown, as usual):

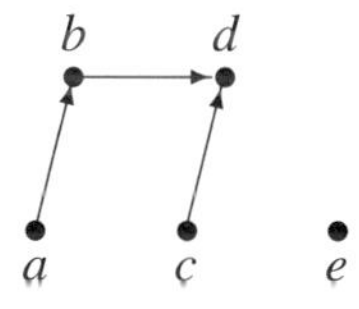

We have consumed three comparisons. The longest chain in the poset is $a \to b \to d$, with b the median of the three numbers. Insert e in this chain, starting with a comparison at the median (binary search on three sorted numbers). There are four "gaps" between numbers in this chain and e will fall in one of the gaps. With two more comparisons we determine the gap, getting one of the four cases in the following figure:

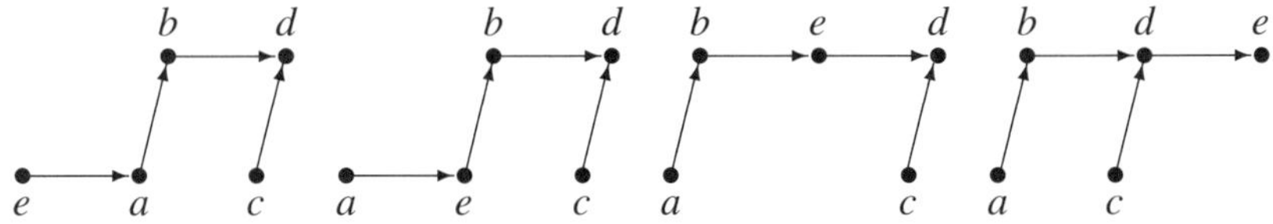

What remains to be done is inserting c in the longest chain of the resulting poset. In all four cases the longest chain has length 4, but d on the chain is comparable to c, $(c \le d)$. So, we insert c in a chain of length 3 by binary search, consuming the last two allowed comparisons.

1.8.4 The decision tree contains two nodes labeled $a \overset{?}{\le} b$. Let the lower (closer to the root) be labeled u and the higher be labeled v. Let w be the parent of v. If v is in the left subtree of u, then $a \le b$. The right subtree rooted at v is completely redundant; on no input will the algorithm follow paths in that subtree (it stands for inputs with $a \le b$ and $a > b$ simultaneously). Eliminate v and its right subtree. Promote the root of the left subtree of v to replace v as a child of w. A symmetric argument applies for the case v is in the right subtree of u. The resulting decision tree corresponds to a new comparison-based sorting algorithm. Obviously, the new tree has a shorter total path length. Hence, the new algorithm has a smaller average running time. Note that this procedure may not reduce the worst-case running time, which may occur on a path not containing v and that path will not be affected by the elimination of redundancy.

1.9.1 An adversary can present an input that always has the champion (largest number) playing and winning the first match. After that comparison, whatever algorithm is given will still have to contend with finding the first k order statistics among $n - 1$ elements.

1.9.2 Let us define the median of $X_1, \ldots, X_n$ as $X_{(n/2)}$ (n even). A leaf ℓ in the decision tree contains the median and thus determines a dichotomy (Lemma 1.1) with $B = \{X_{(n/2+1)}, \ldots, X_n\}$ and its complement. The corresponding dichotomy for two medians bears information about the minimal element of B (the larger of the two medians). Unlike the case of two medians, the dichotomy of the one-median problem does not bear information on the minimal element of B (the case does not have the analog of the larger median among two). An adversary can contrive an input for which no prior information is available at ℓ on B. While the median remains to be known as the maximal element of B^c, all elements of B can be its minimal element, achieving $|\min_\ell(B)| = n/2$. The weight associated with ℓ becomes $2^{1+n/2}$, violating property **P2**.

1.9.3 Directly verify; if $X_2 \leq X_1$,

$$\frac{X_1 + X_2}{2} - \frac{|X_1 - X_2|}{2} = \frac{X_1 + X_2}{2} - \frac{X_1 - X_2}{2} = X_2,$$

otherwise,

$$\frac{X_1 + X_2}{2} - \frac{|X_1 - X_2|}{2} = \frac{X_1 + X_2}{2} - \frac{X_2 - X_1}{2} = X_1.$$

For the maximum, we have

$$\max\{X_1, X_2\} = (X_1 + X_2) - \min\{X_1, X_2\} = \frac{X_1 + X_2}{2} + \frac{|X_1 - X_2|}{2}.$$

1.9.4 Start with

$$\min\{X_1, X_2, X_3\} = \min\big\{\min\{X_1, X_2\}, \min\{X_2, X_3\}\big\};$$

use the formula of Exercise 1.9.3 for the minimum of two numbers to obtain

$$\frac{1}{2}\left(\frac{X_1 + X_2}{2} - \frac{|X_1 - X_2|}{2} + \frac{X_2 + X_3}{2} - \frac{|X_2 - X_3|}{2}\right)$$
$$-\frac{1}{2}\left|\left(\frac{X_1 + X_2}{2} - \frac{|X_1 - X_2|}{2}\right) - \left(\frac{X_2 + X_3}{2} - \frac{|X_2 - X_2|}{2}\right)\right|,$$

which simplifies to the given formula. Similarly, find the maximum from the relation

$$\max\{X_1, X_2, X_3\} = \max\big\{\max\{X_1, X_2\}, \max\{X_2, X_3\}\big\}.$$

After the simplification you find

$$\max\{X_1, X_2, X_3\} = \frac{1}{4}\big(X_1 + 2X_2 + X_3 + |X_1 - X_2| + |X_2 - X_3|$$
$$+\big|X_1 - X_3 + |X_1 - X_2| - |X_2 - X_3|\big|\big).$$

Finding the maxima of the pairs $\{X_1, X_2\}$, $\{X_2, X_3\}$, and $\{X_3, X_1\}$ excludes the minimum of the three. The minimum of the three maxima must be the median:

$$\text{median}\{X_1, X_2, X_3\} = \min\big\{\max\{X_1, X_2\}, \max\{X_2, X_3\}, \max\{X_3, X_1\}\big\}.$$

Simplifying, one obtains

$$\text{median}\{X_1, X_2, X_3\} = \frac{1}{8}\Big(2X_1 + 3X_2 + 3X_3$$
$$+|X_1 - X_2| + 2|X_2 - X_3| + |X_1 - X_3|$$
$$-\big|X_1 - X_3 + |X_1 - X_2| - |X_2 - X_3|\big|$$

$$-\Big|X_2 - X_1 + |X_2 - X_3| - |X_1 - X_3|\Big|$$

$$-\Big|X_2 - X_3 + |X_1 - X_2| - |X_1 - X_3|$$

$$-|X_1 - X_3 + |X_1 - X_2| - |X_2 - X_3||$$

$$+|X_2 - X_1 + |X_2 - X_3| - |X_1 - X_3|| \Big|\Big).$$

1.9.5 The idea is similar to that of Exercise 1.8.3. There we wanted to find a total order. Here we want to jump as quickly as we can to the conclusion that certain elements are excluded from consideration. We focus on resolving order relations among data still qualified to be in the "middle block." Many paths in the decision tree are symmetric. We consider essential cases that are not mirror images. All other cases are argued by relabeling or symmetry. Follow the same algorithm of Exercise 1.8.3 until you construct the first diagrammed poset in Exercise 1.8.3 with three comparisons. The elements with the most uncertainty at this time are c and e; let us compare them. If $e \le c$ we have the poset

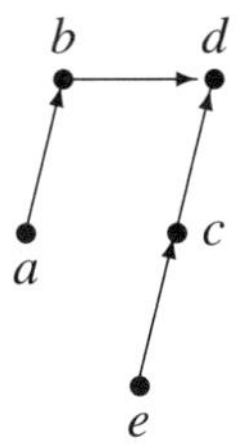

Now compare b and c to get one of the two posets

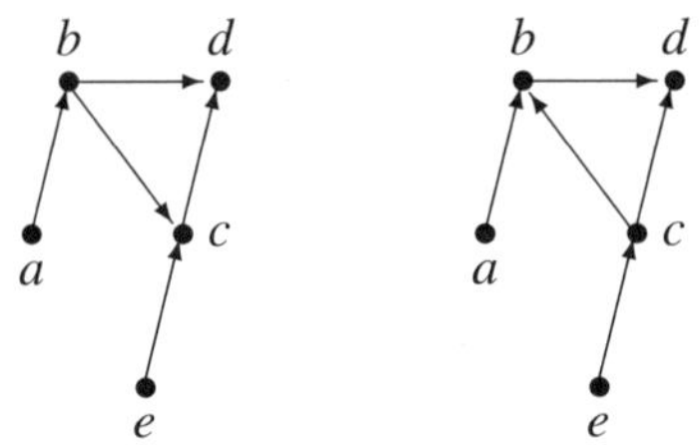

We have gathered enough information to exclude extrema. For example, in the left poset above, we have discovered d is the maximum, and there are three elements above a and three below c. So, a, c and d are out and we need only to resolve the potential middle block $\{b, e\}$ by one additional query. The median is $\max\{b, e\}$.

1.10.1 Let $\Xi_n = (\xi_1, \xi_2, \dots, \xi_n)$ be an arbitrarily fixed permutation. Then

$$\mathbf{Prob}\{\Pi'_n = \Xi_n\} = \mathbf{Prob}\{\Pi_n = (\xi_n, \xi_1, \xi_2 \dots, \xi_{n-1})\} = \frac{1}{n!}.$$

1.10.2 Let $\Lambda_n = (\lambda_1, \lambda_2, \dots, \lambda_n)$ be a fixed permutation of $\{1, 2, \dots, n\}$. By construction, $\mathbf{Prob}\{\xi'_1 = \lambda_1\} = 1/n$, $\mathbf{Prob}\{\xi'_2 = \lambda_2\} = 1/(n-1)$, and so on, until $\mathbf{Prob}\{\xi'_n = \lambda_n\} = 1$. Therefore,

$$\mathbf{Prob}\{\Xi'_n = \Lambda_n\} = \frac{1}{n} \times \frac{1}{n-1} \times \cdots \times 1 = \frac{1}{n!},$$

and Ξ'_n is a random permutation. Note that the probability distribution p_i does not play any role in the construction. We get the same result no matter what distribution is given. The randomization obliterates any nonuniformity in the distribution of permutations.

1.10.3 The nth key ranks j with probability $1/n$, by a fundamental property of random permutations (Proposition 1.6). It was shown in Exercise 1.6.4 that this insertion hits the jth leaf in a left-to-right labeling of the leaves with probability $1/n$.

1.11.1 The given condition is easily satisfied by most probability generating functions. Hold u fixed throughout de-Poissonization. Let $P(z, u)$ be the Poisson transform of $\phi_n(u)$. The given condition on the bivariate generating function gives

$$\begin{aligned}\big|P(z, u)\big| &= \Big|\sum_{n=0}^{\infty} \phi_n(u)\frac{z^n e^{-z}}{n!}\Big| \\ &= \Big|\sum_{n=0}^{\infty} \phi_n(u)\frac{z^n}{n!}\Big|\big|e^{-z}\big| \\ &\le K,\end{aligned}$$

for some constant $K > 0$. Hence, inside the de-Poissonization cone,

$$\big|P(z, u)\big| \le K|z|^0.$$

Outside the de-Poissonization cone

$$\big|P(z, u)e^z\big| \le \Big|\sum_{n=0}^{\infty} \phi_n(u)\frac{z^n}{n!}\Big| < K\big|e^z\big| = Ke^{\Re z} \le Ke^{|z|\cos\theta}.$$

In the de-Poissonization lemma, we can take $c = 0$ and $\alpha = \cos\theta$ to write

$$\phi_n(u) = P(n, u) + O\Big(\frac{\ln n}{\sqrt{n}}\Big).$$

In terms of probabilities, we can approximate fixed-population probabilities by the Poissonized probabilities:

$$\mathbf{Prob}\{X_n = k\} = \mathbf{Prob}\{X_{N(n)} = k\} + O\Big(\frac{\ln n}{\sqrt{n}}\Big),$$

where $N(n) = \mathrm{POISSON}(n)$. This path sheds light on the more general problem of de-Poissonization of doubly indexed arrays and whose Poisson transform is bivariate.

1.11.2 While the analogous relation is true for means (the Poissonized mean is the same as the Poisson transform of fixed-population mean), the relation does not hold for variances basically because means are linear operators and variances are not. It is sufficient to go through an instance. Let $B_i = \mathrm{BERNOULLI}(1/2)$, for $i = 1, 2, \ldots$, be independent. Set $X_n = B_1 + B_2 + \cdots + B_n$. We have

$$\mathbf{E}[B_i] = \frac{1}{2}, \qquad \mathbf{Var}[B_i] = \frac{1}{4}.$$

It follows that

$$\mathbf{E}[X_n] = \frac{n}{2},$$

and by independence

$$\mathbf{Var}[X_n] = \frac{n}{4}.$$

When n is Poissonized into $N = N(z)$, the Poissonized variance is most easily computed from the conditional variance formula

$$\begin{aligned}\mathbf{Var}\big[X_{N(z)}\big] &= \mathbf{Var}\big[\mathbf{E}[X_N \mid N]\big] + \mathbf{E}\big[\mathbf{Var}[X_N \mid N]\big]\\ &= \mathbf{Var}\Big[\frac{1}{2}N\Big] + \mathbf{E}\Big[\frac{1}{4}N\Big]\\ &= \frac{z}{4} + \frac{z}{4}\\ &= \frac{z}{2}.\end{aligned}$$

The Poisson transform of fixed-population variances is

$$\begin{aligned}V(z) &= \sum_{n=0}^{\infty} \mathbf{Var}[X_n]\frac{z^n e^{-z}}{n!}\\ &= \sum_{n=0}^{\infty} \frac{n}{4} \times \frac{z^n e^{-z}}{n!}\\ &= \frac{z}{4}.\end{aligned}$$

In this instance the Poissonized variance and the Poisson transform of fixed-population variances are not the same.

The quantity $\mathbf{Var}[X_{N(n)}]$ does not have a Poissonization interpretation like that of $\mathbf{E}[X_{N(n)}]$. The correct view is to think of the fixed-population variances $\mathbf{Var}[X_n]$ as a sequence, with Poisson transform $V(z)$. Then find by de-Poissonizing that $V(n)$ approximates $\mathbf{Var}[X_n]$, with small quantifiable errors.

1.11.3 If j is even, the binary representation of j has a 0 rightmost bit, and $\nu(j) = \nu(j/2)$; otherwise the binary representation of the odd number j has a 1 as its rightmost bit, and $\nu(j) = \nu(\lfloor j/2 \rfloor) + 1$. Hence

$$\begin{aligned} Z(n) &= \sum_{j=1}^{n-1} \nu(j) \\ &= \sum_{\substack{j=1 \\ j \text{ even}}}^{n-1} \nu\Big(\frac{j}{2}\Big) + \sum_{\substack{j=1 \\ j \text{ odd}}}^{n-1} \Big[\nu\Big(\Big\lfloor \frac{j}{2} \Big\rfloor\Big) + 1\Big] \\ &= \sum_{k=1}^{\lceil n/2 \rceil - 1} \nu(k) + \sum_{k=0}^{\lfloor n/2 \rfloor - 1} [\nu(k) + 1] \\ &= Z\Big(\Big\lceil \frac{n}{2} \Big\rceil\Big) + Z\Big(\Big\lfloor \frac{n}{2} \Big\rfloor\Big) + \Big\lfloor \frac{n}{2} \Big\rfloor. \end{aligned}$$

1.11.4 Let $b_n = \lfloor n/2 \rfloor$. Then $\triangle \bigtriangledown b_n = (-1)^{n+1}$. Apply Theorem 10.2 to the recurrence

$$Z(n) = Z\Big(\Big\lceil \frac{n}{2} \Big\rceil\Big) + Z\Big(\Big\lfloor \frac{n}{2} \Big\rfloor\Big) + b_n,$$

with boundary condition $Z(1) = 0$. The Dirichlet transform of the sequence $\{\triangle \bigtriangledown b_n\}_{n=1}^{\infty}$ is

$$\begin{aligned} \eta(s) &= \sum_{n=1}^{\infty} \frac{(-1)^{n+1}}{n^s} \\ &= \sum_{k=1}^{\infty} \frac{1}{(2k-1)^s} - \sum_{k=1}^{\infty} \frac{1}{(2k)^s} \\ &= \sum_{k=1}^{\infty} \frac{1}{k^s} - \frac{2}{2^s} \sum_{k=1}^{\infty} \frac{1}{k^s} \\ &= \zeta(s)\Big(1 - \frac{1}{2^{s-1}}\Big). \end{aligned}$$

Delange's recurrence equation has as solution

$$Z_n = \frac{n}{2\pi i}\int_{3-i\infty}^{3+i\infty} \frac{\eta(s)n^s}{s(s+1)(1-2^{-s})}\,ds.$$

By choosing an appropriate contour, we evaluate the integral via residues of enclosed poles:

$$\begin{aligned} Z_n &= n\Big[\operatorname*{Res}_{s=-1} + \sum_{k=-\infty}^{\infty} \operatorname*{Res}_{s=2\pi ik/\ln 2}\Big]\frac{\zeta(s)n^s}{s(s+1)(1-2^{-s})}\Big(1-\frac{1}{2^{s-1}}\Big) \\ &= n\Big[\frac{1}{4n} + \Big(\frac{1}{2}\lg\pi + \frac{1}{2}\lg n - \frac{2+\ln 2}{4\ln 2}\Big) + \delta(\lg n)\Big] \\ &= \frac{1}{2}n\lg n + n\Big[\frac{2\ln\pi - 2 - \ln 2}{4\ln 2} + \delta(\lg n)\Big] + \frac{1}{4}, \end{aligned}$$

where δ is the fluctuating function:

$$\delta(u) = -\frac{1}{\ln 2}\sum_{\substack{k=-\infty\\k\neq 0}}^{\infty} \frac{\zeta(2\pi ik/\ln 2)e^{2\pi iku}}{(2\pi ik/\ln 2)(1+2\pi ik/\ln 2)}.$$

1.11.5 Take the Mellin transform of both sides:

$$f^*(s) = 2^{s+1}f^*(s) + \Gamma(s),$$

that is,

$$f^*(s) = \frac{\Gamma(s)}{1-2^{s+1}}.$$

This transform exists in $-2 < \Re s < -1$. (An alternative way to derive the transform is to iterate for an (exact) infinite harmonic sum, then transform it.)

Invert the transform over a vertical line $\Re s = -3/2$; the poles of the integrand are simple poles at $s_k = -1 + 2\pi ik/\ln 2$, for $k = \pm 1, \pm 2, \ldots$, simple poles at $S_j = -j$ (for $j \geq 2$), a simple pole at $s = 0$, and a double pole at $s = -1$. Move the line integral to the right (close the box at $\Re s = d$, $d > 0$). The poles to compensate for are s_k, for $k = \pm 1, \pm 2, \ldots$ and the poles at $s = -1$ and $s = 0$. A residue calculation yields

$$\begin{aligned} f(x) &= -\Big[\operatorname*{Res}_{s=-1} + \operatorname*{Res}_{s=0} + \sum_{\substack{k=-\infty\\k\neq 0}}^{\infty} \operatorname*{Res}_{s=s_k}\Big]\frac{x^{-s}\Gamma(s)}{1-2^{s+1}} \\ &= x\lg x + \big(2\gamma - 2 + \ln 2 + \delta(\lg x)\big)\frac{x}{2\ln 2} + 1 + O\Big(\frac{1}{x^d}\Big), \end{aligned}$$

where δ is the oscillating function

$$\delta(u) = -\frac{1}{\ln 2} \sum_{\substack{k=-\infty \\ k\neq 0}} \Gamma\Big(-1 + \frac{2\pi i k}{\ln 2}\Big) e^{-2\pi i k u}.$$

The magnitude of $\delta(u)$ is bounded:

$$\big|\delta(u)\big| = \frac{1}{\ln 2} \sum_{\substack{k=-\infty \\ k\neq 0}} \Big|\Gamma\Big(-1 + \frac{2\pi i k}{\ln 2}\Big)\Big| \approx 0.2488660506 \times 10^{-6}.$$

1.11.6 For any fixed x, the given function, $f(x)$, is a harmonic sum with base function $B(x) = 1/(e^x - 1)$. Let $g^*(s)$ be the Mellin transform of a function $g(x)$. Then

$$f^*(s) = B^*(s) \sum_{k=1}^{\infty} \frac{1}{k^s} = B^*(s)\zeta(s),$$

with

$$B^*(s) = \mathcal{M}\{e^{-x}\big(1 + e^{-x} + e^{-2x} + \ldots\big); s\} = \mathcal{M}\Big\{\sum_{j=0}^{\infty} e^{-x(j+1)}; s\Big\},$$

which is again the Mellin transform of a harmonic sum with base function e^{-x}. The latter transform is $\mathcal{M}\{e^{-x}; s\}\zeta(s) = \Gamma(s)\zeta(s)$. So,

$$f^*(s) = \Gamma(s)\zeta^2(s),$$

existing in the domain $\Re s > 1$. The poles are at $s = 1, 0, -1, -2, \ldots$. Invert the transform by an integral on the line $\Re s = 2$. Close a contour to the left to enclose poles (say by a large semicircle centered at 2). If the circle intercepts the horizontal axis at d, $-k - 1 < d < -k$, the contour encloses the poles $s = 1, s = 0, -1, \ldots, -k$. The further to the left the circle intercepts the horizontal axis, the more poles are factored in and the higher the accuracy will be. A full series expansion is obtained by taking an infinite radius for the semicircle. One obtains

$$\begin{aligned} f(x) &= \sum_{k=-1}^{\infty} \operatorname*{Res}_{s=-k} x^{-s}\Gamma(s)\zeta^2(s) \\ &= \operatorname*{Res}_{s=1} + \operatorname*{Res}_{s=0} + \operatorname*{Res}_{s=-1} + \cdots \\ &= -\frac{\ln x}{x} + \frac{\gamma}{x} + \frac{1}{4} - \frac{x}{144} - \cdots. \end{aligned}$$

1.11.7 The first part of this problem is a special case of the function of Exercise 1.11.6 (evaluated at ln 2). This gives the general approach to this type of problem. Look at the given series as a function evaluated at a special value. The Mellin transform formulation may give a full asymptotic expansion near $x = 0$. Let $g^*(s)$ denote the Mellin transform of $g(x)$.

(a) Evaluate the function of Exercise 1.11.6 at ln 2:

$$\sum_{k=1}^{\infty} \frac{1}{2^k - 1} = \frac{\gamma - \ln\ln 2}{\ln 2} + \frac{1}{4} - \frac{\ln 2}{144} - \frac{\ln^3 2}{86400} - \cdots \approx 1.606699.$$

(b) The given sum is $f(\ln 2)$, where

$$f(x) = \sum_{k=1}^{\infty} \frac{1}{e^{kx} + 1},$$

defined for $x > 0$. The setup is almost identical to part (a) of this problem. First one gets the Mellin transform

$$f^*(s) = \frac{\Gamma(s)\zeta^2(s)}{1 - 2^{1-s}},$$

by a routine almost identical to the steps in Exercise 1.11.6; the minor deviation here is when we get to expanding the base function $e^x + 1$ as a series; it will have alternating signs. Invert the Mellin transform to get

$$f(x) = \frac{\ln 2}{x} - \frac{1}{4} + \frac{x}{48} + \frac{x^3}{5760} + \cdots .$$

The constant of the problem is approximately

$$f(\ln 2) = \frac{3}{4} + \frac{\ln 2}{48} + \frac{\ln^3 2}{5760} - \cdots \approx 0.764498.$$

(c) Let $f(x)$ be the harmonic sum

$$f(x) = \sum_{k=1}^{\infty} \ln\left(1 + \frac{1}{e^{kx}}\right).$$

The given sum is $f(\ln 2)$. The formal Mellin transform of the harmonic sum is

$$f(x) = B(s) \sum_{k=1}^{\infty} \frac{1}{k^s},$$

where $B(s)$ is the Mellin transform of the base function $\ln(1 + e^{-x})$. Use the expansion

$$\ln(1+y) = \sum_{j=1}^{\infty} \frac{(-1)^{j-1} y^j}{j},$$

for $y > 0$, to obtain a harmonic sum for the base function. Then

$$B(s) = -\mathcal{M}\{e^{-x}; s\} \sum_{j=1}^{\infty} \frac{(-1)^j}{j^{s+1}},$$

which can be reorganized as a combination of Zeta functions. Hence,

$$f^*(s) = -\Gamma(s)\zeta(s)\zeta(s+1)\Big(\frac{1}{2^s} - 1\Big),$$

existing in $\Re s > 1$, and having singularities at $1, 0, -1, -2, \ldots$. Thus,

$$f(x) = -\sum_{k=-1}^{\infty} \operatorname*{Res}_{s=k} \Gamma(s)\zeta(s)\zeta(s+1)\Big(\frac{1}{2^s} - 1\Big)x^{-s}$$

$$= \frac{\pi^2}{12x} + \frac{1}{2}\ln 2 - \frac{1}{24}x + \cdots .$$

Finally,

$$\sum_{k=1}^{\infty} \ln\Big(1 + \frac{1}{2^k}\Big) = f(\ln 2) \approx 0.868876.$$

1.11.8 Let S_n be the given sum. Then

$$S_n = -\frac{1}{2\pi i} \oint_{\Lambda} \frac{\beta(-z, n+1)}{1 - 2^z}\, dz,$$

where Λ is the boundary of the box joining the four points $3/2 \pm i$, and $n+1 \pm i$. Apply Rice's method to enlarge the contour to one with boundaries at $-d \pm iM$ and $n+2 \pm iM$, $d > 0$ arbitrary, with M large and not allowing the boundaries to coincide with any of the poles of the integrand. Any solution of the equation

$$2^z = 1 = e^{2\pi i k}, \qquad k = 0, \pm 1, \pm 2, \ldots,$$

is a pole of the integrand (that is, $z_k = 2\pi i k / \ln 2$), and so are $\tilde{z}_j = j$, for $j = 0, 1, \ldots, n$, the poles of the Beta function. So, all the poles are simple except the double pole at 0. Rice's method gives

$$S_n = \operatorname*{Res}_{z=1} \frac{\beta(-z, n+1)}{1 - 2^z} + \operatorname*{Res}_{z=0} \frac{\beta(-z, n+1)}{1 - 2^z}$$

$$+ \sum_{\substack{k=-\infty \\ k\neq 0}}^{\infty} \operatorname*{Res}_{z=z_k} \frac{\beta(-z, n+1)}{1-2^z} + O\big(n^{-d}\big)$$

$$= -n + \frac{H_n}{\ln 2} - \frac{1}{2} + \delta(n) + O\big(n^{-d}\big),$$

where $\delta(n)$ is the oscillating function

$$\delta(n) = -\frac{1}{\ln 2} \sum_{\substack{k=-\infty \\ k\neq 0}}^{\infty} \beta(-z_k, n+1),$$

whose magnitude is bounded:

$$\big|\delta(n)\big| \le \frac{1}{\ln 2} \sum_{\substack{k=-\infty \\ k\neq 0}}^{\infty} \left|\frac{\Gamma(-z_k)\Gamma(n+1)}{\Gamma(n+1-z_k)}\right|$$

$$\le \frac{1}{\ln 2} \sum_{\substack{k=-\infty \\ k\neq 0}}^{\infty} \big|\Gamma(-z_k)\big|\,\big|n^{z_k}\big|\Big(1 + O\Big(\frac{1}{n}\Big)\Big)$$

$$\le \Big(\frac{1}{\ln 2} \sum_{\substack{k=-\infty \\ k\neq 0}}^{\infty} \big|\Gamma(-z_k)\big|\Big) + O\Big(\frac{1}{n}\Big),$$

by Stirling' approximation. The fixed part is approximately $0.1573158429 \times 10^{-5}$, and of course the $O(1/n)$ can be made arbitrarily small.

CHAPTER 2

2.1 In the sentinel version each new key in the given list will travel all the way back to the sentinel (and be compared with it), giving a total of $1+2+\cdots+10 = 55$ comparisons. BINARY INSERTION SORT makes $0+1+1+2+2+2+2+3+3+3 = 19$ comparisons.

If the list is reversed, each new key is immediately stopped after one comparison in LINEAR INSERTION SEARCH, giving 10 comparisons. BINARY INSERTION SORT still makes $0+1+2+2+3+3+3+3+4+4 = 25$ comparisons.

2.2 Simply remove the read statement **read**(K) from the algorithm (5th line in Figure 2.3).

2.3 We discuss the case of integer data. All other types are sorted similarly. The sorted list is pointed to from the outside by *top*, that points to the largest key inserted so far. The key to be inserted is K. Each entry on the list is assumed to be a record carrying the information field *key*, and the nexus field *nexus*, set up in the following way:

```
type pointer =↑node;
     node = record
          key : integer;
          nexus : pointer;
     end;
```

The nodes appear in decreasing order, each pointing to a node carrying a smaller (or equal) key. A pointer p sets out to find the first element that is less than or equal to K. Another pointer q sets out to find the parent of that node. So, q is made to lag one step behind p. Once a position is found, a new node is allocated (by a call to the built-in procedure **new**) and linked. A sentinel $-\infty$ key is placed at the end of the list to guarantee termination under a simplified condition. In spirit, this algorithm is closest to BACKWARD LINEAR INSERTION SORT (with a sentinel):

```
call new(top);
top↑.key ← −∞;
top↑.nexus ← nil;
for i ← 1 to n do
     begin
          read(K);
          p ← top;
          q ← nil;
          while K > p↑.key do
               begin
                    q ← p;
                    p ← p↑.nexus;
               end;
          call new(r);
          r↑.nexus ← p;
          r↑.key ← K;
          if q = nil then top ← r
          else q↑.nexus ← r;
     end;
```

2.4 The general shape of the (deterministic) insertion tree at the ith stage is a right-oriented "spine" and left-oriented "legs" that dangle from the spinal nodes (each "complete" leg is of length $c - 1$ internal nodes). If n is not an exact multiple of c, one last incomplete leg hangs from last node in the spine. (See the first few JUMP INSERTION trees for $c = 3$ in Exercise 2.8 below.) Let us refer to the leg attached to the root as the first leg, the leg attached to the root's right spinal node as the second, and so on. The jth leg has one leaf at each of the levels $j+1, j+2, \ldots, j+c-2$, and has two leaves on level $j+c-1$. Let X_i be the random number of comparisons used to insert the ith key, in a tree of size $i - 1$. There are $\lfloor (i-1)/c \rfloor$ complete legs. The last (highest) spinal node is at level $\lfloor (i-1)/c \rfloor$. There are $r_i \equiv i - 1 \ \mathbf{mod}\ c$ nodes in the incomplete leg, each associated with a leaf, except the highest in the leg, which has two leaves. One leaf appears at each of the levels $\lfloor (i-1)/c \rfloor + k$, for $k = 1, \dots, r_i - 1$ and two leaves appear at level $\lfloor (i-1)/c \rfloor + r_i$. One computes

$$\mathbf{E}[X_i] = \frac{1}{i} \sum_{j=1}^{\lfloor (i-1)/c \rfloor} \Big[\Big(\sum_{k=1}^{c-2} (j+k) \Big) + 2(j+c-1) \Big] + L_i,$$

where L_i is the contribution of the incomplete leg:

$$L_i = \frac{1}{i} \Big[(r_i + 1) \Big\lfloor \frac{i-1}{c} \Big\rfloor + 1 + 2 + \cdots + (r_i - 1) + 2r_i \Big] = O\Big(\max\Big\{1, \frac{c^2}{i}\Big\}\Big).$$

Hence

$$\begin{aligned} \mathbf{E}[X_i] &= \frac{1}{i} \sum_{j=1}^{\lfloor (i-1)/c \rfloor} \Big[cj + \frac{1}{2}(c-1)(c+2) \Big] + O\Big(\max\Big\{1, \frac{c^2}{i}\Big\}\Big) \\ &= \frac{c}{2i} \Big\lfloor \frac{i-1}{c} \Big\rfloor \Big(\Big\lfloor \frac{i-1}{c} \Big\rfloor + c \Big) + O\Big(\max\Big\{1, \frac{c^2}{i}\Big\}\Big) \\ &\sim \frac{i}{2c} + \frac{c}{2} + O\Big(\max\Big\{1, \frac{c^2}{i}\Big\}\Big). \end{aligned}$$

The number of comparisons over all the stages is

$$C_n = X_1 + X_2 + \cdots + X_n,$$

with average

$$\begin{aligned} \mathbf{E}[C_n] &= \mathbf{E}\big[X_1 + X_2 + \cdots + X_n\big] \\ &= \sum_{i=1}^{n} \Big(\frac{i}{2c} + \frac{c}{2} + O\Big(\max\Big\{1, \frac{c^2}{i}\Big\}\Big)\Big) \\ &= \frac{n^2 + n}{4c} + \frac{cn}{2} + \sum_{i=1}^{n} O\Big(\max\Big\{1, \frac{c^2}{i}\Big\}\Big). \end{aligned}$$

2.5 In Exercise 2.4 we found that the number of comparisons of JUMP INSERTION SORT is

$$\mathbf{E}[C_n] = \frac{n^2 + n}{4c} + \frac{cn}{2} + \sum_{i=1}^{n} O\Big(\max\Big\{1, \frac{c^2}{i}\Big\}\Big).$$

Of course, if c is held fixed, as $n \to \infty$, the term $n^2/(4c)$ dominates and all the other terms are subsumed in the $O(n)$ term. If c is allowed to grow with n other terms may become asymptotically nonnegligible. If $c = O(n^{\frac{1}{2}-\varepsilon})$, the term $n^2/4c$ alone contributes $\Omega(n^{\frac{3}{2}+\varepsilon})$ comparisons. If $c = \Omega(n^{\frac{1}{2}+\varepsilon})$, the term $cn/2$ alone contributes $\Omega(n^{\frac{3}{2}+\varepsilon})$ comparisons. The asymptotically optimal order is $c = \Theta(\sqrt{n}\,)$, as it achieves $\Theta(n^{3/2})$ comparisons, and the contributions of the incomplete legs across the various stages add up to the negligible $O(n \ln n)$. To find the leading asymptotic term, let

$$f(c) = \frac{x^2}{4c} + \frac{cx}{2}.$$

Set $f'(c) = 0$ to find the (asymptotically) optimal choice $c = \lfloor \sqrt{n/2} \rfloor$.

2.6 By the same arguments of Exercise 2.4 we obtain

$$\mathbf{E}[X_i^2] = \frac{1}{i} \sum_{j=1}^{\lfloor (i-1)/c \rfloor} \Big[\sum_{k=1}^{c-2} (j+k)^2 + 2(j+c-1)^2 \Big]$$
$$+ \frac{2}{i}\Big(r_i + \Big\lfloor \frac{i-1}{c} \Big\rfloor\Big)^2 + \frac{1}{i} \sum_{k=1}^{r_i - 1} \Big(k + \Big\lfloor \frac{i-1}{c} \Big\rfloor\Big)^2.$$

At $c = \lfloor \sqrt{n/2} \rfloor$, $\mathbf{E}[C_n] \sim n^{3/2}/\sqrt{2}$, and

$$s_n^2 = \mathbf{Var}[C_n] = \mathbf{Var}[X_1] + \cdots + \mathbf{Var}[X_n] \sim \frac{1}{\sqrt{2}} n^{5/2}.$$

The height of the insetion tree is $O(\sqrt{n}\,)$. The growth rate in s_n is faster than the height's. Theorem 2.1 then asserts that:

$$\frac{C_n - n^{3/2}/\sqrt{2}}{n^{5/4}} \xrightarrow{\mathcal{D}} \mathcal{N}\Big(0, \frac{1}{2}\Big).$$

2.7 BINARY SEARCH chooses position $\lceil n/2 \rceil$ as its first probe into a list of size n. Thus, $\lceil n/2 \rceil$ is the label of the root of BINARY SEARCH's insertion tree. As BINARY SEARCH is recursive, the left and right subtrees are the insertion trees for BINARY SEARCH on $\lceil n/2 \rceil - 1$ nodes, and $n - \lceil n/2 \rceil = \lfloor n/2 \rfloor$ nodes, respectively. By induction, the subtrees are complete on their respective sizes. If n is odd, $\lceil n/2 \rceil$ is the data median, and the two subtrees have the same order; two complete trees of the same height are attached to a root, yielding a complete tree. If n is even, the left subtree has size that is smaller by 1 than the size of the right subtree. The two complete subtrees are of successive order. If the left subtree is not perfect; the right subtree is complete and of the same height. If the left subtree is pefect, the right subtree is complete with height larger by 1. In all cases, attatching the particular patterns of complete trees that arise on sucessive orders as subtrees of a root, yields a complete tree.

2.8 In all the strategies below the tree T_1 is empty.

(a) The insertion trees $T_2, \ldots, T_8$ for BACKWARD LINEAR INSERTION SORT are:

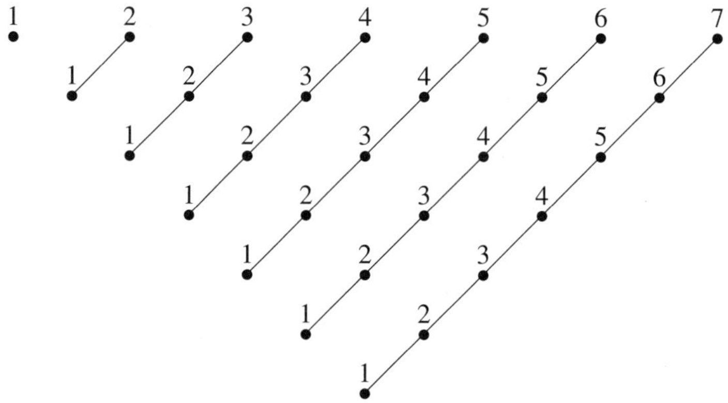

The insertion trees $T_2, \ldots, T_8$ for FORWARD LINEAR INSERTION SORT are:

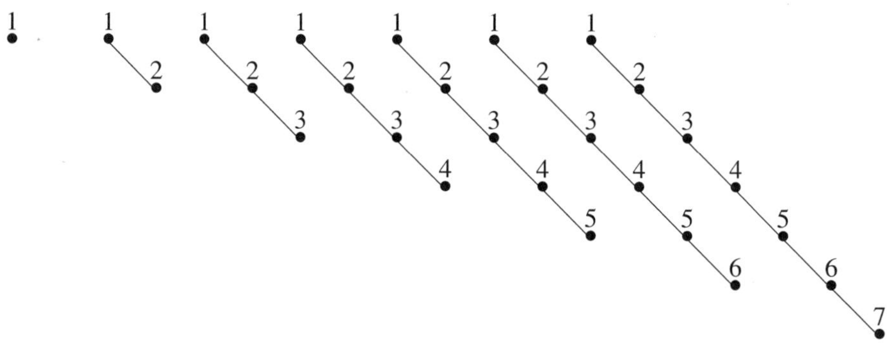

The sentinel versions may probe one additional position. For example, for BACKWARD LINEAR INSERTION SORT, the sentinel sits at position 0 and will be reached in the case of inserting a key less than all they keys that have appeared so far. The insertion tree T_i' to insert the ith key is the same as T_i of BACKWARD LINEAR INSERTION SORT without a sentinel, with a node labeled 0 at the bottom of the drawing. The following figure illustrates T_8':

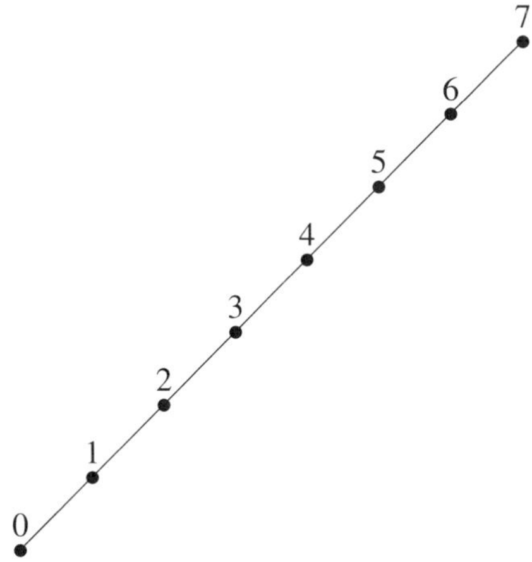

The insertion trees of the sentinel version of FORWARD LINEAR INSERTION SORT are mirror images of insertion trees of the sentinel version of BACKWARD LINEAR INSERTION SORT with the labeling reversed on the path from the root to the highest node (the sentinel in this case is $+\infty$ at the bottom of the array).

The proof of the tree-growing property is the same for any of these linear search strategies. Take, for example, BACKWARD LINEAR INSERTION SORT. The tree T_{i+1} grows from T_i with a new node adjoined as a left child of the highest node. (The nodes have to be relabeled, though, to correspond to the new probe sequences.)

(b) The insertion trees $T_2, \ldots, T_8$ for BINARY INSERTION SORT are:

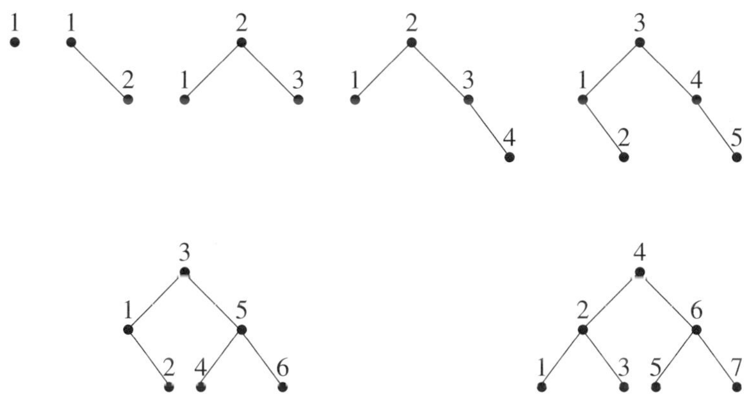

As discussed in Exercise 2.7, T_i is complete. If T_i is not perfect, T_{i+1} is a complete tree of the same height as T_i. The shape of T_{i+1} can therefore be obtained from T_i by adjoining a node at the highest level of T_i as a child of a node at the second highest level. If T_i is perfect, T_{i+1} has one node hanging from the the triangular shape of T_i, a shape that can be obtained from that of T_i by adjoining a node as a child of a node at the highest level of T_i. (In either case the nodes have to be relabeled to correspond to the new probe sequences.)

(c) The insertion trees $T_2, \ldots, T_8$ for 3-JUMP INSERTION SORT are:

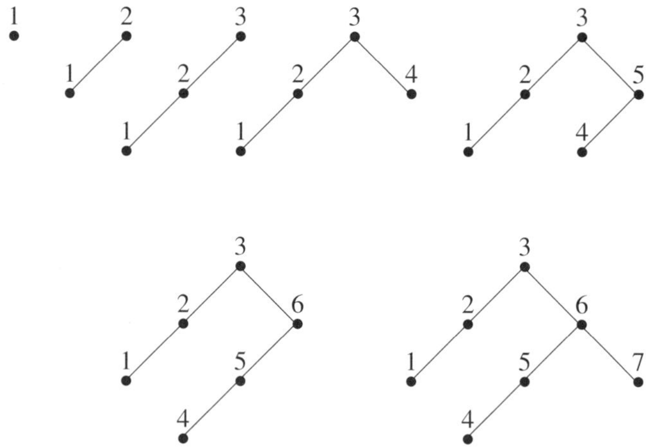

If T_i has an incomplete leg ($n \not\equiv 0$ **mod** c), T_{i+1} grows from T_i by adjoining a new node as a left child of the highest node in the leg. Otherwise, ($n \equiv 0$ **mod** c), T_i grows from T_i by adjoining a node as a right child of the highest node on the spine. (In either case the nodes have to be relabeled to correspond to the new probe sequences.)

2.9 The average cost of inserting the ith key is the average depth of a leaf in the insertion tree (of order $i - 1$). That is,

$$\frac{D_1 + D_2 + \cdots + D_i}{i},$$

where D_j is the depth of the jth leaf (in any systematic labeling of the leaves). The cost is therefore $\mathcal{X}_i / i$, where $\mathcal{X}_i$ is the external path length of the insertion tree. Among all binary tree shapes of order $(i - 1)$, the complete tree has the least external path length (a result of Exercise 1.6.8). A complete tree of order $i - 1$ corresponds to the BINARY INSERTION SORT strategy at the ith stage.

2.10 The insertion tree T_j for the jth key of this randomized algorithm is the random binary search tree (of size $j - 1$) as can been seen inductively from the equal probability of root selection (at the top level of recursion and recursively in the subtrees). Note that T_j is independent of T_{j+1}, as if the algorithm obliterates any information in T_j and constructs a new one at the $(j + 1)$st stage. For example, T_7 may be a linear tree on 6 nodes, and T_8 may be the complete binary tree of size 7. One cannot appeal to Theorem 2.1, which assumes a sequence of insertion trees of *increasing* heights.

Let X_j be the number of comparisons to insert the jth ky, and C_n be the overall number of comparisons to insert n keys. The first random probe $P_j \overset{D}{=}$ UNIFORM$[1 \mathinner{..} j-1]$. Given P_j, there are $P_j - 1$ keys above it, defining P_j gaps. If the next insertion occurs above P_j, it is equally likely to fall in any of the P_j gaps, so it will hit one of the gaps above P_j with probability P_j/j. Insertion below P_j follows similar reasoning. We have

$$C_n = X_1 + X_2 + \ldots + X_n,$$

and X_j's are independent, with

$$X_j = 1 + \begin{cases} X_{P_j}, & \text{with probability } \frac{P_j}{j}; \\ X_{j-P_j}, & \text{with probability } 1 - \frac{P_j}{j}. \end{cases}$$

We have

$$\mathbf{E}[X_j \mid P_j] = 1 + \frac{P_j}{j} X_{P_j} + \frac{j - P_j}{j} X_{j-P_j},$$

and

$$\begin{aligned} \mathbf{E}[X_j] &= \sum_{p=1}^{j-1} \mathbf{E}[X_j \mid P_j = p] \, \mathbf{Prob}\{P_j = p\} \\ &= \frac{1}{j-1} \sum_{p=1}^{j-1} \Big(1 + \frac{p}{j} X_p + \frac{j-p}{j} X_{j-p}\Big) \\ &= 1 + \frac{2}{j(j-1)} \sum_{p=1}^{j-1} p \mathbf{E}[X_p]. \end{aligned}$$

This telescopic sum is easily handled (as was done in the text on several other instances) by differencing a version of the recurrence with $j-1$ from the jth version. The differencing eliminates the sum; one obtains

$$\mathbf{E}[X_j] = \frac{2}{j} + \mathbf{E}[X_{j-1}],$$

which unwinds to $\mathbf{E}[X_j] = 2H_j - 2$. By a similar routine, one obtains the second moment from $\mathbf{E}[X_j^2 \mid P_j]$, etc. One finds $\mathbf{Var}[X_j] = 2H_j - 4H_j^{(2)} + 2$.

For the overall number of comparisons we have

$$\mathbf{E}[C_n] = \sum_{j=1}^{n} (2H_j - 2) \sim 2n \ln n.$$

Compared with BINARY INSERTION SORT's asymptotic average $n \lg n$, this randomized algorithm performs only $2 \ln 2 \approx 138\%$ worse. The variance is

$$s_n^2 \stackrel{\text{def}}{=} \mathbf{Var}[C_n] = \sum_{j=1}^{n} \big(2H_j - 4H_j^{(2)} + 2\big) \sim 2n \ln n.$$

Let $\phi_Y(t)$ be the characteristic function of a random variable Y. So, $\phi_{C_n}(t) = \prod_{j=1}^{n} \phi_{X_j}(t)$. All moments of X_j exist, and the characeristic function of $X_j - (2H_j - 2)$ has an expansion up to any desired moment. We expand up to the second moment (as $t \to 0$):

$$\begin{aligned} e^{\left(C_n - \sum_{j=1}^{n}(2H_j - 2)\right)it} &= \prod_{j=1}^{n} \phi_{X_j - (2H_j - 2)}(t) \\ &= \prod_{j=1}^{n} \Big[1 - \frac{1}{2}\mathbf{Var}[X_j]t^2 + O(t^3)\Big] \\ &= \exp\Big\{\sum_{j=1}^{n} \ln\Big(1 - \big(2H_j - 4H_j^{(2)} + 2\big)\frac{t^2}{2} + O(t^3)\Big)\Big\} \\ &= \exp\Big\{-\sum_{j=1}^{n} \big(2H_j - 4H_j^{(2)} + 2\big)\frac{t^2}{2} + O(nt^3)\Big)\Big\}. \end{aligned}$$

Set $t = v/\sqrt{\sum_{j=1}^{n}\big(2H_j - 4H_j^{(2)} + 2\big)}$, for fixed v. As $n \to \infty$,

$$\exp\Big\{\Big(\frac{C_n - \sum_{j=1}^{n}(2H_j - 2)}{\sqrt{\sum_{j=1}^{n}\big(2H_j - 4H_j^{(2)} + 2\big)}}\Big)iv\Big\} \to e^{-v^2/2}.$$

Use Slutsky's theorem to throw out nonessential factors to get

$$\frac{C_n - 2n \ln n + 2(1 - \gamma)n}{\sqrt{2n \ln n}} \xrightarrow{\mathcal{D}} \mathcal{N}(0, 1).$$

2.11 Regardless of the search algorithm, in the array implementation an insertion position is found and all data are pushed down one slot to create vacancy. The number of moves is the same for all search strategies (including LINEAR INSERTION SORT, where searching and data movement are integrated). Let M_n be the number of moves while inserting n keys by LINEAR INSERTION SORT (and consequently by any INSERTION SORT). The variable M_n is an accumulation of the moves done for the ith insertion, $i = 1, \dots, n$. The number of moves for the ith insertion is: $X_i - 1$, where X_i is LINEAR INSERTION SORT's number of comparisons for the ith key. Thus

$$M_n = (X_1 - 1) + (X_2 - 1) + \cdots + (X_n - 1) = C_n - n,$$

where C_n is the overall number of comparison, and has a Gaussian law. It follows that

$$\frac{M_n - \frac{1}{4}n^2}{n^{3/2}} = \frac{C_n - \frac{1}{4}n^2}{n^{3/2}} - \frac{1}{\sqrt{n}} \xrightarrow{\mathcal{D}} \mathcal{N}\Big(0, \frac{1}{36}\Big).$$

CHAPTER 3

3.1 The distribution funcrion $F(x)$ is strictly increasing. Its inverse $F^{-1}(u)$ is well defined. For $0 \le u \le 1$, the distribution function of the random variable $F(X)$ is

$$\mathbf{Prob}\{F(X) \le u\} = \mathbf{Prob}\{X \le F^{-1}(u)\} = F\big(F^{-1}(u)\big) = u,$$

like the distribution function of UNIFORM(0, 1).

3.2 (a) Any positive integer j can be written as

$$j = \underbrace{1 + 1 + \cdots + 1}_{j \text{ times}}.$$

We take $R(1, s_1, \ldots, s_k) = 0$.

(b) Suppose each of the integers $\mu, \mu + 1, \ldots, \mu + s_i - 1$ can be expressed as a linear combination of $s_1, \ldots, s_k$. For integer $\nu \ge \mu + s_i$, we can write $\nu = \mu + j$, with $j = qs_i + r$, for $q \ge 1$, and $0 \le r < s_i$. Then for $\alpha_1, \ldots, \alpha_k \ge 0$,

$$\begin{aligned}\nu &= (\mu + r) + qs_i \\ &= (\alpha_1 s_1 + \cdots + \alpha_k s_k) + qs_i \\ &= \alpha_1 s_1 + \cdots + \alpha_{i-1}s_{i-1} + (q + \alpha_i)s_i + \alpha_{i+1}s_{i+1} + \cdots + \alpha_k s_k.\end{aligned}$$

3.3 Let i and j be two positions in an array $A[1 \,..\, n]$ that is k ordered and h ordered, with k and h relatively prime, with $j - i > R(h, k) = (h-1)(k-1) - 1$. The hint asserts that $j - i$ can be represented in the form $ah + bk$, for nonnegative integers a and b. This implies that $A[i] \le A[j]$, as we can go back a hops of h (along the h-ordered file to which $A[j]$ belongs) to reach $A[j - ah] \le A[j]$, then go further back b hops of k (along the k-ordered file to which $A[j - ah]$ belongs), to reach $A[i] \le A[j - ah]$.

In the g-stage, we use LINEAR INSERTION SORT along files of g keys apart. Starting at arbitrary j, we reach $i = j - sg$ in s steps. Then, for s sufficiently large, $j - i = sg \ge (h-1)(k-1)$, and $A[i] \le A[j]$. The minimum s sufficient to reach a stopper is therefore $s_{min} = \lceil (k-1)(h-1)/g \rceil$. In the g stage, each of the n keys needs at most s_{min} comparisons to be inserted.

3.4 The h_s-stage orders h_s arrays of sizes $n/h_s + O(1)$ by LINEAR INSERTION SORT. Each such subarray requires $O(n^2/h_s^2)$ compasrisons by the results of Section 2.3; the h_s-stage calls for at most $O(n^2/h_s)$. For the $k_n, k_n - 1, \ldots, \lfloor k_n/2 \rfloor$ stages, a total of $O(n^2/(2^{k_n} - 1) + \cdots + n^2/(2^{\lfloor \frac{1}{2}k_n \rfloor} - 1))$ comparisons are made. The function inside O is bounded by

$$\begin{aligned}\frac{2n^2}{2^{\lfloor \lg n \rfloor}} + \frac{2n^2}{2^{\lfloor \lg n \rfloor - 1}} + \cdots + \frac{2n^2}{2^{\left\lfloor \frac{1}{2} \lfloor \lg n \rfloor \right\rfloor}} &\le \frac{2n^2}{2^{\lfloor \lg n \rfloor}} \left(1 + 2 + \cdots + 2^{\left\lceil \frac{1}{2} \lfloor \lg n \rfloor \right\rceil}\right) \\ &\le 4n \times 2^{\left\lceil \lfloor \frac{1}{2} \lg n \rfloor \right\rceil + 1} \\ &\le 16 n \sqrt{n}.\end{aligned}$$

Each of the h_s-stages with $h_s \in \{2^{\lfloor k_n/2 \rfloor} - 1, \ldots, 1\}$, makes $O(h_{s+2} h_{s+1} n / h_s) = O(2^{s+2} \times 2^{s+1} n/(2^s - 1)) = O(2^s n)$ by Exercise 3.3. Combining the $\lfloor k_n/2 \rfloor - 1$, $\lfloor k_n/2 \rfloor - 2, \ldots, 1$ stages we have the bound

$$O(n\sqrt{n}) + O\big((1 + 2 + 4 + \cdots + 2^{\lfloor \frac{1}{2} \lfloor \lg n \rfloor \rfloor - 1})n\big) = O(n^{3/2}).$$

CHAPTER 4

4.1 The algorithm BUBBLE SORT (as implemented in Figure 4.3) handles repeated data correctly and does not require any modification. Sooner or later, replicated data will come together and will be considered good pairs.

4.2 Yes, BUBBLE SORT is a stable sorting algorithm—a good pair is left unchanged in case of equality on a primary key. If the two keys in a good pair are equal on the primary keys but are ordered according to a secondary key, the two keys stay in the same order after they are compared.

4.3 The permutation

$$(1, 2, 3, \ldots, n - P_n, n, n - 1, n - 2, \ldots, n - P_n + 1)$$

requires P_n passes and corresponds to the sequence of assignments $n, n - 1$, $n - 2, \ldots, n - P_n + 1$.

The permutation

$$(P_n, P_n - 1, \ldots, 1, P_n + 1, P_n + 2, \ldots, n)$$

requires P_n passes and corresponds to the sequence of assignments n, $P_n - 1$, $P_n - 2, \ldots, 2, 1$.

4.4 Implement the sequence $t_1, \ldots, t_k$ as an array $t[1..k]$. Go through $k - 1$ passes, where at the jth pass, the increment t_{k-j+1} is used to compare and exchange keys that are t_{k-j+1} apart. A final pass with $t_1 = 1$ amounts to an execution of regular BUBBLE SORT. We assume that BUBBLE SORT (the algorithm of Figure 4.3) is implemented as a parameterless sort procedure that sets up all its local variables and accesses the subject array $A[1 \,..\, n]$ globally:

```
        for j ← 1 to k − 1 do
            begin
                h := t[k − j + 1];
                for i ← 1 to n − h do
                    if A[i] > A[i + h] then
                        swap(A[i], A[i + h]);
            end;
            call BubbleSort;
```

CHAPTER 5

5.1 The only change needed is to reduce the iterations of the outer loop. Replace the statement

for $i \leftarrow 1$ **to** $n - 1$**do**

by

for $i \leftarrow 1$ **to** k **do**.

The algorithm deterministically makes

$$(n - 1) + (n - 2) + \cdots + (n - k) = nk - \frac{1}{2}k(k + 1)$$

comparisons.

5.2 SELECTION SORT is oblivious to the information gained over the stages. After it has already progressed to the stage where the first $j - 1$ order statistics (ranked $i_1, \ldots, i_{j-1}$) have been found and it is about to embark on finding the i_jth order statistic, the algorithm will still handle the task by finding the $i_j - i_{j-1}$ smallest among the remaining $n - i_{j-1}$ keys in the manner of Exercise 5.1. This can be directly implemented by invoking the adaptation in Exercise 5.1 to find the i_kth order statistic. Such an algorithm will find the first i_k smallest (including the keys ranked $i_1, \ldots, i_k$) in a given set of n keys. At termination, the required order statistics will be respectively at positions $i_1, \ldots, i_k$ of the array. The algorithm makes $ni_k - \frac{1}{2}i_k(i_k + 1)$ comparisons.

5.3 Let $Y_n^{(k)}$ be the random number of comparisons to select the random set. Let the random ranks be $i_1, \ldots, i_k$. The distribution of the maximal rank i_k is then

$$\mathbf{Prob}\{i_k = j\} = \frac{\binom{j-1}{k-1}}{\binom{n}{k}}.$$

According to the result of Exercise 5.2,

$$Y_n^{(k)} = n i_k - \frac{1}{2} i_k (i_k + 1).$$

(a) Averaging:

$$\mathbf{E}\big[Y_n^{(k)}\big] = \left(n - \frac{1}{2}\right) \mathbf{E}[i_k] - \frac{1}{2}\mathbf{E}[i_k^2],$$

involving the first two moments of i_k. These are

$$\mathbf{E}[i_k] = \sum_{j=k}^{n} j \frac{\binom{j-1}{k-1}}{\binom{n}{k}} = \frac{k(n+1)}{k+1},$$

and

$$\mathbf{E}[i_k^2] = \sum_{j=k}^{n} j^2 \frac{\binom{j-1}{k-1}}{\binom{n}{k}} = \frac{k(kn+k+n)(n+1)}{(k+1)(k+2)}.$$

So,

$$\mathbf{E}\big[Y_n^{(k)}\big] = \frac{k(n+1)(3n+kn-2k-2)}{2(k+1)(k+2)}.$$

(b) The second moment of $Y_n^{(k)}$ involves third and fourth moments of i_k. These are

$$\mathbf{E}[i_k^3] = \frac{k(n+1)(k^2n^2 + 2k^2n + k^2 + 3kn - k + 3kn^2 + n + 2n^2)}{(k+1)(2+k)(k+3)},$$

and

$$\begin{aligned}\mathbf{E}[i_k^4] = k(n+1)(&15kn^2 + 12k^2n^2 + 2k^2n + k^3 - kn + 6n^2 \\ &- 5k^2 + 6n^3 + 11kn^3 + 6k^2n^3 + k^3n^3 \\ &+ 3k^3n^2 + 3k^3n)/\big[(k+1)(2+k)(3+k)(k+4)\big].\end{aligned}$$

The variance of $Y_n^{(k)}$ is therefore

$$\begin{aligned}\mathbf{Var}\big[Y_n^{(k)}\big] = k(n+1)(&13k^3n + 5kn^3 + 10n - 41kn^2 - 4k^4 - 10k \\ &+ 11n^3 - 18k^3 + 45kn - 21n^2 - 24k^2 + 48k^2n \\ &- 14k^2n^2)/\big[(2+k)^2(k+1)^2(3+k)(k+4)\big].\end{aligned}$$

(c) If $k = 1$, the order statistic $i_1 \stackrel{D}{=} U_n$ that has a discrete uniform distribution on $\{1, \ldots, n\}$. Distributionally,

$$Y_n^{(1)} \stackrel{D}{=} nU_n - \frac{1}{2}U_n(U_n + 1).$$

So,

$$F_n(x) \stackrel{\text{def}}{=} \mathbf{Prob}\left\{\frac{1}{n^2}Y_n^{(1)} \le x\right\} = \mathbf{Prob}\{U_n(2n - U_n - 1) \le 2xn^2\}.$$

Let $g_n(y) = y(2n - y - 1)$. The following figure illustrates this parabola:

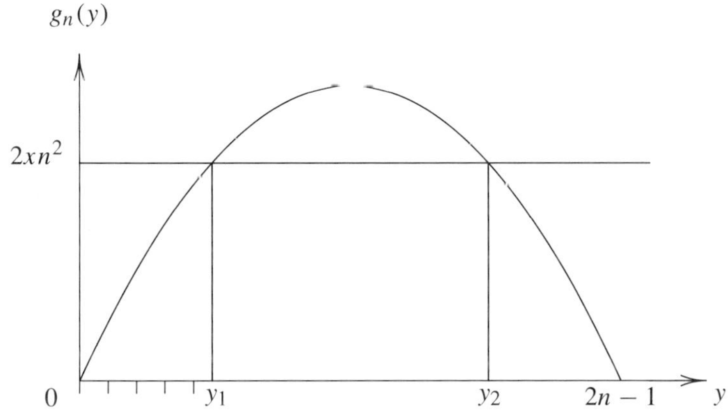

The function $g_n(y)$ is a parabola with apex at $y = \frac{1}{2}(2n - 1)$, and intersects the absissa at 0 and $2n - 1$. So, $g_n(U_n) \le 2xn^2$, if $U_n \le y_1$ or $U_n \ge y_2$, where $y_1 = y_1(n, x)$ and $y_2 = y_2(n, x)$ are solutions of the equation $g_n(y) = 2xn^2$. Of course, U_n is nonnegative, and the solution

$$y_1 = \frac{1}{2}(2n - 1) - \frac{1}{2}\sqrt{(2n - 1)^2 - 8xn^2}$$

corresponds to feasible values ($1 \le U_n \le \lfloor y_1 \rfloor$), whereas the solution

$$y_2 = \frac{1}{2}(2n - 1) + \frac{1}{2}\sqrt{(2n - 1)^2 - 8xn^2}$$

does not correspond to any feasible values of U_n ($n < \lceil y_2 \rceil < U_n \le 2n - 1$). Hence,

$$\begin{aligned} F_n(x) &= \mathbf{Prob}\{U_n \le y_1\} \\ &= \mathbf{Prob}\{U_n \le \lfloor y_1 \rfloor\} \end{aligned}$$

$$= \frac{1}{n}\lfloor y_1 \rfloor$$

$$= \frac{1}{2n}\Big[2n - 1 - \sqrt{(2n-1)^2 - 8xn^2}\,\Big] + O\Big(\frac{1}{n}\Big)$$

$$\to 1 - \sqrt{1 - 2x}.$$

The limiting density of $n^{-2}Y_n^{(1)}$is $1/\sqrt{1-2x}$.

5.4 Consider $n = 3$. Let Π_3 be a a random permutation of $\{1, 2, 3\}$, and let Π_3' be that permutation after it has been subjected to the stage of selecting the smallest element. The following table shows all the possible outcomes:

Π_3			R_{K_3-1}	Π_3'			U_2
1	2	3	0	1	2	3	1
1	3	2	0	1	3	2	2
2	1	3	1	1	2	3	1
2	3	1	1	1	3	2	2
3	1	2	1	1	3	2	2
3	2	1	2	1	2	3	1

One sees from the table that

$$\frac{2}{6} = \mathbf{Prob}\{R_{K_3-1} = 1, U_2 = 2\}$$

$$\neq \mathbf{Prob}\{R_{K_3-1} = 1\} \times \mathbf{Prob}\{U_2 = 2\}$$

$$= \frac{3}{6} \times \frac{3}{6}.$$

5.5 The average cost of one search for a randomly chosen key is $\frac{1}{n}(1 + 2 + \cdots + n) = \frac{1}{2}(n+1)$. On average, m searches cost $m(n+1)/2$. This cost is less than $n(n-1)/2$, the cost of SELECTION SORT, if $m \le n(n-1)/(n+1)$. For all $n \ge 2$, up to $m = \lfloor n(n-1)/(n+1) \rfloor = n - 2$ searches, repeated searching is cheaper. For more than $n - 2$ searches, SELECTION SORT will do better.

CHAPTER 6

6.1 COUNT SORT is stable; it does not move any data—if two keys are ordered according to a secondary key, they remain so ordered after the count.

6.2 The comparison

if $A[j] < A[i]$

accounts for each key $A[j]$ once its relatoin to $A[i]$, favoring the higher index in case of a tie. Let $i_1, i_2, \dots, i_k$ be the positions of all the repeats of a certain value. Suppose there are $c - 1$ keys smaller than $A[i_1]$. At the end of the algorithm $count[i_i] = c$, position i_1 has never been favorite in breaking a tie. Position i_2 has been favored only once (the ties between $A[i_1]$ and $A[i_2]$), and lost all decisions breaking the other ties (the ties between $A[i_2]$ and $A[i_r]$, $r = 3, \dots, k$). The argument extends to positions $i_3, \dots, i_k$: position i_r has been favored only $r - 1$ times (the ties between $A[i_s]$ and $A[i_r]$, $s = 1, \dots, r - 1$)), and lost all decisions breaking the other ties (the tie between $A[i_r]$ and $A[i_s]$, $s = r + 1, \dots, k$). At the end of the count, $count[i_r] = c$ plus the number of times $A[i_r]$ has been favored in tie breaking, that is, $count[i_r] = c + r - 1$. The algorithm assigns these repeats the ranks $c, c + 1, \dots, c + k - 1$. The next value larger than these repeats will be considered the $(c + k)$th largest, as it should be.

6.3 The rate of convergence is $O(1/\sqrt{n}\,)$. This can be found from a result in Section 2.3 for the number of comparisons, C_n, of LINEAR INSERTION SORT on n random keys. In Section 2.3 we established that

$$C_n = n + Y_n,$$

where Y_n is the number of inversions in a random permutation of $\{1, \dots, n\}$. Recall that the number of jumps J_n is the same as Y_n. Thus,

$$\frac{J_n - \frac{1}{4}n^2}{n^{3/2}} \stackrel{D}{=} \frac{C_n - \frac{1}{4}n^2}{n^{3/2}} - \frac{n}{n^{3/2}}.$$

There are two components on the right hand side. The normalized number of comparisons of LINEAR INSERTION SORT converges in distribution to $\mathcal{N}(0, \frac{1}{36})$ at the rate of $O(1/\sqrt{n}\,)$, as discussed in Section 2.3. The deterministic component $n/n^{3/2} = 1/\sqrt{n}$ converges to 0 (of course at the rate of $1/\sqrt{n}\,$). Combined, the two components approach $\mathcal{N}(0, \frac{1}{36})$ in distribution at the rate of $O(1/\sqrt{n}\,)$.

CHAPTER 7

7.1 None.

7.2 Implement the algorithm on $A[1\,..\,n+1]$, with $A[1\,..\,n]$ holding the raw data and $A[n+1]$ initialized to ∞. While handling the stretch $A[i\,..\,j]$, take the pivot as $A[i]$. Let i and $j + 1$ be respectively the initial values for F and B, then let these pointers drift as described:

(a)

```
pivot ← A[i];
F ← i;
B ← j + 1;
repeat F ← F + 1 until A[F] > pivot or F ≥ j;
repeat B ← B − 1 until A[B] ≤ pivot;
```

```
while F < B do
    begin
        swap(A[F], A[B]);
        repeat F ← F + 1 until A[F] > pivot;
        repeat B ← B − 1 until A[B] ≤ pivot;
    end;
swap(A[i], A[B]);
```

(b) When QUICK SORT gets to work recusively on $A[\ell\,..\,u]$, it has already moved a pivot at the previous level of recursion to $A[u+1]$, or $A[u+1]$ is ∞ in the extremal case $u = n$. The key at $A[u+1]$ is larger than all the keys of $A[\ell\,..\,u]$. So, $A[u+1]$ is the sentinel at no extra effort.

(c) Suppose the pointer F moves a total distance f, and B moves a total distance b. Each move corresonds to a comparison. At termination of PARTITION, F and B have just crossed. Regardless of where the crossing point is, we have committed to $f + b = f + (n - f + 1) = n + 1$ comparisons.

(d) In the best case for swaps (data in increasing order), one swap is made (to exchange the pivot with itself). In the worst case for swaps, for every pair $A[F]$ and $A[B]$ considered there will be a swap. One goes half the way always swapping, then one last swap follows to put the pivot where it belongs. The number of swaps in this worst case is $1 + \lfloor\frac{1}{2}(n-1)\rfloor$. An instance of bad data for swaps is $\lceil\frac{1}{2}n\rceil, n, n-1, \dots, \lceil\frac{1}{2}n\rceil + 1, \lceil\frac{1}{2}n\rceil - 1, \lceil\frac{1}{2}n\rceil - 2, \dots, 1$.

Let S_n be the number of swaps that Sedgewick's algorithm performs on a random permutatoin $(\pi_1, \dots, \pi_n)$. Then

$$S_n = 1 + \mathbf{1}_{\{\pi_2 > \pi_n\}} + \mathbf{1}_{\{\pi_3 > \pi_{n-1}\}} + \cdots + \mathbf{1}_{\{\pi_{\lceil \frac{n}{2}\rceil} > \pi_{\lceil \frac{n+1}{2}\rceil + 1}\}};$$

each indicator is a BERNOULLI($\frac{1}{2}$) random variable, and the indicators are i.i.d. The average number of swaps is

$$\mathbf{E}[S_n] = 1 + \frac{1}{2}\left(\left\lceil\frac{n}{2}\right\rceil - 1\right) \sim \frac{1}{4}n.$$

The variance is

$$\mathbf{Var}[S_n] = \frac{1}{4}\left(\left\lceil\frac{n}{2}\right\rceil - 1\right) \sim \frac{1}{8}n.$$

The central limit tendency

$$\frac{S_n - \frac{1}{4}n}{\sqrt{n}} \xrightarrow{\mathcal{D}} \mathcal{N}\Big(0, \frac{1}{8}\Big)$$

follows from the Central Limit Theorem for i.i.d. random variables.

7.3 Let $k_n = n - \lfloor n^\alpha \rfloor$, for $0 \le \alpha \le 1$. Let $(\pi_1, \pi_2, \dots, \pi_{k_n - 1})$ be the best-case permutation on $\{1, \dots, k_n - 1\}$ for QUICK SORT. The permutation $(\pi_1, \pi_2, \dots, \pi_{k_n-1}, k_n, k_n + 1, \dots, n)$ uses the largest elements n, $n-1, \dots, k_n$ as pivots (in this order), accumulating $(n-1) + (n-2) + \cdots + k_n - 1 = \frac{1}{2}[(n^2 - n) - (k_n - 2)(k_n - 1)] = \Theta(n^{\alpha+1})$ comparisons in the first $n - k_n + 1$ stages. Each stage isolates the largest element in what is left, and leaves the rest intact. When $(\pi_1, \pi_2, \dots, \pi_{k_n-1})$ is left, QUICK SORT finishes off with additional $\Theta(k_n \ln k_2) = \Theta(n \ln n)$ comparisons. The total number of comparisons is $\Theta(n^{\alpha+1} + n \ln n)$. For instance, QUICK SORT performs in $\Theta(n^{3/2})$ on the construction when $k_n = n - \lfloor \sqrt{n} \rfloor$.

7.4 The sample space Ω_n consists of $n!$ points (permutations); each is a permutation ω of $\{1, \dots, n\}$. QUICK SORT takes the sample point ω and partitions it into $\omega = \omega' \, p \, \omega''$, where p is the pivot and ω' is a permutation of $\{1, \dots, p-1\}$, and ω'' is a permutation of $\{p+1, \dots, n\}$. The permutation ω' gives rise recursively to C_{p-1} comparisons, which is thus well defined on Ω_n. Similarly, ω'' gives rise to C_{n-p} comparisons, which is also well defined on Ω_n.

7.5 Taking variances

$$\mathbf{Var}[C] = \mathbf{E}\big[C^2\big] = \mathbf{E}\big[U^2 C^2\big] + \mathbf{E}\big[(1-U)^2 \tilde{C}^2\big] + \mathbf{E}\big[G^2(U)\big];$$

cross products disappeared because U, C and $\tilde{C}$ are independent, and $\mathbf{E}[C] = \mathbf{E}[\tilde{C}] = 0$. Hence

$$\begin{aligned}\mathbf{Var}[C] &= \mathbf{E}\big[U^2\big]\mathbf{E}\big[C^2\big] + \mathbf{E}\big[(1-U)^2\big]\mathbf{E}\big[\tilde{C}^2\big] + \mathbf{E}\big[G^2(U)\big] \\ &= 2\mathbf{E}\big[U^2\big]\mathbf{Var}\big[C\big] + \mathbf{E}\big[G^2(U)\big].\end{aligned}$$

Reorganize as

$$\mathbf{Var}[C] = \frac{\int_0^1 G^2(u)\,du}{1 - 2\int_0^1 u^2\,du} = 7 - \frac{2\pi^2}{3}.$$

Convergence of the second-order Wasserstein distance implies

$$\mathbf{Var}\Big[\frac{C_n - 2n\ln n}{n}\Big] \to \mathbf{Var}[C],$$

or

$$\frac{1}{n^2}\mathbf{Var}[C_n] \to 7 - \frac{2\pi^2}{3}.$$

7.6 Let $d_2(F, G) = \inf_{X,Y} ||X - Y||_2 = \inf_{X,Y} \mathbf{E}[(X-Y)^2]$ be the second-order Wasserstein metric between distribution functions F and G, where $\inf_{X,Y}$ is taken over all pairs X, Y of random variables with distribution functions F and G, respectively.

(a) Under any reasonable measure, the distance between an object and itself must be 0. This is the case here—let V and W be two identically distributed random variables with distribution function F; in the infimum, the choice $X = Y$ yields 0, which cannot be reduced as $d_2(F, F)$ is non-negative. So, $d_2(F, F) = 0$.

(b) Let F_Z be the distribution function of a random variable Z. Then

$$\begin{aligned} d_2(F_U, F_{W_n}) &= \inf_{X \overset{D}{=} Y \overset{D}{=} U} \mathbf{E}\Big[\Big(X - Y\Big(1 + \frac{1}{n}\Big)\Big)^2\Big] \\ &\le \inf_{X \overset{D}{=} U} \mathbf{E}\Big[\Big(X - X\Big(1 + \frac{1}{n}\Big)\Big)^2\Big] \\ &= \frac{1}{n^2} \inf_{X \overset{D}{=} U} \mathbf{E}[X^2] \\ &= \frac{1}{n^2} \int_0^1 u^2\, du \\ &= \frac{1}{3n^2} \\ &\to 0, \end{aligned}$$

a sufficient condition for $W_n \xrightarrow{\mathcal{D}} U$.

7.7 Let F_c be the (degenerate) distribution function of the constant c and G be that of an arbitrary random variable with mean μ and second moment s. Then:

$$\begin{aligned} d_2(F_c, G) &= \inf_{X,Y} \mathbf{E}\big[(X - Y)^2\big] \\ &= \inf_{X} \mathbf{E}\big[(X - c)^2\big] \\ &= s - 2\mu c + c^2. \end{aligned}$$

Set to 0 the derivative of $d_2(F_c, G)$ with respect to c, to obtain $c = \mu$. The constant μ is the closeset number in the Wasserstein sense to a random variable with mean μ.

7.8 To identify the minimum, FIND must go through the partitioning stage (requiring $n - 1$ comparisons), then it makes a definte move to the left (to locate the minimum among the first P_n keys), where P_n is the landing position of the pivot. The random number of comparisons to identify the minimum satisfies the functional equation

$$Q_n \overset{D}{=} n - 1 + Q_{P_n - 1}.$$

Taking averages,

$$\mathbf{E}[Q_n] = n - 1 + \mathbf{E}[Q_{P_n - 1}]$$

$$= n - 1 + \sum_{p=1}^{n} \mathbf{E}[Q_{P_n - 1} \mid P_n = p]\,\mathbf{Prob}\{P_n = p\}$$

$$= n - 1 + \frac{1}{n} \sum_{p=1}^{n} \mathbf{E}[Q_{p-1}].$$

The telescoping nature of this recurrence suggests differencing:

$$n\mathbf{E}[Q_n] - (n-1)\mathbf{E}[Q_{n-1}] = 2(n-1) + \mathbf{E}[Q_{n-1}],$$

from which we have

$$\begin{aligned} \mathbf{E}[Q_n] &= \frac{2(n-1)}{n} + \mathbf{E}[Q_{n-1}] \\ &\quad - 2\sum_{j=1}^{n} \frac{j-1}{j} \\ &= 2n - 2H_n \\ &\sim 2n. \end{aligned}$$

The calculation of the variance is similar. We only outline the steps of this longer task. Square the functional equation and take expectation. The expected value of Q_n appears and we can use the value $\mathbf{E}[Q_n]$ already obtained. What we have is a telescoping recurrence, which can be solved by iteration, just as in the case of the mean.

7.9 As established in Exercise 7.8,

$$Q_n \stackrel{D}{=} n - 1 + Q_{P_n - 1}.$$

Scaling with the asymptotic order of the variance:

$$\frac{Q_n}{n} \stackrel{D}{=} \frac{(n - 1 + Q_{P_n-1})}{n} = 1 - \frac{1}{n} + \frac{Q_{P_n-1}}{P_n - 1} \times \frac{P_n - 1}{n}.$$

If Q_n/n converges to a limit Q, then so does $Q_{P_n-1}/(P_n - 1)$, because $P_n \xrightarrow{\text{a.s.}} \infty$. With $(P_n - 1)/n$ converging in distribution to U, a UNIFORM(0, 1) random variable, the functional equation for the limit might be

$$Q \stackrel{D}{=} UQ + 1.$$

This guessed solution can now be established by showing that the second-order Wasserstein distance between the distribution functions of Q_n/n and Q converges to 0, with computations similar to those preceding and included in the proof of Theorem 7.4.

7.10 Let $\phi(t)$ be the characteristic function of Q. According to the functional equation established in Exercise 7.9,

$$\begin{aligned}\phi(t) &= \mathbf{E}\big[e^{itQ}\big]\\ &= \mathbf{E}\big[e^{it(1+UQ)}\big]\\ &= e^{it}\mathbf{E}\big[e^{itUQ}\big]\\ &= e^{it}\int_0^1 \mathbf{E}\big[e^{ituQ}\big]\,du\\ &= e^{it}\int_0^1 \phi(tu)\,du.\end{aligned}$$

Change the variable of integration to $v = tu$, to obtain

$$te^{-it}\phi(t) = \int_0^t \phi(v)\,dv.$$

Take derivatives with respect to t, to get the differential equation

$$\phi'(t) = \frac{e^{it} - 1 - it}{t}\phi(t),$$

whose solution is

$$\phi(t) = \exp\Big(\int_0^t \frac{e^{ix} - 1 - ix}{x}\,dx\Big).$$

The square of this characteristic function is the characteristic function of (randomized) FIND's normalized number of comparisons.

7.11 Suppose Y has bounded support. Then $\mathbf{Prob}\{|Y| > M\} = 0$, for some $M < \infty$. The infinite divisibility of Y allows us to find i.i.d. random variables $A_1, \ldots, A_n$, such that for any n, $Y = A_1 + \cdots + A_n$. The events $A_i > M/n$, $i = 1, \ldots, n$, imply that $Y > M$. So, if $\mathbf{Prob}\{A_1 > M/n\} = p > 0$, the probability

$$\begin{aligned}\mathbf{Prob}\{|Y| > M\} &\ge \mathbf{Prob}\{Y > M\}\\ &\ge \mathbf{Prob}\{A_1 > M/n, \ldots, A_n > M/n\}\\ &= \mathbf{Prob}\{A_1 > M/n\}\ldots\mathbf{Prob}\{A_n > M/n\}\\ &= p^n\\ &> 0,\end{aligned}$$

a contradiction. We must have $A_1 \le M/n$, almost surely. By independence, $\mathbf{Var}[Y] = \mathbf{Var}[A_1] + \cdots + \mathbf{Var}[A_n] = n\mathbf{Var}[A_1] \le n\mathbf{E}[A_1^2] \le n(M^2/n^2)$.

This is true for all n. Take limits to see that $\mathbf{Var}[Y] = 0$. This is impossible because we already know that $\mathbf{Var}[Y] = 7 - \frac{2}{3}\pi^2$.

7.12 When the characteristing function $\phi_X(t)$ is absolutely integrable, the random variable X is absolutely continuous and has a density given by the Fourier inversion formula

$$f_X(x) = \frac{1}{2\pi} \int_{\infty}^{\infty} e^{-itx} \phi_X(t)\, dt.$$

The integrability of $|\phi_Y(t)|$ can be seen from

$$\begin{aligned} \big|\phi_Y(t)\big| &= \left|\exp\left(2 \int_0^1 \frac{e^{itx} - 1 - itx}{x}\, dx\right)\right| \\ &= \exp\left(2\Re \int_0^t \frac{e^{iu} - 1 - iu}{u}\, du\right) \\ &= \exp\left(2 \int_0^t \frac{\cos u - 1}{u}\, du\right); \end{aligned}$$

the cosine integral in the exponentiation is a well known function with asymptotic expansion

$$\int_0^t \frac{\cos u - 1}{u}\, du = -\ln t - \gamma + O\left(\frac{1}{t}\right), \qquad \text{as } t \to \infty.$$

Hence,

$$\big|\phi_Y(t)\big| = \exp\left(-2\ln t - 2\gamma + O\left(\frac{1}{t}\right)\right) = \Theta\left(\frac{1}{t^2}\right),$$

and $\int_{-\infty}^{\infty} \big|\phi_Y(t)\big|\, dt < \infty$.

7.13 Bound the upper tail probability with the moment generating function as follows:

$$\begin{aligned} \mathbf{Prob}\{Y > y\} &= \mathbf{Prob}\{tY > ty\} \\ &= \mathbf{Prob}\big\{e^{tY} > e^{ty}\big\} \\ &\le \frac{\mathbf{E}\big[e^{tY}\big]}{e^{ty}} \qquad \text{(by Markov's inequality).} \end{aligned}$$

Get the required moment generating function from the characteristic function $\exp\left(\int_0^1 \frac{e^{itx}-1-itx}{x}\, dx\right)$ by replacing it with t. It follows from the convexity of $e^{tx} - 1$ that (for fixed t) the equation $e^{tx} - 1 = xe^t$ has two solutions, one of them is $x_1 = 0$, the other is $x_2 > 1$. Consequently, we have the inequality $\frac{e^{tx}-1-tx}{x} \le e^t - t$, in the interval $x \in (0, 1)$. Further,

$$\mathbf{Prob}\{Y > y\} \le \exp(e^t - t - yt).$$

Minimize the exponent by setting to 0 its derivative with respect to t.

7.14 Bound the lower tail probability with the moment generating function as follows:

$$\begin{aligned}\mathbf{Prob}\{Y < -y\} &= \mathbf{Prob}\{-tY > ty\}\\ &\le \frac{\mathbf{E}\big[e^{-tY}\big]}{e^{ty}}\\ &\le \exp\big(t^2 - yt\big).\end{aligned}$$

Minimize the exponent.

7.15 For $Z_n = \text{UNIFORM}[1\,..\,n]$,

$$\begin{aligned}\mathbf{Prob}\left\{\frac{Z_n}{n} \le z\right\} &= \mathbf{Prob}\{Z_n \le nz\}\\ &= \mathbf{Prob}\{Z_n \le \lfloor nz\rfloor\}\\ &= \frac{\lfloor nz\rfloor}{n}\\ &\to z,\end{aligned}$$

if $0 < z < 1$. The distribution function of Z_n/n is that of UNIFORM(0, 1). Hence P_n/n and M_n/n converge in distribution two independent uniforms.

Further, let $Y_n = \mathbf{1}_{\{M_n<P_n\}} P_n/n$, and compute its distribution function

$$\begin{aligned}\mathbf{Prob}\{Y_n \le z\} &= \sum_{1\le m,p\le n} \mathbf{Prob}\left\{\mathbf{1}_{\{M_n<P_n\}}\frac{P_n}{n} \le z \mid P_n = p,\, M_n = m\right\}\\ &\qquad\qquad \times \mathbf{Prob}\{P_n = p,\, M_n = m\}\\ &= \sum_{1\le m,p\le n} \mathbf{Prob}\big\{\mathbf{1}_{\{m<p\}}p \le nz\big\}\\ &\qquad\qquad \times \mathbf{Prob}\{P_n = p\}\mathbf{Prob}\{M_n = m\}\\ &= \frac{1}{n^2}\sum_{1\le m<p\le n} \mathbf{Prob}\{1\times p \le \lfloor nz\rfloor\}\\ &\qquad\qquad + \frac{1}{n^2}\sum_{1\le p\le m\le n} \mathbf{Prob}\{0 \le \lfloor nz\rfloor\}\\ &= \frac{1}{n^2}\sum_{p=1}^{n}\sum_{m=1}^{p-1} \mathbf{Prob}\{p \le \lfloor nz\rfloor\} + \frac{1}{n^2}\sum_{p=1}^{n}\sum_{m=p}^{n} 1\\ &= \frac{1}{n^2}\sum_{p=1}^{n}(p-1)\mathbf{Prob}\{p \le \lfloor nz\rfloor\} + \frac{1}{2n^2}n(n+1)\end{aligned}$$

$$
\begin{aligned}
&= \frac{1}{n^2}\sum_{p=1}^{\lfloor nz \rfloor}(p-1) + \frac{n+1}{2n} \\
&= \frac{\lfloor nz \rfloor\big(\lfloor nz \rfloor - 1\big)}{2n^2} + \frac{n+1}{2n} \\
&\to \frac{z^2+1}{2}.
\end{aligned}
$$

On the other hand, $Y = \mathbf{1}_{\{W<U\}}U$ has the distribution function

$$
\begin{aligned}
\mathbf{Prob}\{Y \le z\} &= \int_0^1 \mathbf{Prob}\big\{\mathbf{1}_{\{W<u\}}u < z\big\}\,du \\
&= \int_0^z \mathbf{Prob}\Big\{\mathbf{1}_{\{W<u\}} < \frac{z}{u}\Big\}\,du \\
&\qquad + \int_z^1 \mathbf{Prob}\Big\{\mathbf{1}_{\{W<u\}} < \frac{z}{u}\Big\}\,du \\
&= \int_0^z du + \int_z^1 \mathbf{Prob}\big\{\mathbf{1}_{\{W<u\}} = 0\big\}\,du \\
&= z + \int_z^1 \mathbf{Prob}\big\{W \ge u\}\,du \\
&= z + \int_z^1 (1-u)\,du \\
&= \frac{z^2+1}{2}.
\end{aligned}
$$

Hence, $Y_n \xrightarrow{\mathcal{D}} Y$.

7.16 Let $a_n^{(p)} = ||Y^{(p)} - Y_n^{(p)}||_2^2$. This upper bound on the second-order Wasserstein distance converges to 0. The case $a_n^{(1)}$ was studied at length. The computation here is similar and we shall only outline its salient points. Write

$$
\begin{aligned}
a_n^{(p)} &\stackrel{\text{def}}{=} ||Y_n^{(p)} - Y^{(p)}||_2^2 \\
&= \mathbf{E}\Big[\Big(\big(Y_n^{(p)} + 2H_p\big) - \big(Y^{(p)} + 2H_p\big)\Big)^2\Big] \\
&= \mathbf{E}\Big[\sum_{r=0}^{p}\Big(\frac{\lceil nU \rceil - 1}{n} Y^{(r)}_{\lceil nU \rceil - 1} - UY^{(r)}\Big)^2 K_r^{(p)} \\
&\qquad + \Big(\frac{n - \lceil nU \rceil}{n}\tilde{Y}^{(r)}_{n-\lceil nU \rceil} - (1-U)\tilde{Y}^{(p-r)}\Big)^2 K_r^{(p)} + o_p(1)\Big],
\end{aligned}
$$

which follows from the mutual exclusion of the indicators $K_r^{(p)}$ (the cross product $K_i^{(p)} K_j^{(p)} = 0$, when $i \neq j$) and the fact that the families $\{Y_n^{(p)}\}$ and $\{Y^{(p)}\}$ have 0 mean. The norm can be computed by conditioning on $\lceil nU \rceil$. One obtains

$$a_n^{(p)} = \frac{2}{n} \sum_{j=1}^{n} \sum_{r=0}^{p} \mathbf{E}\Big[\Big(\frac{j-1}{n} Y_{j-1}^{(r)} - \frac{j-1}{n} Y^{(r)}\Big)^2\Big] \mathbf{E}\big[K_r^{(p)} \mid \lceil nU \rceil = j\big] + o(1)$$

$$\leq \frac{2}{n} \sum_{j=1}^{n} \sum_{r=0}^{p} \Big(\frac{j-1}{n}\Big)^2 a_{j-1}^{(r)} + o(1).$$

It can be shown by a double induction on n and p that $a_n^{(p)} \to 0$, as $n \to \infty$: Assume $a_n^{(r)} \to 0$, as $n \to \infty$, for $r = 1, \ldots, p-1$. Then for p go through an induction on n similar to that in the proof of Theorem 7.4.

7.17 (a) The two harmonics in the first moment are symmetric. When summed over j, we can write

$$\sum_{j=1}^{n} \mathbf{E}[S_j] = 2\Big(\sum_{j=1}^{n} H_j\Big) - 2n = 2\Big(\sum_{j=1}^{n} \sum_{k=1}^{j} \frac{1}{j}\Big) - 2n.$$

Interchange the order of summation.

(b) Similar to (a).

(c) Let $m \geq 3$. The answer to Exercise 1.6.5 is developed for the general moment before obtaining the special cases of first and second moments. From the general expression there, uniformly in $j \leq n/2$ we have

$$\mathbf{E}[S_j^m] = 2 \sum_{k=1}^{j-2} \big[k^{m-2} + O(k^{m-3})\big] + \sum_{\substack{k=j-1 \\ k \neq 0}}^{n-j-1} \big[k^{m-2} + O(k^{m-3})\big]$$

$$+ 2j \sum_{\substack{k=j-1 \\ k \neq 0}}^{n-j-1} \big[k^{m-3} + O(k^{m-4})\big]$$

$$+ 2(n+1) \sum_{k=n-j}^{n-2} \big[k^{m-3} + O(k^{m-4})\big]$$

$$+ n^{m-1} + O(n^{m-2})$$

$$= \Big(\frac{1}{m-1} - \frac{2}{m-2}\Big)\big[j^{m-1} + (n-j)^{m-1}\big] + \Big(\frac{m}{m-2}\Big) n^{m-1} + O(n^{m-2}).$$

For n even,

$$\sum_{j=1}^{n/2} \mathbf{E}[S_j^m] = \Big(\frac{1}{m-1} - \frac{2}{m-2}\Big)\Big[\sum_{j=1}^{n/2} (n-j)^{m-1} + \sum_{j=1}^{n/2} j^{m-1}\Big]$$

$$+ \frac{1}{2}\Big(\frac{m}{m-2}\Big)n^m + O\big(n^{m-1}\big)$$

$$= \Big(\frac{1}{m-1} - \frac{2}{m-2}\Big)\left[\sum_{j=n/2}^{n-1} j^{m-1} + \sum_{j=1}^{n/2} j^{m-1}\right]$$

$$+ \frac{1}{2}\Big(\frac{m}{m-2}\Big)n^m + O\big(n^{m-1}\big)$$

$$= \Big(\frac{1}{m-1} - \frac{1}{m-2}\Big)\frac{n^m}{m} + \frac{1}{2}\Big(\frac{m}{m-2}\Big)n^m + O\big(n^{m-1}\big)$$

$$= \Big(\frac{1}{2} + \frac{1}{m-1}\Big)n^m + O\big(n^{m-1}\big).$$

By symmetry the sum over $j = \frac{1}{2}n + 1, \dots, n$ is the same. For n even the required result follows.

For n odd, verify that the asymptotic result is the same—only few lower-order terms are different. These are hidden in O anyway.

7.18 The reader is assumed familiar with sums of the form $S_k(m) = \sum_{j=1}^{k} j^m$, for small values of m:

(a) Write $H_r = \sum_{j=1}^{r} 1/j$. Multiply out and exchange the order of summation.

(b) Obtain the first sum by multiplying out and using $S_{p-1}(1)$ and $S_{p-1}(2)$. The second and third sums are symmetric and both are obtained as in (a).

The "squared" nature of $A(z)$ indicates that $A(z)$ is the square of some elementary function. One finds

$$A(z) = \Big(\sum_{p=0}^{\infty} (j+1)H_j z^j\Big)^2 = \frac{1}{(1-z)^4}\Big[z + \ln\Big(\frac{1}{1-z}\Big)\Big]^2.$$

The result is obtained upon extraction of coefficients.

7.19 Carefully review the proof of Theorem 7.10 and recall the definition of I_j and S_j from the pragraphs preceding the theorem. Let S_n be the number of partition steps to sort n random keys. Then A_n, the size of the MQS tree, is given by

$$A_n = \sum_{j=1}^{n} I_j,$$

with average

$$\mathbf{E}[A_n] = \sum_{j=1}^{n} \mathbf{E}[I_j] = \sum_{j=1}^{n} \mathbf{Prob}\{I_j = 1\}.$$

We have developed most of the answer already in proving Theorem 7.10. Compute $\mathbf{Prob}\{I_j = 1\}$ by conditioning on S_j:

$$\mathbf{E}[A_n] = \sum_{j=1}^{n}\sum_{k=0}^{n-1}\left[1 - \frac{\binom{n-k-1}{p}}{\binom{n}{p}}\right]\mathbf{Prob}\{S_j = k\}.$$

Note that this expression is the same as a similar expression in the computation of $\mathbf{E}[C_n^{(p)}]$, except that it is free of the multiplier k. Specifically after the manipulation via the identity for signless Stirling number one obtains

$$\mathbf{E}[A_n] = n - \sum_{j=1}^{n}\sum_{k=0}^{n-1}\left[\sum_{r=1}^{p}\frac{\left[{p \atop r}\right](n-p-k)^r}{\langle n-p+1\rangle_p}\right]\mathbf{Prob}\{S_j = k\};$$

compare with (7.16). The strategy then is the same, except that $\mathbf{E}[S_j^t]$ will appear instead of $\mathbf{E}[S_j^{t+1}]$ after expanding $(n-p-k)^{r-t}$ by the binomial theorem. Isolate the terms containing the zeroth and first moments of S_j, we obtain:

$$\begin{aligned}\mathbf{E}[A_n] = n - n\sum_{j=1}^{n}\sum_{r=1}^{p}\left[{p \atop r}\right]\frac{(n-p)^r}{\langle n-p+1\rangle_p}\\ +\sum_{j=1}^{n}\sum_{r=1}^{p}\left[{p \atop r}\right]\frac{r(n-p)^{r-1}}{\langle n-p+1\rangle_p}\mathbf{E}[S_j]\\ +\sum_{j=1}^{n}\sum_{r=1}^{p}\left[{p \atop r}\right]\sum_{t=2}^{r}(-1)^{t+1}\binom{r}{t}\\ \times\frac{(n-p)^{r-t}}{\langle n-p+1\rangle_p}\mathbf{E}[S_j^t].\end{aligned}$$

So,

$$\sum_{j=1}^{n}\sum_{r=1}^{p}\left[{p \atop r}\right]\frac{(n-p)^r}{\langle n-p+1\rangle_p} = 1 - \frac{p}{n}.$$

By the result of Exercise 7.17, the first moments contribute

$$\sum_{j=1}^{n}\sum_{r=1}^{p}\frac{\left[{p \atop r}\right] r(n-p)^{r-1}}{\langle n-p+1\rangle_p}\mathbf{E}[S_j] = \frac{1}{n}\Big(\sum_{j=1}^{n}\mathbf{E}[S_j]\Big)\Big(1-\frac{p}{n}\Big)\Big[p+O\Big(\frac{1}{n}\Big)\Big]$$
$$= \frac{1}{n}\big[2(n+1)H_n - 4n\big]\Big[p+O\Big(\frac{1}{n}\Big)\Big]$$
$$\sim 2pH_n + O(1).$$

By the result of Exercise 7.17, all higher moments contribute $O(1)$, in view of

$$\sum_{j=1}^{n}\frac{\mathbf{E}[S_j^t]}{n^t} = \frac{t+1}{t-1} + O\Big(\frac{1}{n}\Big).$$

Putting it all together,

$$\mathbf{E}[A_n] = 2pH_n + O(1).$$

Every partition step invokes comparisons of the pivot within the array of order statistics sought to search for a splitting position within that array. However, any reasonable search method (even an inefficient one like LINEAR SEARCH) requires at most p comparisons. If index comparisons are to be taken into account (as in the case of integer data) these comparisons may only add at most $2p^2H_n + O(1)$ (recall that p is fixed). This can only alter the second-order term in the total number of comparisons.

CHAPTER 8

8.1 Start with the exact distribution (8.2). Form the distribution function

$$\mathbf{Prob}\{P_n \le k\} = \sum_{p=1}^{k}\frac{(p-1)(n-p)}{\binom{n}{3}}$$
$$= \frac{6}{n(n-1)(n-2))}\sum_{p=1}^{k}(p-1)(n-p)$$
$$= \frac{3nk^2 - 3nk + 2k - 2k^3}{n(n-1)(n-2)}.$$

For $0 \le x \le 1$,

$$\mathbf{Prob}\Big\{\frac{P_n}{n} \le x\Big\} = \mathbf{Prob}\{P_n \le \lfloor xn \rfloor\}$$
$$= \frac{1}{n^3}\Big(1+O\Big(\frac{1}{n}\Big)\Big)\big(3n\lfloor xn\rfloor^2 - 3n\lfloor xn\rfloor + 2\lfloor xn\rfloor - 2\lfloor xn\rfloor^3\big)$$

$$= \frac{1}{n^3}\left(1 + O\left(\frac{1}{n}\right)\right)\left[3n\left(x^2n^2 + O(n)\right) + O(n^2)\right.$$
$$\left. -2\left(x^3n^3 + O(n^2)\right)\right]$$
$$= 3x^2 - 2x^3 + O\left(\frac{1}{n}\right)$$
$$\to 3x^2 - 2x^3.$$

Hence $P_n/n \xrightarrow{\mathcal{D}} L$, where L has the distribution function $3x^2 - 2x^3$, for $0 \le x \le 1$. The density of L is $6x(1-x)$.

8.2 According to PARTITION, deterministically the insertion of $A[1]$ into $A[s+1\,..\,n]$ costs $N_1 = n - s$ comparisons. There are N_2 keys that are at least as large as $A[1]$. The insertion of $A[2]$ into a subarray of N_2 elements will cost N_2 comparisons, and it will leave N_3 keys above $A[2]$, and so on. The insertion component of the sequential scheme makes $N_1 + N_2 + \cdots + N_s$ comparisons. On average, $\mathbf{E}[N_j]$ is about $(n-s) - (j-1)n/(s+1)$. The avearge number of comparisons to insert all the sample under the sequential scheme is therefore

$$s(n-s) - \frac{s(s-1)n}{2(s+1)} \sim \frac{1}{2}ns.$$

The choice $n = s\ln s$ optimizes the quantile scheme. With this choice, asymptotically the sequential scheme makes $\frac{1}{2}s^2\ln s$ comparisons, which is about $s/(2\lg n) \sim s/(2\ln s) \to \infty$ as much as the comparisons that the quantile scheme makes. With this choice of s, the overall number of comparisons is obtained as in Theorem 8.3, with $(n-s)\lg s$ replaced with $s(n-s) - \frac{s(s-1)n}{2(s+1)}$, which gives an expression asymptotic to $\frac{1}{2}ns \sim \frac{s}{2\lg n}n\lg n$, an almost quadratic order of magnitude that is higher than optimal sorting (because $s/\lg n \to \infty$ for this choice of s).

However, the sample size that minimizes the average cost in the insertion stage for the sequential scheme may be different from the best sample size for the quantile scheme. Toawrd a fairer comparison, let us minimize

$$g(s) = 2s\ln s + s(n-s) - \frac{s(s-1)n}{2(s+1)} - 2n(\ln n - \ln s) - 2(n-s),$$

the total number of comparisons in the sequential scheme. This procedure gives $\frac{1}{4}n - o(n)$ as an asymptotic choice of sample size, leading to a quadratic order of magnitude. Even the best choice in the sequential scheme cannot optimize SAMPLE SORT.

CHAPTER 9

9.1 One achieves the worst case in Floyd's algorithm when *every* tree root considered travels all the way to become a highest leaf in the subheap it is a root of. The arrangement

$$21 \quad 25 \quad 38 \quad 44 \quad 59 \quad 53 \quad 72 \quad 80 \quad 67 \quad 49$$

does that. With $1 + 2 + 2 + 4 + 6 = 15$ comparisons, Floyd's algorithm organizes these data into a heap.

9.2 An application of Floyd's algorithm on Π_4, a random permutation of $\{1, 2, 3, 4\}$, results in one of the following three heaps:

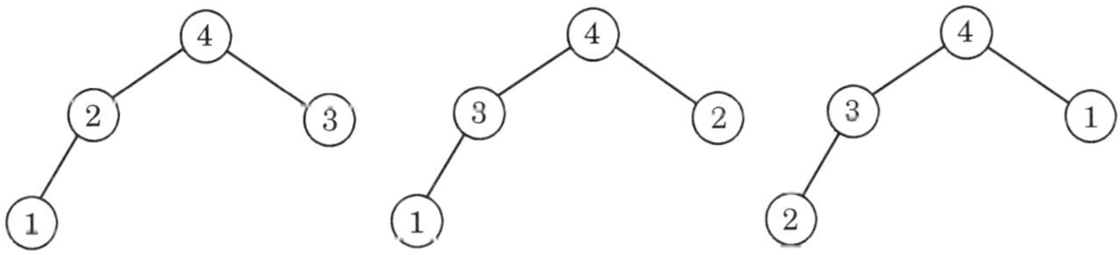

Exchange $A[1] = 4$ with $A[4]$ and let $\Pi'_3 = (A[1], A[2], A[3])$. Prune the highest node in the tree. This results in one of the trees:

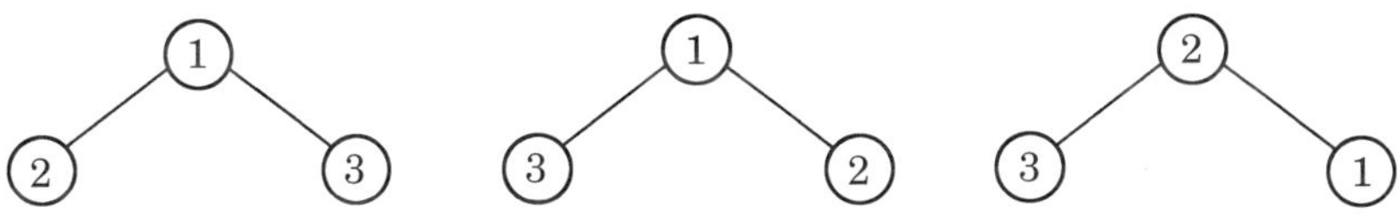

Certain configurations do not appear. For example, $\mathbf{Prob}\{\Pi'_3 = (3, 2, 1)\} = 0$; Π'_3 is *not* a random permutation.

9.3 Let S_k be the size of the subtree rooted at position k in the array. Let $\mathcal{H}_n$ be the number of heaps possible on n distinct keys. All these heaps are different admissible labelings of one complete binary tree. A heap of size n can be made up from a heap in the left subtree, a heap in the right subtree, and the largest element at the root. The labels of the left subheap can be chosen in $\binom{n-1}{S_2}$. Then

$$\mathcal{H}_n = \mathcal{H}_{S_2}\mathcal{H}_{S_3}\binom{n-1}{S_2}.$$

Iterating this recurrence, we get

$$\frac{\mathcal{H}_n}{n!} = \frac{1}{n} \times \frac{\mathcal{H}_{S_2}}{S_2!} \times \frac{\mathcal{H}_{S_3}}{S_3!}$$

$$= \frac{1}{S_1 S_2 S_3}\left[\frac{\mathcal{H}_{S_4}\mathcal{H}_{S_5}}{S_4!\,S_5!}\right]\left[\frac{\mathcal{H}_{S_6}\mathcal{H}_{S_7}}{S_6!\,S_7!}\right]$$

$$\vdots$$

$$= \frac{1}{\prod_{k=1}^{n} S_k}.$$

The number of heaps possible on n distinct keys is $n!$ divided by the product of the sizes of all subtrees in the structure.

CHAPTER 10

10.1 If merged by *LinearMerge*, the two lists

$$\begin{matrix} 1 & 3 & 5 & 7 \\ 2 & 4 & 6 & \end{matrix}$$

leave one key to the transfer stage. *LinearMerge* makes 6 comparisons to merge the two lists.

The two lists

$$\begin{matrix} 4 & 5 & 6 & 7 \\ 1 & 2 & 3 & \end{matrix}$$

leave 4 keys to the transfer stage. *LinearMerge* makes 3 comparisons to merge the two lists.

10.2 For only one key to be left over to the transfer stage, it must be the largest. It can stay at the tail of the n-long list, in which case choose the m-long list from the remaining $m + n - 1$ keys, or it can stay at the tail of the m-long list, in which case choose the n list from the remaining $m + n - 1$ keys. The number of such constructions is

$$\binom{m+n-1}{m} + \binom{m+n-1}{n}.$$

For n keys to be left over to the transfer stage, they must be the n largest. The first m smallest keys must go into the m-long list. Only one pair of lists is possible under this constraint.

10.3 Recall the relation between the leftover L_{mn} for the transfer stage and M_{mn} the number of comparisons for merging. Here we are dealing with the special case of two lists of sizes $\lfloor n/2 \rfloor$ and $\lceil n/2 \rceil$:

$$Y_n = M_{\lfloor n/2 \rfloor, \lceil n/2 \rceil} = \lfloor n/2 \rfloor + \lceil n/2 \rceil - L_{\lfloor n/2 \rfloor, \lceil n/2 \rceil}.$$

For $L_{\lfloor n/2 \rfloor, \lceil n/2 \rceil}$, we have an exact distribution (Equation (10.1)). Hence

$$\begin{aligned}\mathbf{Prob}\{n - Y_n \ge y\} &= \mathbf{Prob}\{L_{\lfloor n/2 \rfloor, \lceil n/2 \rceil} \ge y\} \\ &= \frac{\binom{n-y}{\lfloor n/2 \rfloor} + \binom{n-y}{\lceil n/2 \rceil}}{\binom{n}{\lfloor n/2 \rfloor}}.\end{aligned}$$

Work out factorial cancellations; for any fixed integer y apply Stirling's approximation to obtain

$$\begin{aligned}\mathbf{Prob}\{n - Y_n \ge y\} &= \frac{(n-y)!\,\lceil n/2 \rceil!}{n!\,(\lceil n/2 \rceil - y)!} + \frac{(n-y)!\,\lfloor n/2 \rfloor!}{n!\,(\lfloor n/2 \rfloor - y)!} \\ &\sim n^{-y} \lceil n/2 \rceil^{y} + n^{-y} \lfloor n/2 \rfloor^{y} \\ &\sim \frac{1}{2^{y-1}}.\end{aligned}$$

This is the tail distribution for GEOMETRIC(1/2), for each integer y.

10.4 Let the truth of the assertion for $0, 1, \ldots, n-1$ be our induction hypothesis. If n is even, we can remove the ceils in $\lceil n/2 \rceil$. The recurrence is

$$W_n = 2W_{n/2} + n - 1.$$

By hypothesis

$$\begin{aligned}W_n &= 2\Big(\frac{n}{2}\Big\lceil \lg \frac{n}{2} \Big\rceil - 2^{\lceil \lg(n/2) \rceil} + 1\Big) + n - 1 \\ &= n\lceil \lg n - 1 \rceil - 2 \times 2^{\lceil \lg n - 1 \rceil} + n + 1 \\ &= n\lceil \lg n \rceil - 2 \times 2^{\lceil \lg n \rceil - 1} + 1 \\ &= n\lceil \lg n \rceil - 2^{\lceil \lg n \rceil} + 1.\end{aligned}$$

The case n odd is handled similarly.

10.5 The algorithm divides the given list of n data into two halves of sizes $\lceil n/2 \rceil$ and $\lfloor n/2 \rfloor$. The best case arises in a data set when at any stage any pair of lists to be merged are the best cases for their respective sizes. Thus in the best case the division at the top level costs $\lfloor n/2 \rfloor$ comparisons, the least possible number of comparisons for a pair of sizes $\lceil n/2 \rceil$ and $\lfloor n/2 \rfloor$. After the division the cost in each half is for the best possible cases in these halves, giving the recurrence

$$B_n = B_{\lceil n/2 \rceil} + B_{\lfloor n/2 \rfloor} + \left\lfloor \frac{n}{2} \right\rfloor.$$

CHAPTER 11

11.1 Let U be a UNIFORM(0, 1] random variable and let us consider k to be the bucket number in question. Then

$$\begin{aligned}\mathbf{Prob}\{h(U) = k\} &= \mathbf{Prob}\{\lceil bU \rceil = k\} \\ &= \mathbf{Prob}\{k - 1 < bU \le k\} \\ &= \mathbf{Prob}\left\{\frac{k-1}{b} < U \le \frac{k}{b}\right\} \\ &= \frac{k}{b} - \frac{k-1}{b} \\ &= \frac{1}{b},\end{aligned}$$

a result that can be seen from the symmetry of the buckets. For n independent keys, the probability that they all hit the kth bucket is $1/b^n$.

11.2 Take the case $b = 2$. It may happen (with small probability) that the n uniform keys are all in the interval $(0, 2^{-(n+1)})$. At the top level of recursion, all the keys fall into the first bucket (the interval 1/2), then recursively again they fall together in the first bucket of the rescaled interval (buckets of length 1/4), and so on. After n recursive calls the data have not yet separated; they still fall in the first bucket of that level (of length $1/2^n$). The same can be done for any b.

11.3 (a) The saddle point equation (11.3) becomes:

$$z\Phi'(u, z) = \Phi(u, z).$$

The expansion technique that led to determining the saddle point at $z = 1 + O(1 - u)$ still holds verbatim without any changes.

(b) The uniform bound is sufficient, but not necessary. All we really need in the proof is a uniform bound not on Y_j itself, but rather on its mean and variance. It is an easy induction to show that $\mu_j \le 2j$ and $s_j \le 6j$. Thus both series in μ and σ^2 converge and one can accurately compute their values up to any number of places from exact recurrences.

11.4 The extra operations within buckets are recursively the same as those at the top level of recursion ($\alpha = 1$ in the basic recurrence (11.1)). The extra operations W_n satisfy Theorem 11.1. The total number of operations $C_n = W_n + n$

follows the central limit theorem

$$\frac{C_n - \mu n}{\sqrt{n}} \xrightarrow{\mathcal{D}} \mathcal{N}(0, \sigma^2),$$

with

$$\mu = 1 + e^{-1} \sum_{j=0}^{\infty} \frac{\mu_j}{j!};$$

$$\sigma^2 = e^{-1} \sum_{j=0}^{\infty} \frac{s_j}{j!} - (\mu - 1)^2,$$

and μ_j and s_j are the first and second moments of the number of operations in a bucket containing j keys. To obtain these we have exact recurrences. For example, conditioning on the binomially distributed share of the first bucket, we have

$$\begin{aligned}\mu_n &= n + n \sum_{j=0}^{n} \mathbf{E}[C_j \mid N_1 = j]\, \mathbf{Prob}\{N_1 = j\} \\ &= n + n \sum_{j=0}^{n} \mu_j \binom{n}{j} \frac{1}{n^j} \left(1 - \frac{1}{n}\right)^{n-j}.\end{aligned}$$

One obtains the numerical value $\mu = 2.302023901\ldots$. Similarly, by squaring the basic recurrence (11.1), one obtains a recurrence for the second moments (involving the means μ_j and the cross products $\mu_i \mu_j$) from which one gets the variance $\sigma^2 = 6.456760413\ldots$.

11.5 Find the minimum and maximum (*min* and *max* respectively) of the n data. Divide the interval $max - min$ into n equal intervals and apply the hash function $h(x) = \lceil (x - min)n/(max - min) \rceil$ for bucketing. Two of the points fall in the extreme first and last buckets. Apply the procedure recursively until all data are separated into individual buckets. Collect the sorted data from the buckets.

11.6 Approximate the binomial probability

$$\mathbf{Prob}\{B_n = j\} = \frac{1}{n^j}\left(1 - \frac{1}{n}\right)^{n-j} \binom{n}{j}$$

to the second-order asymptotic by expanding each term to second-order symptotic. We have

$$\binom{n}{j} = \frac{1}{j!} n(n-1)\ldots(n-j+1) = \frac{1}{j!}\Big[n^j - n^{j-1} \sum_{k=1}^{j-1} k + O\big(n^{j-2}\big)\Big].$$

We also have

$$\left(1-\frac{1}{n}\right)^{n-j} = e^{(n-j)\ln(1-1/n)}.$$

Expand the logarithm to get

$$\begin{aligned}\left(1-\frac{1}{n}\right)^{n-j} &= \exp\Big[-(n-j)\Big(\frac{1}{n}+\frac{1}{2n^2}+O\Big(\frac{1}{n^3}\Big)\Big)\Big]\\ &= e^{-1}\Big[1+\frac{2j-1}{2n}+O\Big(\frac{1}{n^2}\Big)\Big].\end{aligned}$$

Compose the binomial probability down to second-order asymptotics:

$$\mathbf{Prob}\{B_n = j\} = \frac{e^{-1}}{j!}\Big[1-\frac{j(j-1)}{2n}+O\Big(\frac{1}{n^2}\Big)\Big]\Big[1+\frac{2j-1}{2n}+O\Big(\frac{1}{n^2}\Big)\Big],$$

which expands to the desired result.

11.7 Each key is hashed twice; the extra cost in the buckets is

$$Z_n = I_1 Y^{(1)}_{N_1} + I_2 Y^{(2)}_{N_2} + \cdots + I_n Y^{(n)}_{N_n}$$

operations (each is in proportion α to hashing). See (11.10), with Y_j being the number of comparisons made by the selection-sort-like algorithm to find a randomly chosen key from among j items. So,

$$C_n = 2n + \alpha Z_n.$$

(a) Let $\mu_j = \frac{1}{3}(j^2-1)$ be the mean of Y_j (as developed in the text). Follow the text's development, but push for the next term in the approximation of binomial probabilities:

$$\begin{aligned}\mathbf{E}[C_n] &= 2n + \alpha\mathbf{E}[Z_n]\\ &= 2n + \alpha\sum_{k=0}^{\infty} k\sum_{j=1}^{\infty} j\mathbf{Prob}\{Y_j = k\}\,\mathbf{Prob}\Big\{\text{BINOMIAL}\Big(n,\frac{1}{n}\Big)=j\Big\}\\ &= 2n + \alpha\sum_{j=0}^{n} j\mu_j\frac{e^{-1}}{j!}\Big[1-\frac{j^2-3j+1}{2n}+O\Big(\frac{1}{n^2}\Big)\Big]\\ &= 2n + \frac{4\alpha}{3} - \frac{2\alpha}{n} + O\Big(\frac{1}{n^2}\Big).\end{aligned}$$

(b) Do the same for the variance to obtain

$$\mathbf{Var}[C_n] = \alpha^2\mathbf{Var}[Z_n] = \frac{589}{180}\alpha^2 - \frac{25}{2n}\alpha + O\Big(\frac{1}{n^2}\Big).$$

11.8 Operations within the bucket containing the required order statistic are hashing operations ($\alpha = 1$). Let $\phi_j(u)$ be the probability generating function of the cost C_n (whose average is μ_n). Then

$$\phi_n(u) = u^n \sum_{j=1}^{n} j\phi_j(u) \times \frac{1}{n^j}\Bigl(1 - \frac{1}{n}\Bigr)^{n-j} \binom{n}{j}.$$

The average satisfies the recurrence

$$\mu_n = \phi_n'(1) = n + \sum_{j=1}^{\infty} j\mu_j \mathbf{Prob}\Bigl\{\text{BINOMIAL}\Bigl(n, \frac{1}{n}\Bigr) = j\Bigr\}.$$

It can be seen from an induction that $\mu_j \le 2j$. The series converges quickly.

Use the finer approximation of the binomial probability in Exercise 11.6. From the recurrence, one can inductively compute $\mu_1, \mu_2, \mu_3, \ldots$ to provide as many terms as necessary to get each series in the expression

$$\mu_n = n + \sum_{j=0}^{n} j\mu_j \frac{e^{-1}}{j!} - \sum_{j=0}^{n} \frac{j\mu_j}{j!} e^{-1} \times \frac{j^2 - 3j + 1}{2n} + O\Bigl(\frac{1}{n^2}\Bigr)$$

accurately for as many digits as required. The series expansion of the average is

$$\mu_n = n + 3.011281835\ldots - \frac{1.75182021\ldots}{n} + O\Bigl(\frac{1}{n^2}\Bigr).$$

Similar work for the variance gives

$$\mathbf{Var}[C_n] \to 11.39004484\ldots.$$

11.9 Given that $Y_n/n \xrightarrow{P} \frac{1}{2}$, which can be expressed as $Y_n/n = \frac{1}{2} + o_P(1)$, with $o_P(1)$ being a quantity approaching 0 (in probability), we have

$$\frac{v(Y_n)}{v(n)} = \frac{Y_n\bigl(4.3500884\ldots + \delta(\lg Y_n)\bigr) + o_P(Y_n)}{n\bigl(4.3500884\ldots + \delta(\lg n)\bigr) + o(n)}.$$

The bounded periodic function δ is continuous, whence $\delta(\lg Y_n) = \delta(\lg(n/2)) + o_P(1)$. Also, by the periodicity of δ, the part $\delta(\lg(n/2))$ is $\delta(\lg n)$. Then,

$$\frac{v(Y_n)}{v(n)} = \frac{Y_n}{n} + o_P(1) \xrightarrow{P} \frac{1}{2}.$$

11.10 Differentiate the probability generating function in (11.12) and evaluate at $u = 1$, to get

$$\phi_n'(1) = \mathbf{E}[C_n] = 2n - 2 - \frac{1}{n}\sum_{k=2}^{n}(-1)^k \frac{k}{2^{1-k}-1}\binom{n}{k}.$$

The sum (let us call it S_n) is a good candidate for an application of Rice's method:

$$S_n = -\frac{1}{2\pi i}\oint_\Lambda f(z)\beta(-z, n+1)\,dz,$$

where Λ is the rectangle with corners at $3/2 \pm i$, and $n + 1 \pm i$; the part $f(k) = k/(2^{1-k} - 1)$ has the analytic continuation $f(z) = z/(2^{1-z} - 1)$, with simple poles at $z_k = 1 \pm 2\pi ik/\ln 2$, $k = 1, 2, \ldots$. The beta function doubles the pole $z_0 = 1$. Shift the contour in the usual way to a larger one whose limit encloses all the poles on the vertical line $\Re z = 1$, as well as the additional pole at 0 (coming from the beta function). Compute

$$S_n = \sum_{k=-\infty}^{\infty} \operatorname*{Res}_{z=z_k} f(z)\beta(-z, n+1) + o(1).$$

For $k \ne 0$, the residue at z_k is $-z_k\Gamma(-z_k)\Gamma(n+1)/(\Gamma(n+1-z_k)\ln 2) = \frac{1}{\ln 2}\Gamma(1-z_k)e^{z_k \ln n}(1+O(1/n))$, by Stirling's approximation. The residue at $z_0 = 1$ is $-\frac{n}{2}(1 + 2H_n/\ln 2)$. Assemble the result from these residues.

CHAPTER 12

12.1 Let $X_n^{(j)}$ be the number of comparisons to insert the jth entry. Then $X_n^{(j)} \overset{\mathcal{D}}{=}$ UNIFORM$[1\,..\,n + j]$, and

$$\frac{X_n^{(j)}}{n+j} \xrightarrow{\mathcal{D}} \text{UNIFORM}(0, 1),$$

for any fixed j, as was shown in Exercise 7.15. In fact, because j is asymptotically negligible relative to n, we have $(n + j)/n \to 1$, and we can write the simpler form

$$\frac{X_n^{(j)}}{n} \xrightarrow{\mathcal{D}} \text{UNIFORM}(0, 1),$$

by Slutsky's theorem. Let $C_n^{(k)} = X_n^{(1)} + \cdots + X_n^{(k)}$ be the cost of inserting k new entries, and let $U_1, \ldots, U_k$ be k independent UNIFORM(0, 1) random variables. Then

$$\frac{C_n^{(k)}}{n} \xrightarrow{\mathcal{D}} U_1 + U_2 + \cdots + U_k \stackrel{\text{def}}{=} A_k.$$

The moment generating function of the convolution A_k is the product of k identical moment generating functions of uniforms. This moment generating function is $[(e^t - 1)/t]^k$.

12.2 Let $(\pi_{i_1}, \ldots, \pi_{i_j})$ be the longest ascending subsequence of Π_n. If elements from $Y_n = \{1, \ldots, n\} - \{\pi_{i_1}, \ldots, \pi_{i_j}\}$ are taken out of Π_n, inversions will remain unless all the elements of Y_n are removed—if not, there is $y \in Y_n$ in the remaining data lying in between π_{i_r} and $\pi_{i_{r+1}}$, and $\pi_{i_r} > y$ or $y > \pi_{i_{r+1}}$. (If neither inequality is satisfied $\pi_{i_r} < y < \pi_{i_{r+1}}$ and $(\pi_{i_1}, \ldots, \pi_{i_r}, y, \pi_{i_{r+1}}, \ldots \pi_{i_j})$ is a longer ascending subsequence, a contradiction). Removal of $Y_n \cup Z$, for any nonempty $Z \subseteq \{\pi_{i_1}, \ldots, \pi_{i_j}\}$ will surely leave an ascending subsequence of Π_n, but it will be shorter than that obtained by removing Y_n only. The length of the ascending subsequence obtained by the removal of Y_n is $n - j$.

12.3 Suppose $(\pi_{i_1}, \ldots, \pi_{i_k}, \ldots, \pi_{i_r})$ and $(\xi_{j_1}, \ldots, \xi_{j_m}, \ldots, \xi_{j_s})$ are two distinct cycles of Π_n. The exchange of π_{i_k} and ξ_{i_m} in Π_n amalgamates the two cycles into the longer cycle

$$(\pi_{i_1}, \ldots, \pi_{i_{k-1}}, \xi_{j_m}, \xi_{j_{m+1}}, \ldots, \xi_{j_s}, \xi_{j_1} \ldots, \xi_{j_{m-1}}, \pi_{i_k}, \ldots, \pi_{i_r}).$$

The exchange of π_{i_μ} and π_{i_ν} in Π_n splits the cycle $(\pi_{i_1}, \ldots, \pi_{i_r})$ into the two cycles $(\pi_{i_1}, \ldots, \pi_{i_{\mu-1}}, \pi_{i_\nu}, \pi_{i_{\nu+1}}, \ldots, \pi_{i_r})$ and $(\pi_{i_\mu}, \pi_{i_{\mu+1}}, \ldots, \pi_{i_{\nu-1}})$.

The goal is to move up from $C(\Pi_n)$ cycles to n cycles (the maximal number, which is that of a sorted permutation) by exchanges. For the exchanges to be minimal, we must restrict them to the splitting type that increases the number of cycles. A cycle of length ℓ needs $\ell - 1$ splits to be decomposed into ℓ unit cycles. Thus $X(\Pi_n) = \sum_{j=1}^{C(\Pi_n)} (\ell_i - 1)$, with $\ell_1, \ldots, \ell_{C(\Pi_n)}$ being the lengths of the cycles of Π_n. Hence $X(\Pi_n) = n - C(\Pi_n)$. Distributionally, $(X_n - n + \ln n)/\sqrt{\ln n} = -(C_n - \ln n)/\sqrt{\ln n} \xrightarrow{\mathcal{D}} -\mathcal{N}(0, 1) \stackrel{D}{=} \mathcal{N}(0, 1)$.

12.4 Let B_n be the number of bits checked. The argument is the same as in the unbiased case to establish the recurrence

$$B_n = n + B_{N_1} + B_{N_2},$$

where N_1 and N_2 are the "shares" of the left and right subfiles after the splitting; the equation is valid for $n \geq 2$. Let $q = 1 - p$. The difference here is that if $p \neq q$, the shares are not symmetric; $N_1 \stackrel{D}{=}$ BINOMIAL(n, p). So, conditioning on N_1, we have

$$\mathbf{E}[B_n] = \sum_{j=0}^{n} \mathbf{E}\big[B_n \mid N_1 = j\big]\,\mathbf{Prob}\{N_1 = j\}$$

$$= n + \sum_{j=0}^{n} \binom{n}{j} p^j q^{n-j}\mathbf{E}[B_j] + \sum_{j=0}^{n} \binom{n}{j} p^j q^{n-j}\mathbf{E}[B_{n-j}].$$

Now go through the Poissonization routine formulating the generating function $B(z) = e^{-z}\sum_{n=0}^{\infty}\mathbf{E}[B_n]z^n/n!$. It is the same routine that led to (11.5); multiply by z^n, sum over $n \ge 2$, etc. You get

$$B(z) = B(pz) + B(qz) + z\big(1 - e^{-z}\big),$$

of which (11.5) is the special case $p = q = \frac{1}{2}$. Take the Mellin transform of the last recurrence to obtain

$$B(s) = \frac{\Gamma(s+1)}{p^{-s} + q^{-s} - 1},$$

existing in $-2 < \Re s < -1$, which has a double pole at -1 that gives the dominant asymptotic. All the other poles are simple and contribute $O(z)$ lower-order terms. Invert the transform:

$$B(z) \sim -\operatorname*{Res}_{s=-1} \frac{z^{-s}\Gamma(s+1)}{p^{-s} + q^{-s} - 1} = \frac{1}{r(p)} z \ln z,$$

where

$$r(p) = -(p \ln p + q \ln q)$$

is the so-called information entropy. De-Poissonization gives $\frac{1}{r(p)} n \ln n$ average for fixed n.

The coefficient $r(p)$ is a continuous function of p. So, as p drifts away slowly from 0.5, a small change will occur in the average. For example, $r(0.48) = -1.444362504\ldots$, which is reasonably close to the ideal unbiased case $r(1/2) = -1/\ln 2 = -1.442695041\ldots$; one may view RADIX SORT as a robust sorting procedure.

12.5 Let $T_n^{(j)}$ be the number of inversions created by the swap $Sw_{i,n-i+1}$. After the k swaps the relative ranks in the inner stretch (positions $k+1, \ldots, n-k$) remain a fresh random permutation on $n - 2k$ keys, with $I_{n-2k}^{(0)}$ inversions, giving rise to the convolution

$$I_n^{(k)} = T_n^{(1)} + T_n^{(2)} + \cdots + T_n^{(k)} + I_{n-2k}^{(0)}.$$

Let the probability generating function of $I^{(0)}_{n-2k}$ be $f_{n-2k}(z)$ and that of $T^{(j)}_n$ be $t^{(j)}_n(z)$. By independence,

$$f^{(j)}_n = t^{(1)}_n(z)t^{(2)}_n(z)\ldots t^{(k)}_n(z)f_{n-2k}(z).$$

The function $f_{n-2k}(z)$ is $(\prod_{j=1}^{n-2k}(1-z^j)/(1-z))/(n-2k)!$, according to (1.13). The swap $Sw_{i,n-i+1}$ acts on a fresh random permutation in the stretch $i, \ldots n-i+1$ and does not unwind the effect of any of the swaps $Sw_{1,n}, \ldots, Sw_{i-1,n-i+2}$. So, $t^{(j)}_n(z)$ is the same as (12.5), with $n-2j+2$ replacing n.

12.6 (a) Take the first and second derivatives at $z = 1$ of the exact probability distribution function (12.5) to show that

$$\mathbf{E}[T_n] = \frac{2}{3}n - \frac{4}{3} \sim \frac{2}{3}n;$$

and

$$\mathbf{Var}[T_n] - \frac{1}{18}n^2 \quad \frac{1}{18}n - \frac{1}{9} \sim \frac{1}{18}n^2,$$

(hence the center and scale factors).

(b) Let $T^*_n = (T_n - 2n/3)/n$. From (12.5), $\mathbf{Prob}\{T_n = k\} = (k+1)/\binom{n}{2}$, for feasible k. Hence, for t in the range specified

$$\begin{aligned}
\mathbf{Prob}\{T^*_n \le t\} &= \mathbf{Prob}\Big\{T_n \le \frac{2}{3}n + tn\Big\} \\
&= \mathbf{Prob}\Big\{T_n \le \Big\lfloor\Big(\frac{2}{3}+t\Big)n\Big\rfloor\Big\} \\
&= \sum_{k=0}^{\lfloor(2/3+t)n\rfloor} \mathbf{Prob}\{T_n = k\} \\
&= \sum_{k=0}^{\lfloor(2/3+t)n\rfloor} \frac{k+1}{\binom{n}{2}} \\
&= \frac{\frac{1}{2}(\lfloor(\frac{2}{3}+t)n\rfloor + 1)(\lfloor(\frac{2}{3}+t)n\rfloor + 2)}{\frac{1}{2}n(n-1)} \\
&\to t^2 + \frac{4}{3}t + \frac{4}{9}.
\end{aligned}$$

The derivative of this limiting distribution function is the stated limit density within the given range for t.

Appendix: Notation and Standard Results from Probability Theory

A.1 LOGARITHMS

The logarithm of x to the base b is denoted by $\log_b x$. Bases in common use are 2, e, and 10, but other bases appear in the book as well. The *natural logarithm* of x, $\log_e x$, is denoted by $\ln n$. Base-2 logarithms appear in almost every aspect of algorithmics; they have been christened with a special symbol: $\log_2 x$ is denoted by $\lg x$.

A.2 ASYMPTOTICS

One says that a function $f(x)$ is $O\big(g(x)\big)$ (pronounced $f(x)$ is big-oh $g(x)$) over the set A, if for some $c > 0$,

$$\left|f(x)\right| \leq c \left|g(x)\right|, \qquad \text{for all } x \in A.$$

The notation is sometimes referred to as Landau's O. The most common application in algorithmics is over a semi-infinite interval $A = [x_0, \infty)$, for some $x_0 \geq 0$. When such a semi-infinite range exists, the O notation is useful in identifying upper bounds on the growth rate of $f(x)$ by a suitable bounding function $g(x)$, for large x.

Likewise, one defines lower bounds—one says that a function $f(x)$ is $\Omega(g(x))$ over the set A, if for some $c > 0$,

$$\left|f(x)\right| \geq c \left|g(x)\right|, \qquad \text{for all } x \in A.$$

Thus, $f(x) = \Omega\big(g(x)\big)$, over A iff $g(x) = O\big(f(x)\big)$ over A.

One says that a function $f(x)$ is *of the same order of magnitude* as $g(x)$ (written $\Theta(g(x))$) over A, if for some $c_1, c_2 > 0$,

$$c_1 \left|g(x)\right| \leq \left|f(x)\right| \leq c_2 \left|g(x)\right|, \qquad \text{for all } x \in A.$$

Thus, $f(x) = \Theta\big(g(x)\big)$ over A, iff $f(x) = \Omega(g(x)\big)$, and $f(x) = O(g(x)\big)$ over A.

To identify orders that can be relatively ignored, one says that a function $f(x)$ is *negligible* in comparison to $g(x)$ (written $o(g(x))$), as x approaches x_0, if

$$\lim_{x \to x_0} \frac{f(x)}{g(x)} = 0.$$

To capture the essence of a function $f(x)$, one says $f(x)$ is *asymptotically equivalent* to $g(x)$, as x approaches x_0, if

$$\lim_{x \to x_0} \frac{f(x)}{g(x)} = 1.$$

Symbolically $f(x) \sim g(x)$, as $x \to 0$, denotes the relation "$f(x)$ is asymptotically equivalent to $g(x)$, as $x \to 0$."

A.3 HARMONIC NUMBERS

Harmonic numbers are ubiquitous in algorithmic analysis. The nth harmonic number of order k is the quantity $\sum_{j=1}^{n} 1/j^k$, and is often denoted by $H_n^{(k)}$; the superscript is usually dropped when it is 1. Harmonic numbers of the first two orders have the asymptotic approximations:

$$H_n = \ln n + \gamma + O\Big(\frac{1}{n}\Big);$$

$$H_n^{(2)} = \frac{\pi^2}{6} - \frac{1}{n} + O\Big(\frac{1}{n^2}\Big),$$

where $\gamma = 0.5772156\ldots$ is Euler's constant.

A.4 PROBABILITY

Many results from standard probability theory are used in the book. For a quick reference, we state in this appendix (without proof) the major results employed. A reader interested in the proofs and other background material may refer to one of the many standard textbooks on probability.

Let $X_1, X_2, \ldots$ be a sequence of random variables (measurable set functions defined on a sample space). We sometimes refer to this sequence or family as $\{X_i\}_{i=1}^{\infty}$ or more simply as $\{X_i\}$. We say that X_n *converges in probability* to X, if for any fixed $\varepsilon > 0$, we have

$$\lim_{n \to \infty} \mathbf{Prob}\{|X_n - X| > \varepsilon\} = 0.$$

When X_n converges in probability to X, we write $X_n \xrightarrow{P} X$.

Theorem A.1 *(The Weak Law of Large Numbers). If $X_1, X_2, \ldots$ are independent, identically distributed, absolutely integrable ($\mathbf{E}|X_1| < \infty$) random variables, with common mean μ, their partial sum $S_n = X_1 + X_2 + \cdots + X_n$ satisfies:*

$$\frac{S_n}{n} \xrightarrow{P} \mu.$$

We say that X_n *converges almost surely* to X, if

$$\mathbf{Prob}\Big\{\lim_{n\to\infty} X_n = X\Big\} = 1.$$

When X_n converges almost surely to X, we write $X_n \xrightarrow{\text{a.s.}} X$. We use the notation a.s. for equality almost surely.

Theorem A.2 *(The Strong Law of Large Numbers). If $X_1, X_2, \ldots$ are independent, identically distributed, absolutely integrable ($\mathbf{E}|X_1| < \infty$) random variables, with common mean μ, their partial sum $S_n = X_1 + X_2 + \cdots + X_n$ satisfies:*

$$\frac{S_n}{n} \xrightarrow{\text{a.s.}} \mu.$$

We say that X_n *converges in distribution* to X, if its distribution $F_{X_n}(x)$ tends to $F_X(x)$, the distribution function of X, at every continuity point x of F_X. When X_n converges in distribution to X, we write $X_n \xrightarrow{\mathcal{D}} X$, we reserve the notation $\overset{\mathcal{D}}{=}$ for exact equality in distribution.

Theorem A.3 *(Continuity Theorem). Let $X_1, X_2, \ldots$ be a sequence of random variables with corresponding characteristic functions $\phi_1(x), \phi_2(x), \ldots$. Let X be a random variable with distribution function $\phi(x)$. Then*

$$X_n \xrightarrow{\mathcal{D}} X \qquad \textit{iff} \qquad \phi_n(x) \to \phi(x).$$

Theorem A.4

$$X_n \xrightarrow{\text{a.s.}} X \qquad \textit{implies} \qquad X_n \xrightarrow{P} X;$$

$$X_n \xrightarrow{P} X \qquad \textit{implies} \qquad X_n \xrightarrow{\mathcal{D}} X;$$

for constant c,

$$X_n \xrightarrow{\mathcal{D}} c \qquad \textit{implies} \qquad X_n \xrightarrow{P} c.$$

The families $\{X_j\}_{j=1}^{\infty}$ and $\{Y_j\}_{j=1}^{\infty}$ are said to be *independent* when the members of one family are independent of each other as well as being independent of all the members of the other family. Formally, the families $\{X_j\}$ and $\{Y_j\}$ are independent if X_i and X_j are independent and Y_i and Y_j are independent for every distinct pair $i \neq j$, and X_i and Y_j are independent for every pair of indexes.

Let $X_1, X_2, \ldots$ be a sequence of random variables. Let the mean and variance of the nth random variable be defined by

$$\mu_n = \mathbf{E}[X_n], \qquad \mathbf{Var}[X_n] = \sigma_n^2.$$

Central limit theorems are concerned with the rate of convergence of the partial sums

$$S_n = X_1 + X_2 + \cdots + X_n$$

to the sum of their means. In fact, central limit theorems apply to doubly indexed arrays of random variables:

$$\begin{array}{lll} X_{11}, X_{12}, & \ldots & , X_{1,r_1} \\ X_{12}, X_{22}, & \ldots & , X_{2,r_2} \\ & \vdots & \end{array}$$

However, all the applications in this book assume that $X_{ij} \equiv X_j$, depending only on j. In these cases a central limit theorems states a result about the sequence $\{X_i\}$; we shall write below central limit theorems in a simpler form that assumes the array to be singly indexed. To state central limit theorems in their simplest forms we assume that the variances are always finite ($\sigma_n^2 < \infty$). Let

$$s_n^2 = \sum_{i=1}^{n} \sigma_i^2.$$

Theorem A.5 *(The i.i.d. Central Limit Theorem). If $X_1, X_2, \ldots$ are independent identically distributed random variables, with common mean μ and common variance $\sigma^2 < \infty$, then their partial sum $S_n = X_1 + X_2 + \cdots + X_n$ converges in distribution:*

$$\frac{S_n - \mu n}{\sqrt{n}} \xrightarrow{\mathcal{D}} \mathcal{N}(0, \sigma^2).$$

Theorem A.6 *(Lindeberg's Central Limit Theorem). Suppose $X_1, X_2, \ldots$ are independent, with corresponding distribution functions $F_1, F_2, \ldots$. If the family $\{X_i\}$ satisfies Lindeberg's condition*

$$\lim_{n \to \infty} \frac{1}{s_n^2} \sum_{i=1}^{n} \int_{|X_i| > \varepsilon s_n} X_i^2 \, dF_i = 0,$$

for every $\varepsilon > 0$, then

$$\frac{\sum_{i=1}^{n} X_i - \sum_{i=1}^{n} \mu_i}{s_n} \xrightarrow{\mathcal{D}} \mathcal{N}(0, 1).$$

Theorem A.7 *(Lyapunov's Central Limit Theorem). Suppose for some $\delta > 0$ the independent random variables $X_1, X_2, \ldots$ have finite absolute moment $\mathbf{E}|X_i|^{2+\delta}$, for all i. If the family $\{X_i\}$ satisfies Lyapunov's condition:*

$$\lim_{n\to\infty} \frac{\sum_{i=1}^{n} \mathbf{E}|X_n - \mu_n|^{\delta+2}}{s_n^{\delta+2}} = 0,$$

then

$$\frac{\sum_{i=1}^{n} X_i - \sum_{i=1}^{n} \mu_i}{s_n} \xrightarrow{\mathcal{D}} \mathcal{N}(0, 1).$$

Lyapunov's theorem uses a condition that is sometimes easier to check than Lindeberg's condition. Lyapunov's condition involves an absolute moment higher than the second. The third absolute moment is typical in applications ($\delta = 1$).

To do "algebra" on converging random variables by performing the algebra on the limits, we have the following theorems for addition and multiplication. Of course subtraction and division can be treated as special cases of addition and multiplication.

Theorem A.8 *If $X_n \xrightarrow{\text{a.s.}} X$, and $Y_n \xrightarrow{\text{a.s.}} Y$ then*

$$X_n + Y_n \xrightarrow{\text{a.s.}} X + Y;$$
$$X_n Y_n \xrightarrow{\text{a.s.}} XY.$$

Theorem A.9 *If $X_n \xrightarrow{P} X$, and $Y_n \xrightarrow{P} Y$ then*

$$X_n + Y_n \xrightarrow{P} X + Y;$$
$$X_n Y_n \xrightarrow{P} XY.$$

Theorem A.10 *Suppose $\{X_n\}_{n=1}^{\infty}$ and $\{Y_n\}_{n=1}^{\infty}$ are two families of random variables. Suppose further that X_i and Y_i are independent, for every i. Let X and Y be independent distributional limits: $X_n \xrightarrow{\mathcal{D}} X$, and $Y_n \xrightarrow{\mathcal{D}} Y$. Then*

$$X_n + Y_n \xrightarrow{\mathcal{D}} X + Y.$$

Theorem A.11 *(Slutsky's Theorem). Suppose $\{X_n\}_{n=1}^{\infty}$ and $\{Y_n\}_{n=1}^{\infty}$ are two families of random variables. If $X_n \xrightarrow{\mathcal{D}} X$, and $Y_n \xrightarrow{\mathcal{D}} c$, for a constant c, then*

$$X_n + Y_n \xrightarrow{\mathcal{D}} X + c;$$
$$X_n Y_n \xrightarrow{\mathcal{D}} cX.$$

A sequence of random variables $\{M_n\}_{n=1}^{\infty}$ is said to be a *martingale* with respect to the sequence of sigma fields $\{\mathcal{F}_n\}_{n=1}^{\infty}$ if:

(i) The sigma fields are increasing, that is, $\mathcal{F}_i \subseteq \mathcal{F}_{i+1}$.

(ii) X_n is a measurable function on $\mathcal{F}_n$.

(iii) For all $n \geq 1$, $\mathbf{E}|X_n| < \infty$.

(iv) $\mathbf{E}\big[X_n \mid \mathcal{F}_{n-1}\big] \overset{\text{a.s.}}{=} X_{n-1}$.

Theorem A.12 *(Martingale Convergence Theorem). A square-integrable Martingale with* $\sup_n \mathbf{E}[M_n^2] = K < \infty$ *converges almost surely (and in* L^2*) to an integrable limit.*

When $\{M_n\}$ is a martingale with respect to the sigma-fields $\{\mathcal{F}_n\}$, the backward differences $\triangledown M_n = M_n - M_{n-1}$ satisfy the property $\mathbf{E}[\triangledown M_n \mid \mathcal{F}_{n-1}] \overset{\text{a.s.}}{=} 0$. Generally, a sequence of random variables Y_n satisfying the property $\mathbf{E}[Y_n \mid \mathcal{F}_{n-1}] \overset{\text{a.s.}}{=} 0$ is called a *martingale difference*.

Analogous to Lindeberg's condition for the ordinary central limit theorems for doubly indexed arrays, there is a conditional Lindeberg's condition which, together with another conditional variance condition, are sufficient to derive a martingale central limit theorem for doubly indexed arrays of martingale differences. For a martingale difference array Y_{kn}, for $k = 1, \ldots, k_n$, for $n \geq 1$ (over the increasing sigma-fields $\mathcal{F}_{kn}$), the *conditional Lindeberg's condition* requires that, for all $\varepsilon > 0$,

$$\sum_{k=1}^{k_n} \mathbf{E}\big[Y_{kn}^2 \mathbf{1}_{\{|Y_{kn}|>\varepsilon\}} \mid \mathcal{F}_{k-1,n}\big] \xrightarrow{P} 0.$$

A *Z-conditional variance condition* requires that

$$\sum_{k=1}^{k_n} \mathbf{E}\big[Y_{kn}^2 \mid \mathcal{F}_{k-1,n}\big] \xrightarrow{P} Z,$$

for the random variable Z.

Theorem A.13 *(Martingale Central Limit Theorem). Let* $\{X_{kn}, \mathcal{F}_{kn}, 1 \leq k \leq k_n\}_{n=1}^{\infty}$ *be a zero-mean, square-integrable martingale array, with differences* Y_{kn} *(over increasing sigma-fields). Let* Z *be an almost-surely finite random variable. If the differences satisfy the conditional Lindeberg's condition and the* Z*-conditional variance condition, the sum of the differences on the* n*th row*

$$\sum_{k=1}^{k_n} Y_{kn} = X_{kn}$$

tends in distribution to a random variable with characteristic function $\mathbf{E}[\exp(-\frac{1}{2}Zt^2)]$.

Bibliography

[1] Aho, A., Hopcroft J. and Ullman, J. (1974). *The Design and Analysis of Computer Algorithms*. Addison-Wesley, Reading, Massachusetts.

[2] Aho, A., Hopcroft J. and Ullman, J. (1983). *Data Structures and Algorithms*. Addison-Wesley, Reading, Massachusetts.

[3] Aigner, M. (1982). Selecting the top three elements. *Discrete Applied Mathematics*, **4**, 247–267.

[4] Ajtai, M., Komlós, J. and Szemérdi, E. (1983). An $O(n \log n)$ sorting network. In *Proceedings of the 15th Annual ACM Symposium on Theory of Computing*, 1–9, Boston.

[5] Ajtai, M., Komlós, J. and Szemérdi, E. (1983). Sorting in $c \log n$ parallel steps. *Combinatorica*, **3**, 1–19.

[6] Aldous, D. (1989). *Probability Approximations via the Poisson Clumping Heuristic*. Springer-Verlag, New York.

[7] Allison, D. and Noga, M. (1979). Usort: An efficient hybrid of distributive partitioning sorting. *BIT*, **22**, 135–139.

[8] Allison, D. and Noga, M. (1980). Selection by distributive partitioning. *Information Processing Letters*, **11**, 7–8.

[9] Anderson, D. and Brown, R. (1992). Combinatorial aspects of C.A.R. Hoare's find algorithm. *Australian Journal of Combinatorics*, **5**, 109–119.

[10] Andrews, D., Bickel, P., Hampel, F., Huber, P., Rogers, W. and Tukey, J. (1972). *Robust Estimates of Location: Surveys and Advances*. Princeton University Press, Princeton, New Jersey.

[11] Apers, P. (1978). Recursive Samplesort. *BIT*, **18**, 125–132.

[12] Arora, S. and Dent, W. (1969). Randomized binary search technique. *Communications of the ACM*, **12**, 77–80.

[13] Arratia, R. and Tavaré, S. (1994). Independent processes approximations for random combinatorial structures. *Advances in Mathematics*, **104**, 90–154.

[14] Baase, S. (1978). *Computer Algorithms: Introduction to Design and Analysis*. Addison-Wesley, Reading, Massachusetts.

[15] Banachowski, L., Kreczmar, A., and Rytter, W. (1991). *Algorithms and Data Structures*. Addison-Wesley, Reading, Massachusetts.

[16] Barbour, A., Holst, L., and Janson, S. (1992). *Poisson Approximation*. Oxford University Press, Oxford, UK.

[17] Bartsow, D. (1980). Remarks on "A synthesis of several sorting algorithms." *Acta Informatica*, **13**, 225–227.

[18] Batcher, K. (1968). Sorting networks and their applications. In *Proceedings of the AFIPS Spring Joint Computer Conference*, **32**, 307–314.

[19] Beck, I. and Krogdahl, S. (1988). A select and insert sorting algorithm. *BIT*, **28**, 726–735.

[20] Bell, D. (1958). The principles of sorting. *Computer Journal*, **1**, 71–77.

[21] Bent, S. and John, J. (1985). Finding the median requires $2n$ comparisons. In *Proceedings of the 17th Annual ACM Symposium on Theory of Computing*, Providence, Rhode Island, 213–216.

[22] Bentley, J. (1985). Programming pearls. *Communications of the ACM*, **28**, 1121–1127.

[23] Bentley, J. and McIlroy, M. (1993). Engineering a sort function. *Software—Practice and Experience*, **23**, 1249–1265.

[24] Berztiss, A. (1975). *Data Structures: Theory and Practice*, 2nd ed. Academic Press, New York.

[25] Billingsley, P. (1986). *Probability and Measure*. Wiley, New York.

[26] Bing-Chao, H. and Knuth, D. (1986). A one-way, stackless Quicksort algorithm. *BIT*, **26**, 127–130.

[27] Blair, C. (1966). Certification of Quickersort. *Communications of the ACM*, **9**, 354.

[28] Blum, M., Floyd, R., Pratt, V., Rivest, R., and Tarjan, R. (1973). Time bounds for selection. *Journal of Computer and System Sciences*, **7**, 448–461.

[29] Boothroyd, J. (1967). Sort of a section of the elements of an array by determining the rank of each element (Algorithm 25). *Computer Journal*, **10**, 308–310.

[30] Bowman, C. (1994). *Algorithms and Data Structures Applications in C*. Harcourt Publishing, Fort Worth, Texas.

[31] Brassard, G. and Bratley, P. (1988). *Algorithmics: Theory & Practice*. Prentice-Hall, Englewood Cliffs, New Jersey.

[32] Brawn, B., Gusatvson, F., and Mankin, E. (1970). Sorting in a paging environment. *Communications of the ACM*, **13**, 483–494.

[33] Breiman, L. (1968). *Probability*. Addison-Wesley, Reading, Massachusetts.

[34] Brown, M. and Tarjan, R. (1979). A fast merging algorithm. *Journal of the ACM*, **26**, 211–226.

[35] Bui, T. and Thanh, M. (1985). Significant improvements to the Ford-Johnson Algorithm for sorting. *BIT*, **25**, 70–70.

[36] Burge, W. (1958). Sorting, trees and measures of order. *Information and Control*, **1**, 181–197.

[37] Burge, W. (1972). An analysis of a tree sorting method and some properties of a set of trees. In *Proceedings of the First USA-Japan Conference*.

[38] Burnetas, A., Solow, D. and Agarwal, R. (1997). An analysis and implementation of an efficient in-place bucket sort. *Acta Informatica*, **34**, 687–700.

[39] Caley, A. (1849). Note on the theory of permutation. *London, Edinburgh and Dublin Philos. Mag. J. Sci.*, **34**, 527–529.

[40] Carlsson, S. (1984). Improving the worst-case behavior of heaps. *BIT*, **24**, 14–18.

[41] Carlsson, S. (1986). Splitmerge–A fast stable merging algorithm. *Information Processing Letters*, **22**, 189–192.

[42] Carlsson, S. (1987a). Average-case results on heapsort. *BIT*, **27**, 2–17.

[43] Carlsson, S. (1987b). A variant of HEAPSORT with almost optimal number of comparisons. *Information Processing Letters*, **24**, 247–250.

[44] Chambers, J. (1971). Partial sorting (Algorithm 410). *Communications of the ACM*, **14**, 357–358.

[45] Chambers, J. (1977). *Computer Methods for Data Analysis*. Wiley, New York.

[46] Chao, C. (1997). A note on applications of the martingale central limit theorem to random permutations. *Random Structures and Algorithms*, **10**, 323–332.

[47] Chao, M. Lin, G. (1993). The asymptotic distributions of the remedian. *Journal of Statistical Planning and Inference*, **37**, 1–11.

[48] Chen, W., Hwang, H., and Chen, G. (1999). The cost distribution of queue mergesort, optimal mergesorts, and power-of-two rules. *Journal of Algorithms*, **30**, 423–448.

[49] Chern, H. and Hwang, H. (2000). Transitional behaviors of the average cost of quicksort with median-of-$(2t+1)$. *Algorithmica*.

[50] Chow, Y. and Teicher, H. (1978). *Probability Theory*. Springer, Berlin.

[51] Chung, K. (1974). *A Course in Probability Theory*. Academic Press, New York.

[52] Comtet, L. (1974). *Advanced Combinatorics*. Reidel, Dordrecht.

[53] Cook, C. and Kim, D. (1980). Best sorting algorithms for nearly sorted lists. *Communications of the ACM*, **23**, 620–624.

[54] Cormen, T., Leiserson, C., and Rivest, R. (1990). *Introduction to algorithms*. McGraw Hill, New York.

[55] Cramer, M. (1997). Stochastic analysis of the merge-sort algorithm. *Random Structures and Algorithms*, **11**, 81–96.

[56] Cunto, W. and Munro, J. (1989). Average case selection. *Journal of the ACM*, **36**, 270–279.

[57] Cunto, W., Munro, J. and Rey, M. (1992). Selecting the median and two quartiles in a set of numbers. *Software—Practice and Experience*, **22**, 439–454.

[58] Cypher, R. (1993). A lower bound on the size of Shellsort sorting networks. *SIAM Journal on Computing*, **22**, 62–71.

[59] Darlington, J. (1978). A synthesis of several sorting algorithms. *Acta Informatica*, **11**, 1–30.

[60] David, F. and Barton, E. (1962). *Combinatorial Chance*. Charles Griffin, London.

[61] David, H. (1980). *Order Statistics*. Wiley, New York.

[62] Davies, E. (1978). *Integral Transforms and Their Applications*. Springer-Verlag, New York.

[63] De Bruijn, N. (1984). *Asymptotic Methods in Analysis*. Reprinted by Dover, New York.

[64] Demuth, D. (1956). *Electronic Data Sorting*. Ph.D. Dissertation. Stanford University, Stanford, California.

[65] Devroye, L. (1984). A probabilistic analysis of the height of tries and of the complexity of triesort. *Acta Informatica*, **21**, 229–237.

[66] Devroye, L. (1984). Exponential bounds for the running time of a selection algorithm. *Journal of Computer and System Sciences*, **29**, 1–7.

[67] Devroye, L. (1986). *Lecture Notes on Bucket Algorithms*. Birkhäuser, Boston.

[68] Devroye, L. (1986). *The Generation of Non-Uniform Random Variates*. Springer-Verlag, New York.

[69] Devroye, L. (1988). Applications of the theory of records in the study of random trees. *Acta Informatica*, **26**, 123–130.

[70] Devroye, L. (1991). Limit laws for local counters in random binary search trees. *Random Structures and Algorithms*, **2**, 303–316.

[71] Devroye, L. and Klincsek, T. (1981). Average time behavior of distributive sorting algorithms. *Computing*, **26**, 1–7.

[72] Dijkstra, E. (1982). Smoothsort, an alternative to sorting in situ. *Scientific and Computational Programming*, **1**, 223–233.

[73] Dinsmore, R. (1965). Longer strings from sorting. *Communications of the ACM*, **8**, 48.

[74] Doberkat, E. (1981). inserting a new element into a heap. *BIT*, **21**, 255–269.

[75] Doberkat, E. (1982a). Asymptotic estimates for the higher moments of the expected behavior of straight insertion sort. *Information Processing Letters*, **14**, 179–182.

[76] Doberkat, E. (1982b). Deleting the root of a heap. *Acta Informatica*, **17**, 245–265.

[77] Doberkat, E. (1984). An average case analysis of Floyd's algorithm to construct heaps. *Information and Control*, **61**, 114–131.

[78] Dobosiewicz, W. (1978a). Sorting by distributive partitioning. *Information Processing Letters*, **7**, 1–6.

[79] Dobosiewicz, W. (1978b). The practical significance of d. p. sort revisited. *Information Processing Letters*, **8**, 170–172.

[80] Dobosiewicz, W. (1980). An efficient variation of Bubble Sort. *Information Processing Letters*, **11**, 5–6.

[81] Dor, D. and Zwick, U. (1995). Selecting the median. In *Proceedings of the 6th Annual ACM-SIAM Symposium on Discrete Algorithms*, 28–37, San Francisco.

[82] Dor, D. and Zwick, U. (1996a). Finding the αn-th largest element. *Combinatorica*, **16**, 41–58.

[83] Dor, D. and Zwick, U. (1996b). Median selection requires $(2 + \varepsilon)n$ comparisons. In *37th Annual Symposium on Foundations of Computer Science*, Burlington, Vermont, 125–134.

[84] Dromey, P. (1984). Exploiting partial order with Quicksort. *Software—Practice and Experience*, **14**, 509–518.

[85] Ducoin, F. (1979). Tri par addressage direct. *RAIRO: Informatique*, **13**, 225–237.

[86] Dudzinski, K. and Dydek, A. (1981). On a stable minimum storage merging algorithm. *Information Processing Letters*, **12**, 5–8.

[87] Eddy, W. and Schervish, M. (1995). How many comparisons does Quicksort use? *Journal of Algorithms*, **19**, 402–431.

[88] Ehrlich, G. (1981). Searching and sorting real numbers. *Journal of Algorithms*, **2**, 1–14.

[89] Ellis, M. and Steele, J. (1981). Fast sorting of Weyl sequences using comparisons. *SIAM Journal on Computing*, **10**, 88–95.

[90] Erdélyi, A. (1953). *Higher Transcendental Functions*. McGraw-Hill, New York.

[91] Erkiö, H. (1980). Speeding Sort algorithms by special instructions. *BIT*, **21**, 2–19.

[92] Erkiö, H. (1981). A heuristic approximation of the worst case of Shellsort. *BIT*, **20**, 130–136.

[93] Espelid, T. (1976). On replacement selection and Dinsmore's improvement. *BIT*, **16**, 133–142.

[94] Estivill-Castro, V. and Wood, D. (1989). A new measure of presortedness. *Information and Computation*, **83**, 111–119.

[95] Estivill-Castro, V. and Wood, D. (1993). Randomized Adaptive Algorithms. *Random Structures and Algorithms*, **4**, 37–57.

[96] Eusterbrock, J. (1993). Errata to "Selecting the top three elements" by M. Aigner: A result of a computer-assisted proof search. *Discrete Applied Mathematics*, **41**, 131–137.

[97] Fagin, R., Nievergelt, N., Pippenger, N., and Strong, H. (1979). Extendible Hashing—a fast access method for dynamic files. *ACM Transactions on Database Systems*, **4**, 315–344.

[98] Feller, W. (1970). *An Introduction to Probability Theory and its Applications*, Vol. 1. Wiley, New York.

[99] Feuzig, W. (1960). Math. sort (Algorithm 23). *Communications of the ACM*, **3**, 601–602.

[100] Flajolet, P. (1983). On the performance evaluation of extendible hashing. *Acta Informatica*, **20**, 345–369.

[101] Flajolet, P. (1988). Mathematical methods in the analysis of algorithms and data structures. In *Trends in Theoretical Computer Science*, E. Börger, ed., 225–304, Computer Science Press, Rockville, MD.

[102] Flajolet, P. (1988). Random tree models in the analysis of algorithms. In *Performance '87*, 171–187.

[103] Flajolet, P. and Golin, M. (1994). Mellin transforms and asymptotics. *Acta Informatica*, **31**, 673–696.

[104] Flajolet, P. and Odlyzko, A. (1990). Singularity analysis of generating functions. *SIAM Journal on Discrete Mathematics*, **3**, 216–240.

[105] Flajolet, P. and Sedgewick, R. (1995). Mellin transforms and asymptotics: finite differences and Rice's integrals. *Theoretical Computer Science*, **144**, 101–124.

[106] Flajolet, P., Gourdon, X., and Dumas, P. (1995). Mellin transforms and asymptotics: harmonic sums. *Theoretical Computer Science*, **144**, 101–124.

[107] Flajolet, P., Grabner, P., Kirschenhofer, P., Prodinger, H., and Tichy, R. (1994). Mellin Transform and asymptotics: Digital sums. *Theoretical Computer Science*, **123**, 291–314.

[108] Flajolet, P., Poblete, P., and Viola, A. (1998). On the analysis of linear probing hashing. INRIA Technical Report 3265. To appear in *Algorithmica*.

[109] Flores, I. (1960). Computer time for address calculation sorting. *Journal of the ACM*, **7**, 389–409.

[110] Floyd, R. (1964). Treesort 3 (Algorithm 245). *Communications of the ACM*, **7**, 701.

[111] Floyd, R. and Rivest, R. (1975). Expected time bounds for selection. *Communications of the ACM*, **18**, 165–172.

[112] Floyd, R. and Rivest, R. (1975). The algorithm SELECT for finding the ith smallest of n elements. *Communications of the ACM*, **18**, 173.

[113] Foata, D. (1965). Problèmes d'analyse combinatoire. *Publication de l'Institut Statistique*, University of Paris, **14**, 81–241.

[114] Ford, L. and Johnson, S. (1959). A tournament problem. *American Mathematical Monthly*, **66**, 387–389.

[115] Frazer, W. and McKellar, A. (1970). Samplesort: a sampling approach to minimal storage tree sorting. *Journal of the ACM*, **17**, 496–507.

[116] Fredman, M. (1976). How good is the information theory bound in sorting. *Theoretical Computer Science*, **1**, 355–361.

[117] Friend, E. (1965). Sorting on electronic computer systems. *Journal of the ACM*, **3**, 134–168.

[118] Fussenegger, F. and Gabow, H. (1978). A counting approach to lower bounds for selection problems. *Journal of the ACM*, **26**, 227–238.

[119] Galambous, J. (1978). *The Asymptotic Theory of Extreme Order Statistics*. Wiley, New York.

[120] Glick, N. (1978). Breaking records and breaking boards. *American Mathematical Monthly*, **85**, 2–26.

[121] Gonnet, G. (1984). On direct addressing sort. *RAIRO: Technique et Science Informatiques*, **3**, 123–127.

[122] Gonnet, G. and Baeza-Yates, R. (1991). *Handbook of Algorithms and Data Structures in Pascal and C*, 2nd ed. Addison-Wesley, Reading, Massachusetts.

[123] Gonnet, G. and Munro, J. (1984). The analysis of linear probing sort by the use of a new mathematical transform. *Journal of Algorithms*, **5**, 451–470.

[124] Gonnet, G. and Munro, J. (1986). Heaps on heaps. *SIAM Journal on Computing*, **15**, 4.

[125] Gotlieb, C. (1963). Sorting on computers. *Communications of the ACM*, **6**, 194–201.

[126] Grabner, P. (1993). Searching for losers. *Random Structures and Algorithms*, **4**, 99–110.

[127] Graham, R., Knuth, D., and Patashnik, O. (1989). *Concrete Mathematics: A Foundation for Computer Science*. Addison-Wesley, Reading, Massachusetts.

[128] Greene, D. and Knuth, D. (1982). *Mathematics for the Analysis of Algorithms*. Birkhäuser, Boston.

[129] Grübel, R. and Rösler, U. (1996). Asymptotic distribution theory for Hoare's selection algorithm. *Advances in Applied Probability*, **28**, 252–269.

[130] Hadian, A (1969). *Optimality Properties of Various Procedures for Ranking n Different Numbers Using only Binary Comparisons.* Ph.D. Dissertation, University of Minnesota, Minneapolis, Minnesota.

[131] Hadian, A. and Sobel, M. (1969). Selecting the t-th largest using binary errorless comparisons. *Colloquia Mathematica Societatis János Bolyai*, **4**, 585–599.

[132] Hall, P. and Heyde, C. (1980). *Martingale Limit Theory and Its Applications*. Academic Press, New York.

[133] Harper, L., Payne, T., Savage, J., and Straus, E. (1975). Sorting $X+Y$. *Communications of the ACM*, **18**, 347–349.

[134] Hennequin, P. (1989). Combinatorial analysis of quicksort algorithm. *RAIRO: Theoretical Informatics and Its Applications*, **23**, 317–333.

[135] Hennequin, P. (1991). *Analyse en Moyenne d'Algorithmes, Tri Rapide, et Arbres de Rechereche*. Ph.D. Dissertation, L'École Polytechnique Palaiseau.

[136] Henrici, P. (1977). *Applied and Computational Complex Analysis*. Wiley, New York.

[137] Hertel, S. (1983). Smoothsort's behavior on presorted sequences. *Information Processing Letters*, **16**, 165–170.

[138] Hibbard, T. (1962). Some combinatorial properties of certain trees with applications to searching and sorting. *Journal of the ACM*, **9**, 13–18.

[139] Hibbard, T. (1963). An empirical study of minimum storage sorting. *Communications of the ACM*, **6**, 206–213.

[140] Hille, R. (1985). Binary trees and permutations. *The Australian Computer Journal*, **17**, 85–87.

[141] Hoare, C. (1961). Partition (Algorithm 63), quicksort (Algorithm 64), and find (Algorithm 65). *Communications of the ACM*, **4**, 321–322.

[142] Hoare, C. (1962). Quicksort. *Computer Journal*, **5**, 10–15.

[143] Hofri, M. (1987). *Probabilistic Analysis of Algorithms*. Springer, Berlin.

[144] Holst, L. (1986). On birthday, collector's, occupancy and other classical urn problems. *International Statistics Review*, **54**, 15–27.

[145] Horowitz, E. and Sahni, S. (1978). *Fundamentals of Computer Algorithms*. Computer Science Press, Potomac, Maryland.

[146] Hu, T. (1982). *Combinatorial Algorithms*. Addison-Wesley, Reading, Massachusetts.

[147] Huits, M. and Kumar, V. (1979). The practical significance of distributive partitioning sort. *Information Processing Letters*, **8**, 168–169.

[148] Hunt, D. (1967). *Partially Ordered Files and the Shell Sorting Algorithm*. Bachelor's Thesis, Princeton University, Pincteon, New Jersey.

[149] Hurley C. and Modarres, R. (1995). Low-Storage quantile estimation. *Computational Statistics*, **10**, 311–325.

[150] Hwang, F. and Deutsch, D. (1973). A class of merging algorithms. *Communications of the ACM*, **20**, 148–159.

[151] Hwang, F. and Lin, S. (1971). Optimal merging of two elements with n elements. *Acta Informatica*, **1**, 145–158.

[152] Hwang, F. and Lin, S. (1972). A simple algorithm for merging two disjoint linearly ordered sets. *SIAM Journal on Computing*, **1**, 31–39.

[153] Hwang, H. (1993). On the solutions of divide and-conquer recurrences. Research Report No. 1439, Ecole Polytechnique, Palaiseau Cedex, France.

[154] Hwang, H. (1997). Asymptotic estimates of elementary probability distributions. *Studies in Applied Mathematics*, **99**, 393–417.

[155] Hwang, H. (1998). Asymptotic expansions of the mergesort recurrences. *Acta Informatica*, **35**, 911–919.

[156] Hwang, H., Yang, B., and Yeh, Y. (1999). *Presorting algorithms: An average-case point of view*. Technical Report, Academia Sinica.

[157] Hyafil, L. (1976). Bounds for selection. *SIAM Journal on Computing*, **5**, 109–114.

[158] Incerpi, J. and Sedgewick, R. (1985). Improved upper bounds on Shellsort. *Journal of Computer and System Sciences*, **31**, 210–224.

[159] Incerpi, J. and Sedgewick, R. (1987). Practical variations of Shellsort. *Information Processing Letters*, **26**, 37–43.

[160] Isaac, E. and Singleton, R. (1956). Sorting by address calculation. *Journal of the ACM*, **3**, 169–174.

[161] Jackowski, B., Lubiak, R. and Sokolowski, S. (1979). Complexity of sorting by distributive partitioning. *Information Processing Letters*, **9**, 100.

[162] Jacquet, P. and Régnier, M. (1986). Trie Partitioning process: limiting distributions. *Lecture Notes in Computer Science*, **214**, 196–210.

[163] Jacquet, P. and Szpankowski, W. (1995). Asymptotic behavior of the Lempel-Ziv parsing scheme and digital search trees. *Theoretical Computer Science*, **144**, 161–197.

[164] Jacquet, P. and Szpankowski, W. (1998). Analytical depoissonization and its applications. *Theoretical Computer Science*, **201**, 1–62.

[165] Janko, W. (1976). A list insertion sort for keys with arbitrary key distributions. *ACM Transactions on Mathematical Software*, **2**, 143–153.

[166] Janson, S. and Knuth, D. (1997). Shellsort with three increments. *Random Structures and Algorithms*, **10**, 125–142.

[167] John, J. (1988). A new lower bound for the set partition problem. *SIAM Journal on Computing*, **17**, 640–647.

[168] Johnson, B. and Killeen, T. (1983). An explicit formula for the C.D.F. of the L_1 norm of the Brownian bridge. *Annals of Probability*, **11**, 807–808.

[169] Jones, B. (1970). A variation on sorting by address calculation. *Communications of the ACM*, **13**, 105–107.

[170] Kac, M. (1949). On deviations between theoretical and empirical distributions. *Proc. N. A. S.*, **35**, 252–257.

[171] Karp, R. (1985). The probabilistic analysis of some combinatorial algorithms. In *Algorithms and Complexity*, J. Traub, ed.

[172] Karp, R. (1986). Combinatorics, complexity and randomness. *Communications of the ACM*, **29**, 98–109.

[173] Khinchin, A. (1957). *Mathematical Foundations of Information Theory*. Dover, New York.

[174] Kingston, J. (1991). *Algorithms and Data Structures: Design, Correctness and Analysis*. Addison-Wesley, Reading, Massachusetts.

[175] Kirkpatrick, D. (1981). A unified lower bound for selection and set partitioning problems. *Journal of the ACM*, **28**, 150–165.

[176] Kirschenhofer, P. and Prodinger, H. (1986). Two Selection Problems Revisited. *Journal of Combinatorial Theory Series*, **A 42**, 310–316.

[177] Kirschenhofer, P. and Prodinger, H. (1998). Comparisons in Hoare's Find algorithm. *Combinatorics, Probability, and Computing*, **7**, 111–120.

[178] Kirschenhofer, P., Prodinger, H., and Martínez, C. (1997). Analysis of Hoare's Find algorithm with median-of-three partition. *Random Structure and Algorithms*, **10**, 143–156.

[179] Kislitsyn, S. (1964). On the selection of the k-th element of an ordered set by pairwise comparisons. *Sibirisk. Mat. Zh.*, **5**, 557–564.

[180] Knuth, D. (1963). Length of strings for a merge sort. *Communications of the ACM*, **6**, 685–687.

[181] Knuth, D. (1969). *The Art of Computer Programming*, Vol. 2: *Seminumerical Algorithms*. Addison-Wesley, Reading, Massachusetts.

[182] Knuth, D. (1973). *The Art of Computer Programming*, Vol. 1: *Fundamental Algorithms*, 2nd ed. Addison-Wesley, Reading, Massachusetts.

[183] Knuth, D. (1976). Big omicron and big omega and big theta. *ACM SIGACT News*, **8**, 18–24.

[184] Knuth, D. (1998). *The Design and Analysis of Algorithms*. Springer-verlag, New York.

[185] Kozen, D. (1992). *Data Structures and Program Design*, 2nd ed. Prentice-Hall, Englewood Cliffs, New Jersey.

[186] Kruse, R. (1987). *Data Structures and Program Design*, 2nd ed. Prentice-Hall, Englewood Cliffs, New Jersey.

[187] Kundu, S. (1977). Sorting tree, nestling tree and inverse permutation. *Information Processing Letters*, **6**, 94–96.

[188] Kwalic, J. and Yoo, B. (1981). Implementing a distributive sort program. *Journal of Information and Optimization Sciences*, **1**, 28–33.

[189] Lazarus, R. and Frank, R. (1960). A high speed sorting procedure. *Communications of the ACM*, **3**, 20–22.

[190] Lent, J. and Mahmoud, H. (1996a). Average-case analysis of multiple Quickselect: An algorithm for finding order statistics. *Statistics and Probability Letters*, **28**, 299–310.

[191] Lent, J. and Mahmoud, H. (1996b). On tree-growing search strategies. *Annals of Applied Probability*, **6**, 20–22.

[192] Levcopoulos, C. and Petersson, O. (1989). Heapsort—Adapted for presorted files. In *Lecture Notes in Computer Science*, F. Dehne, J. Sack, and N. Santoro, eds., **382**, 499–509. Springer-Verlag, New York.

[193] Levcopoulos, C. and Petersson, O. (1990). Sorting shuffled monotone sequences. In *Lecture Notes in Computer Science*, J. Gilbert and R. Karlsson, eds., **447**, 181–191, Springer-Verlag, New York.

[194] Li, M. and Vitányi, P. (1997). *An Introduction to Kolmogorov Complexity and Its Applications*. Springer, New York.

[195] Lifschitz, V. and Pittel, B. (1981). The number of increasing subsequences of the random permutation. *Bell Systems Technical Journal*, **63**, 1827–1843.

[196] Linderman, J. (1984). Theory and practice in construction of a working sort routine. *Journal of Combinatorial Theory, Series A*, **31**, 1–20.

[197] Lindstrom, E. and Vitter, J. (1985). The design and analysis of Bucketsort for bubble memory secondary storage. *IEEE Transactions on Computers*, **C-34**, 218–233.

[198] Loeser, R. (1974). Some Performance tests of "Quicksort" and descendants. *Communications of the ACM*, **17**, 143–152.

[199] Lorin, H. (1975). *Sorting and sort systems*. Addison-Wesley, Massachusetts.

[200] Louchard, G. (1983). The Brownian motion: A neglected tool for the complexity analysis of sorted tables manipulations. *RAIRO: Informatique Théorique*, **17**, 365–385.

[201] Louchard, G. (1986). Brownian motion and algorithm complexity. *BIT*, **26**, 17–34.

[202] MacLaren, M. (1966). Internal sorting by radix plus sifting. *Journal of the ACM*, **13**, 404–411.

[203] Mahmoud, H. (1991). Limiting distributions for path lengths in recursive trees. *Probability in the Engineering and Informational Sciences*, **5**, 53–59.

[204] Mahmoud, H. (1992). *Evolution of Random Search Trees*. Wiley, New York.

[205] Mahmoud, H. and Smythe, R. (1998). Probabilistic analysis of MULTIPLE QUICK SELECT. *Algorithmica*, **22**, 569–584.

[206] Mahmoud, H., Flajolet, P., Jacquet, P., and Régnier, M. (2000). Analytic variations on bucket selection and sorting. *Acta Informatica*.

[207] Mahmoud, H., Modarres, R. and Smythe, R. (1995). Analysis of quickselect: An algorithm for order statistics. *RAIRO: Theoretical Informatics and Its Applications*, **29**, 255–276.

[208] Manacher, G. (1979a). Significant improvements to the Hwang-Lin merging algorithm. *Journal of the ACM*, **26**, 434–440.

[209] Manacher, G. (1979b). The Ford-Johnson sorting algorithm is not optimal. *Journal of the ACM*, **26**, 441–456.

[210] Manber, U. (1989). *Introduction to Algorithms: A Creative Approach*. Addison-Wesley, Reading, Massachusetts.

[211] Mannila, H. (1985). Measures of presortedness and optimal sorting algorithms. *IEEE Transactions on Computing*, **C-34**, 318–325.

[212] Martin, W. and Ness, D. (1972). Optimizing binary tree growth with a sorting algorithm. *Communications of the ACM*, **15**, 88–93.

[213] McDiarmid, C. and Hayward, R. (1992). Strong concentration for Quicksort. In *Proceedings of the 3rd Annual ACM-SIAM Symposium on Discrete Algorithms*. Orlando, Florida, 414–421.

[214] McDiarmid, C. and Hayward, R. (1996). Large Deviation inequality for Quicksort. *Journal of Algorithms*, **21**, 476–507.

[215] McGeoch, C. and Tygar, J. (1995). Optimal sampling strategies for quicksort. *Random Structures and Algorithms*, **7**, 287–300.

[216] McIlroy, P., Bostic, K. and McIlroy, M. (1993). Engineering radix sort. *Computing Systems*, **6**, 5–27.

[217] Mehlhorn, K. (1984). *Data Structures and Algorithms, Vol. 1: Sorting and Searching*. Springer-Verlag, Berlin.

[218] Meijer, H. and Akl, S. (1980). The design and analysis of a new hybrid sorting algorithm. *Information Processing Letters*, **10**, 213–218.

[219] Mellin, H. (1902). Über den zusammenhang zwischen den linearen differential und differnzengleichungen. *Acta Mathematica*, **25**, 139–164.

[220] Melville, R. and Gries, D. (1980). Controlled-density sorting. *Information Processing Letters*, **10**, 169–172.

[221] Meyer, R. (1969). The kth largest coordinate of an ordered n-tuple. *American Statistician*, **23**, 27.

[222] Moon, J. (1968). *Topics on Tournaments*. Holt, Rinehart and Winston, New York.

[223] Morris, R. (1969). Some theorems on sorting. *SIAM Journal on Applied Mathematics*, **17**, 1–6.

[224] Motoki, T. (1982). A note on upper bounds for selection problems. *Information Processing Letters*, **15**, 214–219.

[225] Nijenhuis, A. and Wilf, H. (1978). *Combinatorial Algorithms*, 2nd ed. Academic Press, New York.

[226] Panholzer, A. and Prodinger, H. (1998). A generating functions approach for the analysis of grand averages for Multiple Quickselect. *Random Structures and Algorithms*, **13**, 189–209.

[227] Panny, W. (1986). A note on the higher moments of the expected behavior of straight insertion sort. *Information Processing Letters*, **22**, 175–177.

[228] Panny, W. and Prodinger, H. (1995). Bottom-up mergesort: A detailed Study. *Algorithmica*, **14**, 340–354.

[229] Papadimitriou, C. and Steiglitz, K. (1982). Combinatorial Optimization: *Algorithms and Complexity*. Prentice-Hall, Englewood Cliffs, New Jersey.

[230] Papernov, A. and Stasevich, G. (1965). A method for information sorting in computer memories. *Problems of Information Transmission*, **1**, 63–75.

[231] Pearl, J. (1981) A space Efficient on-line method of computing quantile estimates. *Journal of Algorithms*, **2**, 164–177.

[232] Peltola, E. and Erkiö, H. (1978). Insertion merge sorting. *Information Processing Letters*, **7**, 92–99.

[233] Peters, J. and Kritzinger (1975). Implementation of Samplesort: A minimal storage tree sort. *BIT*, **15**, 85–93.

[234] Pfaltz, J. (1977). *Computer Data Structures*. McGraw-Hill, New York.

[235] Plaxton, C. and Suel, T. (1997). Lower bounds for Shellsort. *Journal of Algorithms*, **23**, 221–240.

[236] Poblete, P. (1987). Approximating functions by their Poisson transform. *Information Processing Letters*, **23**, 127–130.

[237] Pohl, I. (1972). A sorting problem and its complexity. *Communications of the ACM*, **15**, 462–464.

[238] Poonen, B. (1993). The worst case in Shellsort and related algorithms. *Journal of Algorithms*, **15**, 101–124.

[239] Postmus, J., Rinnooy Kan, A., and Timmer, G. (1983). An efficient dynamic selection method. *Communications of the ACM*, **26**, 878–881.

[240] Pratt, V. (1971). *Shellsort and Sorting Networks*. Garland, New York.

[241] Prodinger, H. (1995). Multiple quickselect–Hoare's Find algorithm for several elements. *Information Processing Letters*, **56**, 123–129.

[242] Purdom, P. and Brown, C. (1985). *The Analysis of Algorithms*. Holt, Rinehart and Winston, New York.

[243] Rachev, S. (1991). *Probability metrics and the Stability of Stochastic Models*. Wiley, New York.

[244] Rachev, S. and Rüschendorf, L. (1995). Probability metrics and recursive algorithms. *Advances in Applied Probability*, **27**, 770–799.

[245] Raghavan, P. and Motwani, R. (1995). *Randomized Algorithms*. Cambridge university Press, Cambridge, UK.

[246] Rais, B., Jacquet, P., and Szpankowski, W. (1993). Limiting distribution for the depth in PATRICIA tries. *SIAM Journal on Discrete Mathematics*, **3**, 355–362.

[247] Ramanan, P. and Hyafil, L. (1984). New algorithms for selection. *Journal of Algorithms*, **5**, 557–578.

[248] Ramanujan, S. and Berndt, B. (1985). *Ramanujan's Notebook*. Springer-Verlag, New York.

[249] Rawlins, G. (1992). *Compared to What: An Introduction to the Analysis of Algorithms*. Computer Science Press (an imprint of Freeman and Company, New York).

[250] Régnier, M. (1989). A limiting distribution for quicksort. *RAIRO: Theoretical Informatics and Its Applications*, **23**, 335–343.

[251] Reingold, E. and Hansen, W. (1986). *Data Structures in PASCAL*. Little, Brown Series, Boston, Massachusetts.

[252] Reingold, E., Nievergelt, J., and Deo, N. (1977). *Combinatorial Algorithms*. Prentice-Hall, Englewood Cliffs, New Jersey.

[253] Rényi, A. (1970). *Probability Theory*. Prentice-Hall, Englewood Cliffs, New Jersey.

[254] Rich, R. (1972). *Internal Sorting Methods Illustrated with PL/I Programs*. North-Holland, Amsterdam.

[255] Riordan, J. (1958). *An Introduction to Combinatorial Analysis*. Wiley, New York.

[256] Riordan, J. (1968). *Combinatorial Identities*. Wiley, New York.

[257] Rösler, U. (1991). A limit theorem for quicksort. *RAIRO: Theoretical Informatics and Its Applications*, **25**, 85–100.

[258] Rösler, U. (1992). A fixed point theorem for distributions. *Stochastic Processes and Applications*, **42**, 195–214.

[259] Ross, S. (1983). *Stochastic Processes*. Wiley, New York.

[260] Rousseeuw, P. and Bassett, G. (1990). The remedian: A robust averaging method for large data sets. *Journal of the American Statistical Association*, **409**, 97–104.

[261] Schaffer, R. and Sedgewick, R. (1993). The analysis of heapsort. *Journal of Algorithms*, **15**, 76–100.

[262] Schönhage, A., Paterson, M., and Pippenger, N. (1976). Finding the median. *Journal of Computer and System Sciences*, **13**, 184–199.

[263] Schreier, J. (1932). On tournament elimination systems. *Mathesis Polska*, **7**, 154–160.

[264] Scowen, R. (1965). Quickersort (Algorithm 271). *Communications of the ACM*, **8**, 669–670.

[265] Sedgewick, R. (1977). Quicksort with equal keys. *SIAM Journal on Computing*, **6**, 240–267.

[266] Sedgewick, R. (1977). The analysis of quicksort programs. *Acta Informatica*, **7**, 327–355.

[267] Sedgewick, R. (1978). Data movement in odd-even merging. *SIAM Journal on Computing*, **7**, 239–272.

[268] Sedgewick, R. (1978). Implementing quicksort programs. *Communications of the ACM*, **21**, 847–856.

[269] Sedgewick, R. (1980). *Quicksort*. Garland, New York.

[270] Sedgewick, R. (1983). Mathematical analysis of combinatorial algorithms. In *Probability Theory and Computer Science*, G. Louchard and G. Latouche, eds. Academic Press, London, 123–205.

[271] Sedgewick, R. (1986). A new upper bound for Shellsort. *Journal of Algorithms*, **7**, 159–173.

[272] Sedgewick, R. (1988). *Algorithms*, 2nd ed. Addison-Wesley, Reading, Massachusetts.

[273] Sedgewick, R. (1996). Analysis of Shellsort and related algorithms. Presented at *Fourth Annual European Symposium on Algorithms*, Barcelona.

[274] Sedgewick, R. and Flajolet, P. (1996). *An Introduction to the Analysis of Algorithms*. Addison-Wesley, Reading, Massachusetts.

[275] Selmer, E. (1989). On Shellsort and the Frobenius problem. *BIT*, **29**, 37–40.

[276] Sharp, W. and Curran, J. (1884). Solution to Problem 7382 (Mathematics) proposed by J. J. Sylvester, ed., *Times*, London, **41**.

[277] Shell, D. (1959). A high-speed sorting procedure. *Communications of the ACM*, **2**, 30–32.

[278] Shepp, L. (1966). Ordered cycle lengths in a random permutation. *Transactions of the American Mathematical Society*, **121**, 340–357.

[279] Shepp, L. (1982). On the integral of the absolute value of the pinned Wiener process. *Annals of Probability*, **10**, 234–239.

[280] Shorack, R. and Wellner, A. (1986). *Empirical Processes with Applications to Statistics*. Wiley, New York.

[281] Singleton, R. (1969). An efficient algorithm for sorting with minimum storage (Algorithm 347). *Communications of the ACM*, **12**, 185–187.

[282] Smythe, R. and Wellner, A. (2000). Stochastic analysis of Shellsort. *Algorithmica* (submitted).

[283] Sobel, M. (1968a). On an optimal search for the t best using binary errorless comparisons: The ordering problem. Technical Report No. 113, Department of Statistics, University of Minnesota.

[284] Sobel, M. (1968b). On an optimal search for the t best using binary errorless comparisons: The selection problem. Technical Report No. 114, Department of Statistics, University of Minnesota.

[285] Standish, T. (1980). *Data Structure Techniques*. Addison-Wesley, Reading, Massachusetts.

[286] Stockmeyer, P. and Yao, F. (1980). On the optimality of linear merge. *SIAM Journal on Computing*, **9**, 85–90.

[287] Suraweera, F. and Al-Anzy, J. (1988). Analysis of a modified address calculation algorithm. *Computer Journal*, **31**, 561–563.

[288] Sussenguth, E. (1963). Use of tree structures for processing files. *Communications of the ACM*, **6**, 272–279.

[289] Szpankowski, W. (1987). Solution of a linear recurrence equation arising in the analysis of some algorithms. *SIAM Journal on Algebraic and Discrete Methods*, **8**, 233–250.

[290] Szpankowski, W. (1988). The evaluation of an alternative sum with applications to the analysis of some data structures. *Information Processing Letters*, **28**, 13–19.

[291] Tan, K. (1993). *An Asymptotic Analysis of the Number of Comparisons in Multipartition Quicksort*. Ph.D. Dissertation, Carnegie Mellon University, Pittsburgh, Pennsylvania.

[292] Tan, K. and Hadjicostas, P. (1995). Some properties of a limiting distribution in Quicksort. *Statistics and Probability Letters*, **25**, 87–94.

[293] Tanner, M. (1978). Minimean merging and sorting: an algorithm. *SIAM Journal on computing*, **7**, 18–38.

[294] Tanner, R. (1978). Some properties of a limiting distribution in Quicksort. *Statistics and Probability Letters*, **25**, 87–94.

[295] Tarjan, R. (1983). *Data Structures and Network Algorithms*. SIAM, Philadelphia, Pennsylvania.

[296] Tarjan, R. (1985). Amortized computational complexity. *SIAM Journal on Algebraic and Discrete Methods*, **6**, 306–318.

[297] Thanh, M. and Bui, T. (1982). An improvement of the binary merge algorithm. *BIT*, **22**, 454–461.

[298] Trabb-Pardo, L. and Lin, S. (1977). Stable sorting and merging with optimal space and time. *SIAM Journal on Computing*, **6**, 351–372.

[299] Tranter, C. (1962). *Integral Transforms in Mathematical Physics*. Wiley, New York.

[300] Trotter, W. (1992). *Combinatorics and Partially Ordered Sets: Dimension Theory*. Johns Hopkins University Press, Baltimore.

[301] Vallander, S. (1973). Calculation of the Wasserstein distance between distributions on the line. *Theory of Probability and Its Applications*, **18**, 784–786.

[302] Van der Nat, M. (1979). Binary merging by partitioning. *Information Processing Letters*, **8**, 72–75.

[303] Van der Nat, M. (1980). A fast sorting algorithm, a hybrid of distributive and merge sorting. *Information Processing Letters*, **10**, 163–167.

[304] Van Edmen, M. (1970). Increasing the efficency of quicksort (Algorithm 402). *Communications of the ACM*, **13**, 693–694. (See also an article by the same title in *Communications of the ACM*, **13**, 563–567.)

[305] Verma, R. (1997). A general method and a master theorem for divide-and-conquer algorithms. *Journal of Algorithms*, **16**, 67–79.

[306] Verma, R. (1997). General techniques for analyzing recursive algorithms with applications. *SIAM Journal on Computing*, **26**, 568–581.

[307] Vitter, J. and Flajolet, P. (1990). Analysis of algorithms and data structures. In *Handbook of Theoretical Computer Science*, J. Van Leween, ed. Elsevier, Amsterdam, 431–524.

[308] Wang, X. and Fu, Q. (1996). A frame for divide-and-conquer recurrences. *Information Processing Letters*, **59**, 45–51.

[309] Wegener, I. (1985). Quicksort for equal keys. *IEEE Transactions on Computers*, **C-34**, 362–367.

[310] Wegener, I. (1993). BOTTOM-UP-HEAPSORT: A new variant of heapsort beating, on average, quick sort (if n is not very small). *Theoretical Computer Science*, **118**, 81–98.

[311] Weide, B. (1977). A survey of analysis techniques for discrete algorithms. *ACM Computing Surveys*, **9**, 291–313.

[312] Weide, B. (1978). Space efficient on-line selection algorithm. *Proceedings of the 11th symposium on the interface of Computer Science and Statistics*, 308–311.

[313] Weiss, M. (1989). A good case for Shellsort. *Congressus Numerantium*, **73**, 59–62.

[314] Weiss, M. (1991). Empirical study of the expected running time of Shellsort. *Computer Journal*, **34**, 88–91.

[315] Weiss, M. and Sedgewick, R. (1988). Bad cases for Shaker-sort. *Information Processing Letters*, **28**, 133–136.

[316] Weiss, M. and Sedgewick, R. (1990a). Tight lower bounds for Shellsort. *Journal of Algorithms*, **11**, 242–251.

[317] Weiss, M. and Sedgewick, R. (1990b). More on Shellsort increment sequences. *Information Processing Letters*, **34**, 267–270.

[318] Wells, M. (1965). Applications of a language for computing in combinatorics. *Information Processing Congress 65*. North Holland Publishing Company, Amsterdam. Pergamon Press, Oxford.

[319] Wells, M. (1971). *Elements of Combinatorial Computing*. Pergamon Press, Oxford.

[320] Whittaker, E. and Watson, G. (1935). *A course in Modern Analysis*. Cambridge Press, London.

[321] Wilf, H. (1986). *Algorithms and Complexity*. Prentice-Hall, Englewood Cliffs, New Jersey.

[322] Wilf, H. (1989). *Combinatorial Algorithms: An Update*. SIAM, Philadelphia, Pennsylvania.

[323] Wilf, H. (1990). *Generatingfunctionology*. Academic Press, New York.

[324] Williams, D. (1991). *Probability with Martingales*. Cambridge University Press, Cambridge, UK.

[325] Williams, J. (1964). HEAPSORT (Algorithm 232). *Communications of the ACM*, **7**, 347–348.

[326] Windley, P. (1960). Trees, forests and rearranging. *Computer Journal*, **3**, 84–88.

[327] Wirth, N. (1976). *Algorithms + Data Structures = Programs*. Prentice-Hall, Englewood Cliffs, New Jersey.

[328] Wirth, N. (1986). *Algorithms & Data Structures*. Prentice-Hall, Englewood Cliffs, New Jersey.

[329] Wong, R. (1989). *Asymptotic Approximations of Integrals*. Academic Press, Orlando, Florida.

[330] Woodall, A. (1970). Notes on "Sort of a section of the elements of an array by determining the rank of each element (Algorithm 25)." *Computer Journal*, **13**, 295–296.

[331] Wu, J. and Zhu, H. (1994). Least basic operations on heap and improved heapsort. *Journal of Computer Sciences and Technology*, **9**, 261–266.

[332] Wulf, W., Shaw, M., Hilfinger, P., and Flon, L. (1981). *Fundamental Structures of Computer Science*. Addison-Wesley, Reading, Massachusetts.

[333] Yao, A. (1988). On straight selection sort. Technical Report No. CS-TR-185-88, Department of Computer Science, Princeton University, Princeton, New Jersey.

[334] Yao, A. (1980). An analysis of $(h, k, 1)$ Shellsort. *Journal of Algorithms*, **1**, 14–50.

[335] Yao, A. and Yao, F. (1982). On the average-case complexity of selecting the kth best. *SIAM Journal on Computing*, **11**, 428–447.

[336] Yap, F. (1976). New upper bounds for selection. *Communications of the ACM*, **19**, 501–508.

Index

D

E

F

G

H

WILEY-INTERSCIENCE SERIES IN DISCRETE MATHEMATICS AND OPTIMIZATION

AARTS AND KORST • Simulated Annealing and Boltzmann Machines: A Stochastic Approach to Combinatorial Optimization and Neural Computing
AARTS AND LENSTRA • Local Search in Combinatorial Optimization
ALON, SPENCER, AND ERDŐS • The Probabilistic Method, Second Edition
ANDERSON AND NASH • Linear Programming in Infinite-Dimensional Spaces: Theory and Application
AZENCOTT • Simulated Annealing: Parallelization Techniques
BARTHÉLEMY AND GUÉNOCHE • Trees and Proximity Representations
BAZARRA, JARVIS, AND SHERALI • Linear Programming and Network Flows
CHANDRU AND HOOKER • Optimization Methods for Logical Inference
CHONG AND ZAK • An Introduction to Optimization
COFFMAN AND LUEKER • Probabilistic Analysis of Packing and Partitioning Algorithms
COOK, CUNNINGHAM, PULLEYBLANK, AND SCHRIJVER • Combinatorial Optimization
DASKIN • Network and Discrete Location: Modes, Algorithms and Applications
DINITZ AND STINSON • Contemporary Design Theory: A Collection of Surveys
DU AND KO • Theory of Computational Complexity
ERICKSON • Introduction to Combinatorics
GLOVER, KLINGHAM, AND PHILLIPS • Network Models in Optimization and Their Practical Problems
GOLSHTEIN AND TRETYAKOV • Modified Lagrangians and Monotone Maps in Optimization
GONDRAN AND MINOUX • Graphs and Algorithms *(Translated by S. Vajdā)*
GRAHAM, ROTHSCHILD, AND SPENCER • Ramsey Theory, Second Edition
GROSS AND TUCKER • Topological Graph Theory
HALL • Combinatorial Theory, Second Edition
HOOKER • Logic-Based Methods for Optimization: Combining Optimization and Constraint Satisfaction
IMRICH AND KLAVŽAR • Product Graphs: Structure and Recognition
JANSON, LUCZAK, AND RUCINSKI • Random Graphs
JENSEN AND TOFT • Graph Coloring Problems
KAPLAN • Maxima and Minima with Applications: Practical Optimization and Duality
LAWLER, LENSTRA, RINNOOY KAN, AND SHMOYS, Editors • The Traveling Salesman Problem: A Guided Tour of Combinatorial Optimization
LAYWINE AND MULLEN • Discrete Mathematics Using Latin Squares
LEVITIN • Perturbation Theory in Mathematical Programming Applications
MAHMOUD • Evolution of Random Search Trees
MAHMOUD • Sorting: A Distribution Theory
MARTELLI • Introduction to Discrete Dynamical Systems and Chaos
MARTELLO AND TOTH • Knapsack Problems: Algorithms and Computer Implementations
McALOON AND TRETKOFF • Optimization and Computational Logic
MINC • Nonnegative Matrices
MINOUX • Mathematical Programming: Theory and Algorithms *(Translated by S. Vajdā)*
MIRCHANDANI AND FRANCIS, Editors • Discrete Location Theory
NEMHAUSER AND WOLSEY • Integer and Combinatorial Optimization
NEMIROVSKY AND YUDIN • Problem Complexity and Method Efficiency in Optimization *(Translated bv E. R. Dawson)*

PACH AND AGARWAL • Combinatorial Geometry
PLESS • Introduction to the Theory of Error-Correcting Codes, Third Edition
ROOS AND VIAL • Ph. Theory and Algorithms for Linear Optimization: An Interior Point Approach
SCHEINERMAN AND ULLMAN • Fractional Graph Theory: A Rational Approach to the Theory of Graphs
SCHRIJVER • Theory of Linear and Integer Programming
TOMESCU • Problems in Combinatorics and Graph Theory *(Translated by R. A. Melter)*
TUCKER • Applied Combinatorics, Second Edition
WOLSEY • Integer Programming
YE • Interior Point Algorithms: Theory and Analysis